Studies in International Political Economy

Stephen D. Krasner, Editor
Ernst B. Haas, Consulting Editor

The Power of Ideology

Emanuel Adler

The Power
of Ideology

The Quest for Technological Autonomy
in Argentina and Brazil

University of California Press
Berkeley   Los Angeles   London

University of California Press
Berkeley and Los Angeles, California

University of California Press, Ltd.
London, England

Library of Congress Cataloging-in-Publication Data

Adler, Emanuel.
  The power of ideology.

  Includes index.
    1. Technology and state—Argentina.   2. Technology
and state—Brazil.   I. Title.
T25.A7A35   1987   338.98206   85–20845

ISBN 978-0-520-30116-0 (pbk. : alk. paper)

Contents

List of Tables

List of Figures

Acronyms

ABICOMP	Brazilian Association for Computer and Peripheral Equipment Industries	Brazil
APPD	Association of Data Processing Professionals	Brazil
BID	See IDB	
BNDE	National Economic Development Bank	Brazil
CACEX	Department of Foreign Commerce of the Bank of Brazil	Brazil
CADIE	Argentine Association of Electronic Industries	Argentina
CAPES	Company for the Improvement of Higher Education Personnel	Brazil
CAPRE	Commission for the Coordination of Electronic Data-Processing Activities	Brazil
CBPF	Brazilian Center for Physics Research	Brazil
CCNAI	Coordinating Commission of Articulation with Industry	Brazil
CCT	Scientific and Technological Council	Brazil
CDE	Economic Development Council	Brazil
CDI	Industrial Development Council	Brazil

CDTN	Center for the Development of Nuclear Technology	Brazil
CEDINI	Coordination of Entities for the Defense of an Informatics National Industry	Brazil
CEPAL	See ECLA	
CGE	General Economic Confederation	Argentina
CGT	General Confederation of Workers	Argentina
CNEA	National Atomic Energy Commission	Argentina
CNEN	National Nuclear Energy Commission	Brazil
CNI	National Informatics Commission	Brazil
CNPq	National Research Council, later National Council of Scientific and Technological Development	Brazil
CONACYT	National Council of Science and Technology	Argentina
CONADE	National Development Council	Argentina
CONASE	National Security Council	Argentina
CONCEX	National Foreign Trade Council	Brazil
CONICET	National Council of Scientific and Technical Research	Argentina
CONIN	National Council on Informatics and Automation	Brazil
CONMETRO	National Council for Metrology, Norms, and Industrial Quality	Brazil
COPPE	Coordination of Graduate Programs in Engineering	Brazil
CTI	Industrial Technology Company	Brazil
Digibrás	Brazilian Digital Enterprises SA	Brazil
ECLA	Economic Commission for Latin America	
Eletrobrás	Brazilian Electric Power Facilities Holding Co.	Brazil
EMBRAMEC	Brazilian Mechanics, Inc.	Brazil
ESG	Higher War College	Brazil
FATE	Argentine Tire Mfg. Co.	Argentina
FIBASE	Basic Goods, Inc.	Brazil
FINAME	Special Agency for Industrial Investment	Brazil

FINEP	Studies and Projects Financing Agency	Brazil
FNDCT	National Science and Technology Development Fund	Brazil
FUNAT	Fund for the Support of Technology	Brazil
FUNTEC	Scientific and Technical Development Fund	Brazil
GDP	Gross Domestic Product	
GIN	National Industries Group	Argentina
GNP	Gross National Product	
GTE	Special Working Group	Brazil
HWR	Heavy Water Reactor	
IBGE	Brazilian Institute of Geography and Statistics	Brazil
IBICT	Brazilian Institute for Scientific and Technological Information	Brazil
IBRASA	Brazilian Investments	Brazil
IDB	Interamerican Development Bank	
IEA	Atomic Energy Institute	Brazil
IEN	Nuclear Engineering Institute	Brazil
INMETRO	National Institute for Metrology, Standardization, and Industrial Quality	Brazil
INPI	National Institute of Industrial Property	Brazil
INT	National Institute of Technology	Brazil
INTA	National Institute of Agricultural Technology	Argentina
INTI	National Institute of Industrial Technology	Argentina
INVAP	Applied Research	Argentina
IPEA	Economic and Social Planning Institute	Brazil
IPEN	Energy and Nuclear Research Institute	Brazil
IPT	Technological Research Institute	Brazil
ISEB	Higher Institute of Brazilian Studies	Brazil
ISI	Import Substitution Industrialization	
IUPERJ	University Research Institute of Rio de Janeiro	Brazil
KWU	Kraftwerk Union	West Germany
LWR	Low Water Reactor	
MEC	Ministry of Education and Culture	Brazil

MIC	Ministry of Industry and Commerce	Brazil
MME	Ministry of Mines and Energy	Brazil
NAI	Nuclei of Articulation with Industry	Brazil
Nuclebrás	Brazilian Nuclear Enterprises SA	Brazil
OAS	Organization of American States	
OECD	Organization of Economic Cooperation and Development	
PBDCT	Basic Plan for Scientific and Technological Development	Brazil
Petrobrás	Brazilian Petroleum Co. SA	Brazil
PND	National Development Plan	Brazil
PUC/RJ	Pontifical Catholic University of Rio de Janeiro	Brazil
R&D	Research and Development	
SATI	Technical Assistance Service for Industry	Argentina
SBC	Brazilian Computation Society	Brazil
SBPC	Brazilian Society for the Progress of Science	Brazil
SCI	Superintendency for International Cooperation	Brazil
SECYT	Secretariat of Science and Technology	Argentina
SEGBA	Electrical Services of Greater Buenos Aires	Argentina
SEI	Special Secretariat of Informatics	Brazil
SEPLAN	Planning Secretariat	Brazil
SNDCT	National System for the Development of Science and Technology	Brazil
SNICT	National Science and Technology Information System	Brazil
STI	Industrial Technology Secretariat	Brazil
SUBCYT	Subsecretariat of Science and Technology	Argentina
SUMOC	Superintendency of Money and Credit	Brazil
Telebrás	Brazilian Telecommunications Co. SA	Brazil
UFRJ	Federal University of Rio de Janeiro	Brazil
UIA	Argentine Industrial Union	Argentina
UNCTAD	United Nations Conference for Trade and Development	

UNDP	United Nations Development Program
UNESCO	United Nations Educational, Scientific, and Cultural Organization
UNIDO	United Nations Industrial Development Organization

Acknowledgments

My first wish is to acknowledge and thank my wife, Sylvia, and my children, Shirli, Nadav, and Jonathan, for bearing with me the travail involved in writing and preparing a book manuscript and for showing support and understanding beyond what I could ask or expect. This book is a tribute to them. To my parents, my thanks for believing in and supporting me and my work.

Among the many knowledgeable people who provided inspiration, support, and assistance my thanks go first and foremost to Ernst B. Haas, teacher, mentor, and friend, not only for his constant help—reading, advising, instructing, and guiding my work—but also for his profound intellectual legacy. He provided me with an understanding of human and social relations, the processes of cognitive change, and the importance of designing international relations theory with the values of peace and human welfare in mind.

I am grateful also to David Collier and Pranab Bardhan for their valuable comments on my dissertation, on which this study is partially based.

One of my greatest debts is to the people I interviewed in Washington, D.C., Cambridge, Massachusetts, and five Latin American countries. They were all extremely helpful and kind, sometimes taking long hours from their work to talk to this "stranger."

The support of some of those people exceeded all expectations, and it is difficult to find the right words to thank them. They were Francisco (Paco) Sagasti, Gustavo Flores, and Fernando González Vigil of Peru; Carlos Paredes and Ramón Schulczewski of Bolivia; Alberto Aráoz, Jorge Katz, Jorge Kitroser, Rafael Kohanoff, Jorge Martínez Favini, Jorge Sabato, and Roberto Zubieta of Argentina; Andrea Sandro Calabi, Ivan de Costa Marques, Suelí Mendes dos Santos, Arthur Pereira Nunes, Mário Ripper, José Pelúcio, and Simon Schwartzman of Brazil; and Miguel Wionczek of Mexico. I would also like to express appreciation to the staff of the BID/ECLA science and technology secretariat in Argentina and IUPERJ in Brazil and to all those who provided me with help and hospitality during my trips, especially Miriam and Rafi Barak, Anne and Peter Levine, and Sara and Adolfo Zadunaisky.

Many colleagues and friends have provided information and advice, for which I am indeed grateful. Edson and Marcia Nunes have been especially informative about their home country, Brazil; Van R. Whiting and Yaacov Vertzberger generously offered advice; and I wish Beverly Crawford and Stephan Haggard could know how useful our frequent conversations turned out to be.

I am deeply indebted to the Institute of International Studies at Berkeley, Professor Carl Rosberg, and the staff for financial and administrative support, first as part of the "International Conflict and International Regimes Project," and later as part of my postdoctoral fellowship. I am also indebted to the Institute for the Study of World Politics in New York, the Center for Latin American Studies at Berkeley, the Tinker Foundation, and the Leonard Davis Institute for International Relations at the Hebrew University of Jerusalem for financial assistance.

William K. Muir, Chairman of the Political Science Department at the University of California, Berkeley, and the department staff also provided valuable support, for which I thank them.

I am also indebted to Mary Renaud and Naomi Schneider of the University of California Press, who made my job of bringing the manuscript to publication much easier; to Anne G. Canright for her excellent copyediting; and to Nancy Herington for the preparation of the index.

Last but not least, thanks to Florence Myer, my "English teacher," who guided me in the ways and styles of the English language while—

incidentally—typing the many chapters and drafts. Her professionalism, quality work, and positive attitude were always a source of confidence. My joy in writing lengthy chapters was her pain, and we shared the pain when I had to rewrite them. Nevertheless she was always ready to help, as a good friend usually is. Gracias, Florencia!

I

Continuity of change, preservation of the past in the present, real duration—the living being seems, then, to share these attributes with consciousness. Can we go further and say that life, like conscious activity, is invention, is increasing creation?
—*Henri Bergson,* Creative Evolution

Introduction 1

"My purpose," wrote Ovid two thousand years ago, "is to tell of bodies which have been transformed into shapes of a different kind."[1] My purpose is to tell of shapes that have been transformed into bodies of a different kind; it is to study the role of ideology in shaping change in international relations and, more specifically, in the international political economy.

Despite many years of inquiry into and theorizing about international relations, we know little and agree on less about the nature, roots, and modes of change.[2] This situation owes in part to a methodological inclination in favor of positivism/behaviorism—which made us forget history—and to the paradigmatic dominance of neorealism—which too often emphasized structural change.[3] As a result, international relations theories of the past forty years have been static and mechanistic, aimed at explaining stability, efficiency, and hierarchy rather than the emergent and new in social systems.

Most of these theories have also been deterministic, in at least two senses. They have tended either to deduce and predict political behavior from national and international political-economic structures (for example, contemporary behavioral international political economy theories, which are strongly influenced by neoclassical economic images) or to state that structural change occurs, and in fact history develops, in one direction only (for example, Marxist theories). Although international relations theories differ profoundly according to which deterministic view is being advanced, both explain and predict change as resulting from "structures."

From a paradigmatic and epistemological perspective, structural determinist international relations theories, mainly of the first type, are still considered state-of-the-art.[4] Whether these theories deal with national security or with international political economy, the argument is generally the same: power and economic structures constrain and determine behavior. Causality is assumed to be linear, cybernetic processes are disregarded, and reflection, learning, and adaptation are considered irrelevant.

The determinist character of these theories has led political scientists and economists to aim at prediction and to assume that the fallibility of the theories has less to do with new or emerging conditions and unintended or unexpected events than with the lack of complete information. To be sure, some successes in prediction have been achieved in studies of security and international systems where military power plays a major determining role.[5] But even in security affairs, structural determinists have been unable to deal with the unexpected results of force or threat, because these are shaped at the process level. Futhermore, the insufficiency of structural theories is nowhere better demonstrated than in the political-economic area, in which power is only one, albeit an important, variable and in which processes such as policy making and economic development play a fundamental role.

The inability of the discipline to face the shortcomings of structuralism and determinism even in the presence of so much unexplained change and so many prediction failures has to do, I believe, with the very epistemological and metaphysical basis of these theories—which I call "Being." The image of Being looks for the recurrent, treats human beings as calculators possessing a single system of preferred responses to incentives and constraints, and assumes that the physical and social realms are subject to quasi-mathematical universal laws.

This study will instead be based on the image of "Becoming," which takes everything in nature and society to be in a permanent process of change and emergence, even that which appears to be stable and static. Accordingly, international relations must be understood as a human endeavor, based not only on mechanistic and structural "realities" such as gross national product (GNP) growth and military power but also on knowledge, information, ideas, ideologies, creativity, and will.

"The reason that human politics, economics, and warfare present such a mixed group of behaviors," writes Robert North, "is simple, but fundamental: people are not robots, countries are not merely big machines, and neither domestic nor international affairs are mechanistic. Politics, economics, warfare, and all other essentially social activities are moved by human beings who have minds, emotions, values, preferences, ambitions, and expectations and are thus partly rational, partly irrational—and always subject to change."[6] As we learn about what we consider reality, our perceptions and—most important—our images of the future change. It follows, then, that human beings and their actions will frequently surprise us.

But change is also conditioned by our environment. To quote North, "The growth, expansion, competitions and conflicts of society are all human undertakings which would not take place if people did not make them take place. But it is also true that changes in both the natural and social environment affect the way people think, feel, and act, just as the way people think, feel, and act affects the natural and social environment."[7] Therefore, social understanding requires both that human beings be placed at center stage in the drama of international political-economic change and that we "come to terms with the developing relationship between Human Ideas and a Natural World, neither of which is invariant."[8]

A clarification is warranted to close these epistemological remarks. The growth of knowledge in our discipline, the "state of the art," can be seen as gradual and continuous[9] or as revolutionary.[10] Thus, the structuralist social scientist, whether Marxist or non-Marxist, follows the mathematician in dealing with a "world that dies and is reborn at every instant."[11] Evolutionists, however, have a sense of duration, a sense that T_2 doesn't merely replace T_1, a sense of "real persistence of the past in the present, a duration which is, as it were, a hyphen, a connecting link."[12] Scientific revolutions do not take place when a group of "radical scientists" one day sits down in a smoke-filled room and decides to "overthrow" the old paradigm. A paradigm emerges gradually, usually very slowly, and evolves into maturity. New ideas overlap and eventually supersede the old, which in turn simply fade away.

Paradigms and theories viewed in this way "are not 'true' or 'false,' in any naive sense. Rather they take us farther (or less far) and are theoretically more or less fruitful."[13] Unless they are completely useless, they will leave a mark, and the evolution of knowledge will be affected, even if only in an insignificant way. In more practical terms, the philosophy of science from which my indeterministic theoretical ideas take their cue overcomes the naiveté of logical positivism and behaviorism by assuming that no fully demonstrable "truth" of a theory exists. Rather, a theory should be able to be falsified, should be plausible and based on common sense, should definitely advance the knowledge of the field, and the data used should be subject to scientific verification. Finally, when theories are based on metaphysical claims, as they usually are, these claims should be subject to rational criticism. It is because of such metaphysical bases as Being and Becoming that we cannot entirely disprove structural determinist theories any more than we can prove the truth of indeterminist ones.

There is an asymmetry, though, between structural determinist and process-based indeterminist theories. If structural determinists believe their theories to be true, they should be able to explain and anticipate political behavior and to demonstrate how in fact structures are immune to learning and creativity, to sudden changes of mind, and to the combination of factors that can lead to unintended consequences. In the words of Karl Popper: "He who proposes the stronger theory accepts the burden of proof: he must produce arguments in favor of his theory—mainly by exhibiting its explanatory power."[14]

The Problem

A basic epistemological premise of this study, borrowed from Popper, is that human beings in general, and political decisionmakers in particular, are problem solvers. Humans constantly anticipate the future in order to meet desired goals, attempting to close the gap between what they perceive as reality and what they want to be or achieve. Not only do they react to conditions, but they also continually change political and economic interests through their knowledge, imagination, and beliefs. From this perspective, decision making does not mean simply making choices; it is "essentially an integrative action, . . . it relates the behaviour at different instances of time to expectations; or in other words, it relates present behaviour to impending or future behaviour. And it directs *attention*, by selecting what are relevant objects, and what is to be ignored." [15]

The problem I have chosen to study has been commonly defined in terms of "journeys toward progress," [16] as Albert Hirschman calls the process leading to material progress, scientific and technological development, and modernization. Social science so far has treated the technological and industrial development of the less developed countries mainly from a microeconomic technological perspective [17] or from the point of view of scientific and technological dependency. [18] Few, though, have linked scientific and technological development to ideas and ideologies of progress, development, and modernization.

The countries under scrutiny are the two largest in Latin America, Argentina and Brazil, whose journeys toward progress in the last two decades have contrasted dramatically. Specifically, I will examine the choices these countries made between a strategy of foreign technological acquisition—mainly from multinational corporations—and one of selected technological autonomy. I will look also at some of their successes and failures in overcoming technological dependency. My analysis of the science and technology policies of Argentina and Brazil in the last twenty years and their development of computer and nuclear energy industries will raise what seems, at least at first, to be a paradox: technological development has occurred in those cases where structural indicators would have shown only a small potential for it.

In explaining this apparent paradox, I will illustrate how the decision to follow the road of selected technological autonomy and the successes achieved are related to the ideologies of the actors involved and to their perceptions of their country's ability to set its own economic and technological objectives. I will then study the role played by ideologically motivated intellectuals—scientists, technologists, and economists, whom I call the pragmatic antidependency guerrillas—and their institutions in establishing viable goals, raising public awareness, and influencing the policymakers.

Case studies showing the achievement of some measure of technological autonomy neither prove its desirability or inevitability nor offer a formula for achieving it. But they might shed some light on how the change occurred: where the ideas came from, how they evolved, how they penetrated the political realm, how and why it happened.

Cognitive change in Latin America, not among the sellers of technology, is my concern; I will look in depth at the national actor and then at the adaptive (or nonadaptive) behavior of the foreign technology suppliers. Ideas and ideologies of progress and development matter not only because their evolution means changing policies, institutions, and power at the domestic level but because they may lead to changes at the international level as well. Therefore, I will attempt to trace a causal chain starting from ideological and institutional innovations at the domestic level and ending in new and different relations between domestic buyers and foreign sellers of technology. The latter point will be specifically illustrated by the Brazilian computer development case.

The Case Studies

Argentina reached a high level of both scientific and institutional development by the late 1950s. At that time Argentine science was held in high esteem. The scientific and technological communities had some influence on what resources were to be devoted to research activities and considerable influence on where they were to be invested. But Argentina has failed, with a few—though relevant—exceptions, to turn scientific and technological development and its linkages to industrialization into a national goal. Brazil, in contrast, has done so, even though it started much later and at a considerable disadvantage.

Why has Brazil supported a consistent science and technology policy in the last fifteen years and increasingly applied its growing scientific and technological potential to industrial production? And why has Argentina not been able to do the same? After all, as of the mid-1960s, when our description of these two countries begins, Argentina was significantly more developed scientifically than was Brazil. Its per capita GNP was almost three times that of Brazil,[19] and its manufacturing share of the gross domestic product (GDP) was substantially higher.[20]

It could be argued that although the above is true, Brazil since the late 1960s has had much higher levels of GDP growth than Argentina, and while it more than quadrupled its manufacturing value added between 1960 and 1977, Argentina's only slightly more than doubled.[21] Furthermore, whereas after 1964 Brazil experienced relative political stability, Argentina suffered continuous economic and political turmoil.

These are unquestionably important and valid arguments, but they are insufficient. Argentina's manufacturing share of the GDP in 1978 was still substantially higher than Brazil's.[22] And even though the political-economic emphasis in Argentina after 1976 was on finance and agricultural exports, still Argentina did not cease to be a modernizing and semi-industrialized country.

Dependency arguments, which I will review in depth later, are not only insufficient but also inappropriate, because Brazil as a later industrializer became more dependent on multinational corporations in its industrialization process than did Argentina. Even if we were to agree that both countries were dependent, using mostly foreign technology, dependency theory could not explain the differences between these two countries, only their similarities.

Clearly something else must be involved, for why has Argentina had a consistent and successful nuclear policy since the 1950s and developed its nuclear power potential into a somewhat self-sufficient enterprise, while Brazil has failed to do so? Certainly budget allocations cannot explain this contrast, in view of the mammoth Brazilian investments in nuclear energy in the mid-1970s. And how do we account for the successes of the National Atomic Energy Commission (CNEA) in Argentina at a time of extreme domestic turmoil, when most other scientific and technological enterprises were being suffocated by negligence and inadequate action? And why, although the two countries both became interested in developing computer technology at a time when Argentina was more advanced than Brazil in sophisticated electronics technology, did Brazil end up with a growing domestic computer industry and Argentina with none?

Why has Brazil been able to maintain its science and technology policy course in spite of the economic and financial crises of the late 1970s and early 1980s, while Argentina could not develop a systematic policy even in the years when it was growing at relatively high rates and was involved in advanced processes of industrialization and import substitution? The answers to these questions are not unique and unidimensional, as we will see.

Summary of Findings

Argentina's leaders in the years under study disagreed about what progress is and what the means to achieve it should be. Whereas a few aimed at eventual technological autonomy through industrialization and scientific and technological development, others concentrated on commodity exports, treating science as a cultural enterprise and hoping for economic development without economic planning and modernization without application of science and technology to industrial production.

Brazil's nationalist ideology of development and industrialization together with its quest for international status became the engine for certain groups, among them many intellectuals and military officers, to put Brazil on the road to modernization. Brazilian planning was much more systematic than was Argentine planning, and although they were only indicative and were not fully implemented, Brazilian plans nevertheless did become part of the learning and policy-making processes. In Argentina, on the contrary, the plans were just documents, formal policy that affected hardly anyone or anything beyond their authors. Although consensus in Brazil about autonomous scientific and technological development was not total, it was nevertheless much stronger than in Argentina. And more important, active lobbying by groups of technocrats and intellectuals succeeded in convincing policymakers and then creating the bureaucratic apparatus and the financial devices to enable the idea to survive.

In Argentina the group that fought to bring about scientific and technological development was only temporarily in government. Ideological and political conflicts did not allow any one group enough time to bring about an irreversible change. While Argentina emphasized basic science, Brazil emphasized technology. While Argentina stressed research as a career and let the researcher follow his own scientific instincts, Brazil stressed engineering graduate studies and, increasingly, a link between research and the industrial productive sector. Similarly, information management and data banks were indispensable to achieving Brazil's goals, but in Argentina the few attempts to set up an information system between 1970 and 1974 were unsuccessful.

In Brazil the state guided development by means of indicative planning and a growing budget for science and technology. Planning created an awareness of and a basic set of expectations about the national role of science and technology. In Argentina, in contrast, the state changed its approach four times in fifteen years.

To sum up the marked difference between the science and technology policies of the two countries: Brazil opened doors and pursued choices between alternative policies along the interdependence/autonomy spectrum, whereas Argentina opened and closed them.

In the computer field, Brazil made some improbable choices. In fact, analysts considered that the Brazilian domestic computer industry arose at least in part by chance and wrote its obituary before it actually was born. Nevertheless, today it represents approximately one billion dollars in sales, thousands of jobs and substantial domestic benefits in terms of both scientific and technological development and national security. And it materialized in part from the imagination, ingenuity, and voluntarism of some political actors who succeeded in creating irreversible processes that led to political and economic adaptations and initiatives, in spite of the "rational" implausibilities of the industry. In Argentina, meanwhile, a group of scientists in a rapidly growing domestic private industry developed a micro- and a mini-computer. But the venture lacked the support necessary to become competitive and so did not prosper. When the managers of the enterprise were replaced, because of changing political and ideological conditions, the new management considered the operation inefficient and dismantled it.

Nuclear power policy in Argentina was centered in the CNEA, where de facto all the political, as well as scientific and technological, decisions relating to nuclear power were concentrated. These processes were influenced by an ideology that satisfied the military and conservative—as well as the populist nationalist—sectors and thus allowed the institution to sail across the murky waters of political and economic turmoil. But in Brazil something went wrong in the evolution of nuclear power. Despite the large investments in the 1970s, Brazil still has no nuclear plant working at full capacity, and the domestic technological and industrial capacity of the sector has not been developed, to the annoyance of both the military leaders, who want military capabilities, and the scientists, who seek the development of an autonomous technological capacity.

To summarize, domestic and international economic and political constraints and opportunities played a very important role in the processes to be described in this study. But these alone are not sufficient to explain scientific and technological policy, actions, and outcomes, which sometimes did not follow a structuralist-determinist logic. Something was mediating and catalytically intervening in the process. That something I have found to be related to institutions, and to ideologies of progress and development.

Ideology and Change:
Are Ideologies Real?

"It is one of the minor ironies of modern intellectual history," wrote Clifford Geertz, "that the term 'ideology' has itself become thoroughly ideologized."[23] The confusion and, I believe, misinterpretation of ideology owes a great deal to Karl Mannheim and his sociology of knowledge. Mannheim's principal thesis was that "there are modes of thought which cannot be adequately understood as long as their social origins are obscured."[24] Thus he distinguished between ideologies, as defensive intellectual justifications for the existing social order, and utopias, erected by social progressives to critique and attack the status quo.[25] The sociology of knowledge thus became a relativist enterprise of differentiating the social facts that lead to truth from the cynical ideas, or ideology, used to cover up for the status quo and its interests.

This naive interpretation of ideology as some obscure force opposed to social reality is typically a given in structural determinist Marxist and non-Marxist theories of international relations. From the structural determinist point of view, the "battlefield image of society," as Geertz called it, takes ideology as part of the "strategy" of dealing with the enemy in the universal struggle for advantage. Thus, a clash of principles is nothing but a disguise for a clash of interests.[26] For Marxists, ideology, as an object of the superstructure, usually means "false consciousness." Marx even turned mental processes into "ideology." For him, "the notion that ideas as such govern the history of mankind and direct its evolution appeared an aberration—a professorial construction."[27] Thus, ideology has come to represent everything at the "wrong" end of the spectrum, and all beliefs and preferences have been robbed of any legitimate explanatory role. I argue, rather, that ideas and ideologies, which are specific types of ideas, do matter, that they are real and causally relevant because they have real consequences. As they change, both attention to problems and sectors and the allocation of resources change as well, in turn causing policies to change and institutional developments to occur. However, ideologies are only one powerful factor influencing the policy of technological development and the management of relations with foreign technology producers. Our task will be to understand how this factor interacts with structural constraints and opportunities to produce particular outcomes.

Ideologies are powerful because they tell actors, including institutions and groups within institutions, what their goals are, how important these goals are compared to others, and how to seek them, as well as who their friends and enemies are and why. Ideologies are important for political-economic behavior because they "have origins that cannot be reduced to material developments, . . . [they] can have substantial and independent effects,"[28] and they can have the "obvious potential to develop into potent political forces. This happens when a set of political doctrines is adopted by a group of people, assumes a critical position in their belief systems, and then becomes a guiding force behind their actions."[29]

A cognitive explanation here should by no means be seen as the alternative to an institutional explanation, for the actors I describe succeeded or failed within and through institutions. Indeed, since institutions are carriers for certain ideologies, constellations of collective understanding or consciousness cannot help but become integrated into institutional designs, thereby becoming preconditions for institutional change, even if the institutions later cease to depend directly on the original constellations or if the constellations continue in new or different institutional designs.

An explanation based on ideas and ideologies may involve significant epistemological dangers. For example, the claim that ideas matter may be taken as obvious, or, if stated too strongly, it can become a truism.[30] Therefore, I must first explain the theory and logic behind my assertions. In the rest of this chapter I will explain why ideology should be considered one important causal social variable and how some international political-economic change can be traced to domestic ideological and institutional innovation. This discussion is philosophical and theoretical in nature and may be of lesser interest to those concerned primarily with the empirical material and the application of the most general aspects of the theory to case studies. Nevertheless, I believe that the theoretical discussion is essential to an understanding of the assumptions behind the catalytic, process-oriented, and indeterminist nature of international political-economic change. The remaining three theoretical chapters will describe and explain the choices, ideologies, and actors involved in the processes under scrutiny.

In the empirical portion of the study, parts II and III, I will examine whether ideology indeed was or was not applied in the subject cases.

Indeterminism

Yehuda Elkana has likened determinism to Greek tragedy, where fate is identified with the order of nature, events are seen as inevitable,[31] and human beings have only minor influence. Indeterminism is likened instead to Epic theater. "It [the event] can happen this way, but it can also happen quite a different way." The historical question "is not what were the sufficient and necessary conditions for an event that took place, but rather, what were the necessary conditions for the way things happened, although they could have happened otherwise."[32] In social life, certain events may well have happened differently or not at all. This can certainly be said about Brazil's surprising achievement in establishing its own computer industry and about Argentina's success with nuclear energy, both against great odds.

Popper dismisses determinism on three grounds: scientific knowledge, he argues, is not absolute but approximate; there is an asymmetry of past and future; and the effects of knowledge cannot be anticipated. With regard to the asymmetry of past and future, Popper draws a parallel between scientific determinism and a motion picture film: The "picture or still which is just being projected is *the present.* Those parts of the film which have already been shown constitute *the past.* And those which have not yet been shown constitute *the future.*"[33] This analogy indicates a belief by scientific determinists that future reality already exists somehow and that it can be unveiled by means of scientific methods. This belief, according to Popper, destroys a fundamental asymmetry in our experiences and is in striking conflict with common sense. "All our lives, all our activities are occupied by attempts to affect the future. . . . In contrast to the past which is closed, as it were, the future is still open to influence; it is not yet completely determined." Thus Popper's third argument for dismissing determinism: we cannot predict scientifically results that we will obtain in the course of the growth of our own knowledge.[34]

Popper's propensity theory, though insufficient and controversial,[35] most nearly satisfies the needs of indeterminism and provides us with a research program in which change is the object of study. The main thrust of the propensity theory is that change consists of the realization of some actual propensities; such realizations in turn consist of new, and different, potentialities.[36] I distinguish, as Popper did in the realm of science, between propensity and potential or probability in order to drive home the distinction between determinism and indeterminism. While the determinist sees the future as already determined in a structural reality (to which actors react), the lack of perfect knowledge compels him to make predictions about future events in terms of a range of probabilities. The indeterminist, however, sees the future as actively created and discovered, and this can happen only at the process level.

The premises of indeterminism and propensity support the view that what happens at the level of process is of fundamental importance. Furthermore, these premises help to explain that this study is not merely trying to substitute one determinism for another. Ideological resources are taken not as determinants of change but as real propensities for change, whether or not it occurs. Realization of change is not necessary but depends, among other things, on the existence or absence of other resources—political, institutional, or economic—and on their interaction at the process level. It is this interaction and human intervention in these processes that make outcomes so difficult to predict, even when using statistical probabilities.

Body and Mind: Popper's Worlds 1, 2, and 3

A theory that holds ideologies to be real and that takes them as causal variables requires that we raise the "body–mind" question. Is ideology just a mental factor, and if so, why should we consider it as real? Can ideology or other cognitive factors such as scientific knowledge have consequences that were not *thought* or perceived by the individuals who devised and sustained them?

The body–mind dilemma has not been solved by the greatest philosophers, and I make no claim to have solved it. Rather, I intend to show that social scientists cannot keep apart the "body" and "mind" that Descartes so artificially separated centuries ago and that integrating them does not require us to reduce social phenomena to behavioral psychology or psychoanalysis. I study the beliefs and perceptions of political actors not to assess their mental health but because they have real, sometimes intended and other times unintended, consequences and are thus an inseparable part of the social phenomena we explore. The human mind has the capacity not only to adapt itself to a structural situation but also to change it by affecting the future through choices, policies, and institutions that result from awareness and reflection. "The self-reflective human mind," as Erich Jantsch wrote, "is capable of conceiving many visions of reality and even of inventing the tools for the corresponding transformations of reality."[37]

About twenty years ago, to tackle the body–mind dilemma, Popper introduced a theory that divided the universe into what he called World 1, World 2, and World 3. Popper's theory is useful for my analytical account at this point because World 3 helps explain how ideologies, in interaction with institutions, create propensities for change.

World 1 is the world of physical objects, forces, organisms, brains, rocks, and trees. World 2 is the psychological world—the subjective realm of our minds. "It is the world of feelings, of fear and hope, of dispositions to act, and of all kinds of subjective experiences, including subconscious and unconscious experiences."[38] Beliefs and perceptions are thus an intrinsic part of World 2. World 3 encompasses the products of the human mind, such as language, books, theories, policies, and plans, and is to a remarkable extent autonomous.[39]

Books are one example of World 3 objects. The thoughts contained in a book transcend the physical reality (World 1) of the paper on which they are printed, and because a book can convey information and formulate and communicate a theory, it also transcends the subjective (World 2) mind of the author. Thus a World 3 book can have an autonomous effect on the physical environment and its inhabitants (which are World 1 objects) well beyond the physical death of the individual who conceived it.

What is Popper's philosophical basis for accepting the real existence of World 3? He argues after Alfred Landé that something is real if it can be kicked and can, in principle, kick back.[40] According to Popper, World 3 kicks back. Its objects are abstract, but real, because they have real intended and unintended consequences. When World 2—subjective—beliefs or interpretations of reality are communicated and affect or effect some action, they can acquire a World 3 status and kick back—that is, they can have real consequences. This frequently happens within institutional frameworks.

Social institutions can therefore be seen as World 3 objects. Although institutions belong to the physical world (they are composed of people and can be found in the telephone book) as much as to the subjective worlds of individuals who devise and constitute them, their essence lies also in the formulated and communicated outcomes of thought, such as institutional ideologies, roles, and functions. Once these are created, they may have a life of their own. They may not be visible, but they are intelligible and may have important real consequences. For example, as institutional ideologies become the collective consciousness of a group, they help select the goals and means for social action and may even help produce the means. In this case, institutions are but "carriers" for a particular collective understanding that has consequences of its own.

Social theories may play an important role in producing collective understandings within a group, community, or institution and in this way shape the social reality they come to explain. While theories may not be reality itself,[41] as long as they are believed to be true they may become instruments for social action that identify the problems and their solutions.

Therefore, social theories, when used as models *for* something rather than *of* something, may become reasons for behavior—ideological mind products in the World 3 sense whose consequences may not be foreseen at the time of their conception. Such a transformation happened with dependency theory, especially with what I call pragmatic antidependency. The theory was a "descriptive model of" truth, but it also became a strategy, or "model for," the shaping of reality and thus had real social and material consequences. As for the scientists, engineers, and economists who became involved in the policy-making processes, they provided their institutions with not only the know-how and "know-what" but also the "know-where-to"; in other words, they provided the beliefs, expectations, and goals that showed the way.

Ideology as Collective Understanding

To end this section, I should explain how I have arrived at a World 3 definition of ideology. But first we must go back to the World 2 classification. I define ideology at the individual, and thus subjective, level as a set of beliefs and expectations about politics, economics, and society. Beliefs and expectations are based on the perception of reality, not on reality itself. Because people have different ways of understanding, evaluating, decoding, and giving meaning to their surroundings, their reactions, experience, and knowledge will be characteristically different.[42] From the individual viewpoint, then, the actor attaches the stigma "real" to social situations that are both *perceived* and *interpreted,* and these situations are then real in their consequences for what people do.[43]

The above subjective definition of ideology is consistent with Peter Berger's "reality constructionist" approach. Reality, Berger believes, is socially constructed, meaning that "real" experiences—for example, economic dependency—occur in socially allocated situations and are perceived and interpreted in terms of socially derived, validated, and differentiated meanings.[44] Thus, certain beliefs of what reality is may with the passage of time be accepted as objective reality.[45]

When defining ideology in its World 3 meaning, we have to deal with the concept of collective consciousness. My use of this concept should be distinguished from Émile Durkheim's, which described how a structure, or "whole," constrains the individual and argued that individual mentality is a reflection of collective modes of thought.[46] I argue the opposite: that actors socially and institutionally coordinate their beliefs for some purpose.

I take collective consciousness to be collective understanding. More formally, I use Stephen Toulmin's fourth meaning of consciousness, or *con-scientia* (etymologically: joint [con] knowing [sci]). Beyond "sensibility, attentiveness, and articulateness of individuals [the latter meaning the translation of attentiveness into an explicit motive or intention], we should have recognized that the concerted plans of multiple agents manifest yet another, fourth aspect of consciousness. Agents who act as partners in a shared project, carried out jointly, with the intention of collaborating, and with each having full knowledge of the other's role in the project, are engaged in a 'conscious' collaboration. They act as they do 'consciously'—i.e., in the light of their mutual understandings."[47]

From this perspective, ideology transcends the subjective world by becoming part of a collective set of group beliefs. Applying this fourth aspect of consciousness to ideological, political, and economic change, joint understanding can occur only between individuals who are in direct interaction and are aware of each other's roles. Therefore, again following Toulmin, we can say that political-ideological change occurs not in the opinions of individuals but in the collectivity of beliefs, traditions, and concepts that are responsible for the concentrated action of an institution.[48]

I thus arrive at my World 3 definition of ideology. Political ideology is not a justification, or even an explanation, of political behavior. Rather, it is the collective understanding of individuals who, being conscious of each other's roles, beliefs, expectations, and purposes, offer strategies for action or solutions to problems that can be used to change reality. Ideology thus ceases to be a mental phenomenon and becomes a collective product of the mind, a real blueprint for action that anybody can use, that tells people something should be done, and that can have real consequences.

The Idea and Ideology of Progress

Progress as Ideology

The idea of progress is ideological. A particular set of beliefs leads some people to believe in technology as a precondition for improvement of human conditions; another set of beliefs leads others to ask: improvement for whom? what kind of improvement?

The ideology of material progress has caught the attention of both the elites and the masses for over two centuries in Europe and the United States, and for much of this century in the rest of the world. Its carriers, mainly bureaucratic, technocratic, and intellectual elites, actively nourish it with scientific and technological knowledge. Political leaders take the ideology for granted and work hard to cultivate it, pulled by its magnetic force. The masses also take the ideology for granted. They may not be aware of the ideological content of progress, but they do know what it means to them personally, and they put pressure on political elites to further the ideology. In these ways the ideology of progress affects action and becomes a political legitimizing criterion.

The material and physical world interacts with the ideology as follows: objective physical factors, such as energy, pollution, scientific discovery, raw materials, and military strength, establish the parameters of the possible; but at the same time, within these parameters the kind, scope, and pace of progress are a matter of choice. Choices, in turn, will tend to affect the physical world and subsequent choices.

While the developed countries are beginning to experience the costs and risks of the "journeys toward progress," developing countries are still actively pursuing material progress. Progress from their perspective means only going forward, and anyone in the rich countries who suggests that such movement should be slowed down or redefined is accused of wanting to prevent the developing countries from achieving a measure of equality. Thus progress is viewed by the developing countries not only as modernization and economic and technological development but as a matter of autonomy and equality as well. This is why their nationalist ideology is so strongly linked to development and equality. Liberation, cultural self-affirmation, development, science and technology: these are the core dimensions of the idea of progress in the Third World.[49]

Progress as Modernization

Many Third World countries have tried to follow the journeys toward progress through modernization. These nations' leaders have perceived modernization as something the European countries and later the United States and Japan achieved through rationalization. The aim, according to Edward Shils, has been

> to "mobilize" resources, including the labor of their subjects or citizens, to bring them into a national economy which would produce for national and international markets, to increase productivity by rational technology in industry and agriculture, to arrange the educational system so that it would be rationally articulated with the economic system, inculcating the knowledge and skills necessary for the rationalized economy. The promotion of the sense of nationality was intended to foster attitudes of law-abidingness, respect for national authority, and other dispositions necessary for participation in the grand effort of the rationalization of society. Science was to be developed because it was central in the conception of rationalization; it was to be directed toward "use." In a few instances, the idiom of "planned science" was introduced. Regardless of whether the political elite was democratically elected or a self-selected elite of a regime of a single party, its instrument was to be primarily a rationalized and rationalizing bureaucracy which would saturate the society, control, plan, regulate, and initiate in a rational way the activities and organizational arrangements requisite for a modern society.[50]

What does it take for a society, or, sharpening our focus, for certain elites and groups of individuals within the society, to become identified with this rationalist conception? Those closely involved with technological development know that although technology belongs to the realm of the material, perceiving a task for it is an act of imagination. Thus it takes imagination and knowledge to use or to reproduce a technological artifact—a tool or machine—especially if it is based on science.[51] In the same way, modernization based on science and technology is not only an economic and social act but also an act at the level of ideas, scientific theories, and social institutions. However, the identification with rationalist theories and philosophies and the understanding of scientific and technological development and of its connection with economic development are not immanent in Third World societies. Awareness of these factors has rather been created through the diffusion of ideas, models, theories, and ideologies from Europe and the United States to the Third World, and in the past forty years through the experience of Third World countries with transnationalism—that is, the multinational corporations.

Two fundamental questions therefore present themselves: First, who becomes the carrier of these ideas and ideologies, the force behind the change? Not private industry, because in most cases it is too weak—given its level of development and the poor support it gets—to consider taking on an "implicit policy" of indigenous science and technology development. If they are interested in profits, foreign technologies are preferred. The alternative, then, lies at the public level.

Second, how does political change take place, how do political elites become aware of these things, how do they come to accept and support a science and technology policy, either implicitly or explicitly? A preliminary answer to these questions can be found in remarks made by Leo Hamon, a French professor and politician:

Innovation and reception are furthermore encouraged by the action decided on by energetic groups, even when these are minority ones and the speed of progress then depends on the organization of these groups and also on the resistance they encounter. After a certain time, it is true, science and technology acquire what is called their independence and what could be more aptly named their autonomy. . . . It is, however, only when a specific way of thinking has been reached in intellectual evolution that the way is clear for progress. . . . In any event it is not possible for other societies to attain [and for governments to become aware of] a similar technological boom without undergoing the change in mental outlook—or at least its essential features—which has in fact permitted the success of their forebearers.[52]

Catalytic Interdependence

The catalytic interdependence concept is not meant to describe a new phenomenon or a new kind of interdependence. It does not have the status of law, nor does it represent a fixed or unchanging condition. I am also not reifying my personal preferences. Instead, it portrays cybernetic processes that take place in international political economy and that reflect the nature of the present international political-economic interactions: characterized by sovereign states and nonstate actors such as multinational corporations, acting without a supranational authority or an enforceable set of norms equally applied to all actors, and interacting in what appears to be an increasingly interdependent and complex situation.

I call these processes catalytic because their distinctive origin lies in creative ideological and institutional acts that have the propensity of affecting international interactions and bargaining positions. The aim is to show processes that continually and historically change without predetermined direction, teleology, or ultimate reason. Instead, the direction of change emerges from the purposive and adaptive behavior of actors—which can be limited but not totally determined by political and economic environments—and the unintended consequences of their interaction.

International interaction itself, then, follows a historical pattern of changes and adaptations to changes rather than an ahistorical homeostatic pattern of equilibrium or stability maintenance. From a catalytic interdependence perspective, what needs to be described and explained are not the functional requirements of a system or the structural constraints leading to systemic stability or instability but the changes and adaptations themselves.[53] A new idea, scientific knowledge, and a preference may be catalytic for bringing about economic adaptations and for affecting other actors' adaptive moves. Thus, it is in the process and by the process that political and economic "structures" are changed.

What makes this approach different from others, structuralist and nonstructuralist, is my reliance on propensities rather than on determinants for change and my insistence on the relevance of ideological and institutional creativity in the interaction process.[54] But purposive actions aimed specifically at changing something are not the only creative actions; adaptations can be creative as well. Adaptations should be seen, therefore, not only as responses to constraints, or as functional requirements of some sort, but also as *creative* propensities for more change.

Creative adaptations transcend the structuralist view of human beings as calculators. As Popper put it, "We are not merely calculators. . . . But we are constructors of calculators. We make them because we are interested in problems whose solutions are beyond our limited calculating powers."[55] Hence, from a catalytic interdependence perspective actors do not merely react to structures like machines, rather, they act purposefully according to what they believe to be rational and in their interest. A move that may seem very irrational to a multinational corporation or to a developed country's government, then, may be rational from the perspective of Argentina or Brazil.

A change provoked by actor A and affecting actor B has the propensity of eliciting a response by B; but the response and its nature are not predetermined. B may decide not to react, may also change and adapt, or may refuse to adapt and try to reverse the change. The actors' induced changes are thus dynamic catalysts in a situation of interdependence. Structural factors are very important, for they set the parameters of the possible, but while they may generate the need for adaptation, they may not be able alone to determine the type and nature of the adaptive move. The cognitive factors define the problem—that is, they define and interpret reality—and inject into the processes the actors' beliefs, meanings, imagination, preferences, and expectations.

From this point of view, bargaining in the international political-economic context is nothing more than the process of each of the actors adapting to some changes, which engenders propensities for further changes. Its outcome, therefore, cannot be accurately predicted. What one group of people may perceive as resources for bargaining, others may overlook; what one group of people may consider worth fighting for, others may not. A well-timed, creative, adaptive move that because of its improbability catches the other actors by surprise can change the power balance of the bargaining situation significantly.

Once even partial success is achieved, lessons will be drawn from it. The actors involved in the bargaining process, as well as interested observers, may learn from the outcome or from its unexpected results. The partial success of a weak actor may lead to imitation by other weak actors and to more interaction changes.

Complexity is another reason why bargaining outcomes cannot be predicted. As policymakers focus on a particular constellation of facts that they call reality and as groups develop a collective image of that reality, other facts are being disregarded. This nonvoluntary selection-among-variables process is, of course, relative to how complex an international situation is perceived to be. The more actors involved in a political-economic international situation and the more differentiated and interdependent they are,[56] the more complexity they will perceive. As complexity increases, choices may become more difficult as more events and alternatives are left out and predictability of choices and subsequent outcomes decreases.

*

I conclude this section with an imaginary conversation between myself and two experts on multinational corporations whose "statements" are taken from some of their writings. This conversation illustrates the nature and type of some catalytic changes that have taken place between nation-states and multinational corporations.

Adler (A): My study of change in international political economy started, in part, as a reaction to the determinist arguments and theories that assume that multinational corporations are either absolutely good or absolutely bad for developing countries. I wanted to show that the relations between these countries and the transnationals are basically open, indeterministic, and continually changing; also that the changes originating within the nation-state have to do with both economic and political conditions (constraints) and their interaction with ideas, ideologies, institutions, and choices.

At the ideological level my intention is to show that the coming of the multinational corporations to Latin America triggered or enhanced ideological goals of reduced dependency. The "discovery" of science and technology came about with the realization that knowledge was increasingly important for economic development and that the multinational corporations almost entirely controlled that knowledge. The developing countries challenged the multinationals and then took measures to acquire decision power over their own science and technology and industrialization processes.

Vernon (V): It "appears that the numerous threats to the multinationals that were launched in the 1970s—the spate of nationalizations, the codes of conduct, the U.S. legislation against bribery, the demands and resolutions of the General Assembly—were fueled by . . . a pervasive revulsion in much of the world against the effects of industrialization, against the symbols of entrenched authority, and against the impersonal tyranny of big bureaucracies." [57]

A: The threat that Latin Americans felt about not having control over multinationals in general, and science and technology in particular, was not related to the fact that the multinationals were corporations or that they were identified with industrialization, but to the fact that they were *foreign*. Once dependency began to be perceived as coercive, an attempt to reduce or eliminate it followed. It is only at the level of nationalist images that we can understand the threat posed by multinational corporations.

V: Anti-multinational feelings "of those eager to develop a strong national identity free of outside influence, those repelled by the costs of industrialization, those at war with capitalism as a system, and those distrustful of the politics of the rich industrialized states, especially the United States . . . [have] been strongest among those to whom these issues are most important—the leaders of various national groups. . . . At the upper levels, this shared feeling has created a common cause in which government and business leaders in the developing countries can join with the philosophers and revolutionaries." [58]

A: Precisely. This feeling is what made the work of the pragmatic antidependency guerrillas possible and made room for a science and technology policy of autonomy. We cannot understand what has been achieved in countries like Brazil without focusing on the strong nationalist autonomy-bound ideological elements within intellectual and political elites.

Now, I know that you consider the drive toward autonomy as

an old phenomenon. But isn't there anything new in this phenomenon as related to the present and to the Third World?

V: "What is new . . . is the drastic change in the degree of exposure
of the developing countries both to the opportunities and to the
threats of the outside world. A country determined to shape its
own future no longer has much opportunity for choosing a splen-
did isolation, except at a cost most countries would reject. All the
forces of contact have grown more powerful than they were a few
decades ago: examples are the power of multinational enterprises
to thrust their way into a developing country on the basis of a
unique technology or mastery of scale or capacity for research;
and the power of some developing countries to slough off a multinational enterprise by reaching out to an alternative source of
capital or technology or customers."[59]

A: True, the degree of interdependence has increased, but its nature
is also changing. These changes arise, in part, because of the
persistent search by elites in developing countries for autonomy
and choice, which leads to development of domestic industry and
to discovery of alternative sources of technology. In Brazil it led
to the discovery that scientific and technological knowledge is a
key element of autonomy, an engine toward modernization and
equality at the international level. In Argentina it led to the CNEA's
success. A question that follows from the dynamic characteristic
of these processes is, What happens when countries such as Brazil
cross the science and technology threshold and are able to generate industrial growth and industrial exports? Do they abandon
the nationalist and egalitarian drive toward autonomy?

V: "Some countries that currently place themselves in the developing
category—Brazil, Mexico, . . . India and a half dozen others—are
changing rapidly. They may find themselves supporting in their
national interest policies not unlike those of the industrialized nations. That shift will begin as local entrepreneurs in such countries increase their capacity for assembling capital and technology
and for building up effective access to foreign markets."[60]

A: I think this is already occurring in Brazil. But in the science and
technology field there is such a long way to go and so much to
catch up on that the goal of autonomy is not going to be dropped.
It will gradually change according to the circumstances.

Frank (F): "As developing countries experience rapid economic growth, they gradually become aware of their increased power in bargaining with multinationals. Indigenous firms may master the existing standard technology and thereby undermine the basis for the incentives and priorities accorded to foreign firms. . . . [However,] high-technology manufacturing firms serving the local market are likely to retain their bargaining power longer than firms using standard technology because the contribution of high-technology companies depends on a continuing inflow of resources from the parent in the form of new technology."[61]

A: The awareness you speak of can arise from expectations and so precede rather than follow rapid economic growth, as was the case in Brazil. Furthermore, whereas the view that access to high-technology sectors is almost closed to industrializing countries is widely accepted,[62] I hope to be able to show that even in such high-technology sectors as computers there are no deterministic conditions. The Brazilian computer industry was developed not because it had to happen but because there were individuals and institutions with the determination to make it happen. Thus, many of the models and theories that portray multinationals as evils or as benefactors, the obsolescing bargain excluded,[63] are insufficient and often misleading, because in most cases they disregard both the pattern of change and adaptation to change that takes place between nation-states and multinationals and the ideological commitment of actors in developing countries to increase the rate of change.

V: Yes, the "added evidence [in *Storm over the Multinationals*] . . . went some way to confirm the fact that simpleminded propositions about the effects of multinational enterprises were as a rule highly vulnerable."[64]

F: It "is apparent . . . that an important evolution has taken place in the attitude of transnationals toward the growth process in the developing countries and toward their relationship to that process. To a much greater extent than in the fifties and sixties, the corporations recognize both the diversity of circumstances in the Third World and their own need for flexibility in their approach to individual countries. . . . With this new perspective on the development process has come a greater willingness on the part of many multinational corporations to accept some of the constraints imposed on their activities and mode of operation in the countries of the Third World. Moreover, they have discovered that the consequences for the firm have rarely been catastrophic and in some instances have indeed been beneficial." [65]

V: This seems to be true; "U.S.–based enterprises as a class have grown somewhat less reluctant to enter into joint-ventures with foreign partners than had been the case in earlier decades. Multinational enterprises from all countries have proved increasingly flexible in taking on management contracts, acceding to so-called fade-out clauses, entering into partnerships with state-owned enterprises, and involving themselves in other ambiguous arrangements." [66]

F: Furthermore, just "as transnationals have been increasingly accommodating to the changing relations in the Third World, a significant evolution has been taking place in the attitudes of governments of developing countries. Longer contact and experience with multinationals have given host-country governments a better understanding of how the corporations operate and an appreciation that the relationship need not be of the zero-sum variety, but can be one of mutual gain. In many developing countries, stronger economies and better-trained individuals have led to greater competence and confidence in dealing with the multinationals and a growing tendency to take a pragmatic approach to the relationship." [67]

A: You are right about multinational corporations becoming more flexible and willing to enter into joint ventures with local companies, as was the case with multinational participation in the Brazilian computer sector. I am also glad you recognize that the attitudes of developing countries have evolved. The Brazilian government now has a better understanding of the multinationals and of mutual benefits in the computer sector.

Not all individuals and organizations in developing countries have come to believe in a pragmatic course, but you are right: with more knowledge comes more confidence in the use of capabilities. Third World countries can understand the multinationals better, fear them less, and come to a close working relationship. But in order for all this to happen both sides must continue to adapt to the changing conditions.

Political Science and Historical Processes: Toward Integration

In the course of historical processes people and ideas die, as do institutions, policies, and industries. But new people are born, and aspirations, traditions, ideologies, and plans for action are constantly being introduced into the process and selectively maintained and passed from one generation to the next. Given enough time, all political institutions will be transformed into something else.

The changes, processes, outcomes, and situations that arise from the selection of a perceived reality and from creative adaptation at the national and international level are time-bound and concern not only individuals but also social groups—collectivities of individuals sharing ideologies, knowledge, and plans for action. In other words, they are historical and social at the same time.

Innovations, for example, are not only signs of adaptation but also historical criteria by which to compare the actions of national actors in the context of changing political-economic conditions.[68] It is therefore imperative to introduce to international political economy the study of "how institutional . . . variants make their appearance in the first place; how they show their merits as solutions to outstanding social problems; and how they succeed—given favorable conditions—in spreading and establishing themselves more widely."[69]

When focusing on change, one can see each political institution within a nation as developing somewhat autonomously and having its own history. Institutions do not work necessarily and solely for some common "good" or against some common "ill," but rather as their histories and their standard operating procedures dictate. Their interaction is neither "functional" nor "dysfunctional." Even when we consider these institutions as belonging to some hierarchical notion of state, each would have come into existence at a different point in history to meet some particular set of needs or demands and would be subject to variation at different times for different motives.

To understand change and the sources of creative behavior in political economy, we must therefore disaggregate the concept of state. A traditional argument of "state intervention" or some kind of "statist model" to compare and explain the differences between Argentina and Brazil in the areas of science and technology, computer development, and nuclear energy will not do. Both countries have had strong interventionist states but took very different journeys toward progress. Only when we look at the processes themselves—the creative impact of ideologies and institutions, the causal interaction between them and with physical and human resources, and most of all the human choices that emerged from a varied range of interpretations of reality—can we arrive at a satisfactory answer for the differences highlighted above.

Institutional innovations, creative adaptations, catalytic interdependence—in sum, an approach that integrates systematic scientific with historical analyses—must aim at understanding, and explaining, the dynamic interactions between changing material conditions and the evolution of ideas, ideologies, and institutions.

Domestic and International Choices 2

Science, Technology, and Modernization

For most of the post–World War II period, when Latin American countries were engaged in policies leading to industrialization— mainly of the import substitution type—science and technology were not seen as real problems. Acquisition of Western knowledge proceeded without concern about the source of the technology, how the transfer was made, who made it and toward what sectors, and so on. The fruits of the resulting economic development would, it was assumed, trickle down and eventually be enjoyed by all the population. The literature of the 1950s and 1960s reflected this belief, encouraging rational economic decisions about technological purchases and applications as well as tariff agreements and integration efforts among Latin American countries to bring about increased wealth.[1] Interdependence was perceived as beneficial, and its costs were either ignored or downgraded.

Studies during the last twenty years, however, have increasingly criticized Latin American development programs. Domestic factors emphasized in the literature of the 1950s and early 1960s were now minimized, and dependency on the industrialized world became an all-encompassing theory explaining Latin American development ills.[2] Some writers singled out scientific and technological dependency as the cause of unemployment and the lack of wealth redistribution and basic human well-being—in short, of the failure to achieve "integral development."[3] The inappropriateness of Western technology, its high price, the foreign values it implies, the alienation that accompanies it, and the underdevelopment of a scientific and technological infrastructure are just some of the problems caused by dependent development.

The critics have therefore concluded that only a strategy of self-reliance can lead to indigenous scientific and technological development—and this means autonomy: to be in control of whatever technology is introduced and to depend as much as possible on domestically developed and culturally appropriate technology. Third World organizations such as the United Nations Conference for Trade and Development (UNCTAD) have adopted this view and, through the New International Economic Order (NIEO)—a package of demands by the Third World—they are advancing these goals in the international arena.

Some of the authors and policymakers who adhere to the dependency theory are less prone to dogmatically single out structural problems. Instead, they think better results can be achieved by altering the processes at work. Although they see dependency as a real problem, they also believe that scientific and technological "self-determination" can be achieved and dependency reduced through technological planning, appropriate controls of foreign investment and technology transfer, and constructive development of the indigenous research infrastructure. Because this would not require dispensing with the benefits of modern Western technology, the positive aspects of interdependence would remain.

At the heart of these matters lie fundamental questions about the evolution of scientific and technological knowledge, the process of modernization and the ideologies it triggered, the social mechanisms for the allocation of values, and the nature of interdependence.

The scientific, and later the industrial, revolution grew hand in hand with the idea and then the reality of the nation-state. It would not be difficult for a student of modernization and political economy to realize that these two historical developments are in fact related.

But so are day and night, and neither is the cause of the other. Did the slowly growing awareness that the application of scientific knowledge to the organization, utilization, and manufacturing of materials condition and make possible the rise and growth of the nation-state? Or did the nationalist ideals, and later the reality of the nation-state, bolster scientific discovery and technological development for application to production?

The causal linkage, I suspect, has been a two-way, mutual-influence process, the one reinforcing the other. To see these related developments clearly we have to turn to the period between 1650 and 1700, when the scientific revolution began to take shape. As Geoffrey Barraclough put it, the then-greatest scientific discovery was science itself.[4] The scientific method that resulted is based on factual observation as a means of corroborating abstractions about the surrounding world—physical phenomena were linked with theories and philosophies.

> Philosophers began to worry about machines, systematic astronomical observations helped navigation, and the renewed respect for manual labor . . . reached its culmination with the works of great artists such as da Vinci. The contributions of Copernicus and Galileo on the celestial order led to the triumph of reason over dogma, and constituted a milestone in the transition from religion to science as a way of explaining natural phenomena. Finally, the contribution of Newton, who introduced the idea that the universe was predictable and obeyed certain laws that could be known and tested, radically changed man's conception of the world, giving sense to the Baconian statement that man can master and control nature through understanding.[5]

The impact of this realization cannot be underrated: the history of Europe, and later, through spillover effects, of the whole world, was to be different. This was definitively a turning point in the evolution of world history.[6]

Human awareness that nature can be controlled, together with development of the scientific method, led to the industrial revolution. Materials were meshed and know-how and energy (mechanical and human) applied to the invention of tools and machines. These developments led to a new awareness: on the one hand, individuals realized that the domination of nature and its laws made possible the domination of other societies by force. The linkage of these processes with the idea of nationalism and the nation-state is relevant but will not be studied here. On the other hand, it also became clear that the human capacity to dominate nature meant that nature itself could be exploited to improve the human condition, welfare, and well-being. This is the line to be pursued in this study: the idea of production and growth of productivity.

When people began to realize that there was a link between the organization of knowledge (science and technology) and their welfare, the modernization process was set in motion. Today there is hardly any nation-state or culture that is immune to the modernization process of putting science and technology to work for production, for growth and progress, or immune to the insight that through the utilization of knowledge for production human welfare can be improved.[7] But this process would not have occurred the way it did without the parallel development of the nation-state and nationalism.

Nationalism as we know it today is related to the questions of who gets the fruits of progress and modernization, and at what cost, and whether the route to progress forces one nation-state to stand against others in the anarchical international society, or whether progress can be achieved faster and more securely if many states collaborate. The questions are ideological, and as such they influence policy and action.

Three ideological branches sprang from the awareness of the human capacity to control nature and the onset of the modernization process: *liberalism, Marxism,* and *nationalism.* A fourth, evolved from these three, I call *egalitarian nationalism.* The first two are internationalist in nature; the latter two are based on the belief that the route to progress is a struggle among nation-states. The evolution of these ideologies has led actors to encourage or to curtail interaction, to trade or to protect domestic industry, to call for international cooperation or for collective self-reliance. As the three parent ideologies have been widely defined, described, and explained,[8] I will describe them only briefly.

Progress Without the State: Liberalism

Liberalism was the answer to mercantilism—the dominant political-economic ideology of the seventeenth and eighteenth centuries that induced nations to augment their power by accumulating precious metals. At the heart of the liberal critique was the argument that mercantilism was the road not to progress but to wars and domination.

Liberalism proposed instead a system based on the freedom of the individual. As Adam Smith argued, "The system of individual liberty in a free market would enable the society as a whole to achieve a higher level of national benefit than could be achieved under the system of mercantilism."[9] In further contrast to mercantilism, liberalism was based on the separation of political and economic matters, and the conviction that their affiliation led to power-seeking activities, war, and lack of progress. An awareness of the connection between science and technology and welfare led liberals to the conclusion that the international society was not zero-sum: wealth could be increased and everybody could have more of it, but only through the creation of free markets, not only within societies but also among them. This idea about progress is fundamentally different from the one that results from stocking precious metals.

More practically, liberals argued that patterns of domestic consumption and production differ and that in the absence of trade, domestic relative prices would differ from country to country. The liberal solution, then, was for countries to exchange commodities. No country's condition would be made worse by trade, and all would benefit if goods were exchanged at some intermediate price ratio.[10]

Progress Without Classes and States: Marxism

Marxism, like liberalism, sprang from the spark of the modernization process and reacted with disdain to the static mercantilist idea of gold accumulation. But Marxism criticized liberalism even more strongly, and, because it developed later, it reflected the evolution of ideas throughout another century. In contrast to liberalism, the problem for Marxism was not one of fixed wealth versus growing wealth but one of the uneven distribution of wealth, which in turn was seen as the cause of domestic and international conflicts.[11] Marxism's unit of analysis was the social class, as opposed to liberalism's individual. When perceiving social phenomena as stemming from class inequalities, the natural conclusion was class conflict. At the heart of Marxism was the idea that these conflicts would lead through a dialectic process to a renewal of the society, as happened with the evolution from feudalism to capitalism. The next evolution, then, would be the triumph of the working class, which would eliminate all classes and lead to progress without classes. For Marxism, progress meant doing away with inequality.

Industrialization was the key to the Marxist idea, for it would allow progressive improvement of bad conditions, liberate workers physically and psychologically, and lead eventually to a classless society. Because of class conflict the state would fulfill a role until a classless society was established, but after that there would be no need for a state.

Progress Without Interdependence: Nationalism

A nation, wrote Haas, is "a socially mobilized body of individuals, believing themselves to be united by some set of characteristics that differentiate them (in their own minds) from 'outsiders,' and striving to create or maintain their own state. 'Nationalism' is the body of beliefs held by these people as legitimating their search for uniqueness and autonomy; nationalism is the myth of the successful nation." [12]

Modern nationalism is not only a development of but also an alternative to liberalism and Marxism. At the core of the modern nationalist idea is a perception of threat and vulnerability within an international society, which prevents the nation from achieving power and wealth. Moreover, industrialization is seen as creating new stakes in the struggle for power and wealth: as technology confers new means of domination and production, it influences the staging and the intensity of the conflict. Nationalism can thus be defined as the mechanism by which those who perceive themselves as a nation deal with the consequences of industrialization and internationalization that are so damaging to the nation-state. Nationalists, then, like the mercantilists before them, see the reinforcement of the connections between the economic and political realms as the key to the development of a strong state.

Progress Without Interdependence and Inequality: Egalitarian Nationalism

A variant of the nationalist idea is a mixture of the nationalist rejection of internationalism, both liberal and Marxist, with the Marxist beliefs about equality. The result is a kind of nationalism based not just on the search for uniqueness and autonomy but also on the belief that a nation is inevitably imperfect and unequal, to a great extent because of international factors. Interdependence thus not only constitutes a roadblock to progress for the nation as a whole but also is the cause of inequalities both among and within nations.

Egalitarian nationalism takes uniqueness and autonomy as allowing modernization, which in turn is nurtured by an equitable distribution of its fruits. The linkage between modernization and the nation-state is axiomatic: nationhood derives from the common development toward progress (modernization), where progress necessarily involves the values of equality.

Science and Technology for Development: Domestic Choices

The ideas of development and progress are linked. Development is a process: it is the evolution toward progress and our way of reaching it. Thus development may include an increase in power and economic growth; it may include the ability to choose and change goals and the capacity for adaptation through organization.[13] But it also includes the price we are willing to pay for progress.

From this point of view, it is ideological economic and political choices—beliefs and values—that determine the kind and amount of productivity and the pace of development toward progress and that lead to a modernization process. The questions then become, *why* develop a scientific and technological capacity, and *how*?

Figure 1 provides a taxonomy to order and illustrate the linkages that determine the modernization process, some of whose elements I have already referred to, and some of which I will deal with next.[14]

The ideology of progress creates an image of the future that inspires decisionmakers to start the modernization process at different points. It is progress itself (not its image) that brings about more education and the development of a scientific and technological infrastructure, and these then lead to the adaptation and production of technology, to productivity, and thus to more progress.

Most "late late industrializers"[15] entered the process of modernization at point D. Because they lacked the necessary infrastructure and productivity was low, they bought foreign technology in order to increase productivity, which was supposed to lead to progress, which in turn would lead to more education, to a scientific and technological infrastructure, and to more technology. But with it they also bought the disadvantages of foreign technology and the uneasiness and sometimes damaging effects that excessive dependency on foreign technology may bring. The liberal idea of interdependence favors this choice, but in many cases there was no alternative for developing countries.

Nationalists and egalitarian nationalists offer the alternative of getting into the process at points C and D while technology is under the control of the state. Thus they call for strengthening education and building a national scientific and technological infrastructure that will lead to development of domestic technologies, to increased production and productivity, to progress as desired by the "nation-state," and so on.

Marxists, although few have shown how this could be done in practice, advocate getting in at point E—that is, by changing the ownership of the means of production and therefore the way values are allocated to the society. From this point of view, a social revolution, or at least a broad structural change of the society and its economy, would "guarantee" a ticket on the journey toward progress and thus to the "right" kind of education, the "appropriate" scientific and technological infrastructure, and the socially relevant and culturally suitable technology.

Figure 1.　The Modernization Process

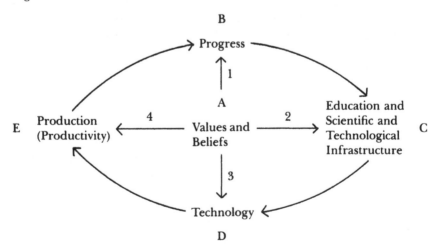

Several choices are involved at each of the above links in the modernization process chain. The first involves the kind of progress desired; all the others depend on this one. The second concerns the kind of education and scientific/technological infrastructure the nation-state would have. For some, scientific inquiry is something to be left to the free choice of the scientists.[16] For others, scientific inquiry has to meet the developmental goals of the nation-state, because science is considered to have a social function and to be connected to social structures and processes. The third choice is about technology and is multiple: one alternative is between foreign and domestic technology, another between "modern" and "appropriate" technology, which usually but not always coincides with the question of capital-intensive versus labor-intensive technology. The question here is, Who will benefit from the technology, and will it fit the culture and the values of the society? The other alternative is between technology and no technology at all. The rapid pace of industrialization and its environmental effects, such as pollution, nuclear accidents, and the depletion of natural resources, have led different schools to argue for the halting of technological development, an idea that envisions the suspension of growth and development, or even a regression to lower levels of production.[17]

The fourth and last choice involves production: who decides on what production to help whom and why? There are three basic alternatives: leaving it to the market to decide automatically the questions and allocations of society's values; letting the state decide in the interest of society through planning; or, the most common today, a mixture of market and planning.

Indeed, this last question transcends production, for all other elements in the taxonomy pose the same question and alternatives: should the problems be solved by individuals, who "know" what is best for themselves, or by the state, which "knows" what is best for the nation? It is the basic question of the free market versus planning.

The Market

The market is based on the principles of barter and exchange and developed naturally from people's tendency to exchange things they have for things they do not have.[18] A market economy is a self-regulating system that seeks to maintain order in the production and the distribution of goods based on the expectation that human beings seek to maximize money gains. Liberalism is the ideology that calls for a market economy: freedom, the individual as the focus of analysis, and the market are all elements of one essential idea about modernization.

Planning

Planning is not based on just one political philosophy, perhaps because in contrast to the market it is a less ingenuous and more common mechanism for dealing with complex matters and with the allocation of values in the society. Marxism, nationalism, and egalitarian nationalism all advocate planning but differ on their reasons for preferring it and on the comprehensiveness of planning.

For Marxists, the state is responsible for allocating values to the society, but because it must be of proletarian direction, it creates plans to facilitate the apportionment of gains and losses equally throughout the society. For nationalists, competition with other states for power and wealth requires a clear set of priorities in the goals and means, and only the state is believed to be able to perform this—through planning. Egalitarian nationalists advocate planning for reasons that combine Marxist and nationalist motives.

Nationalists, like mercantilists before the modernization era, accept the market as a domestic mechanism and in many cases even encourage it. However, they resent the tendency toward free barter among states and the linkages between domestic and international markets. In contrast, egalitarian nationalists, because they are also moved by the value of equality, oppose not only a free international market but also a free domestic market, which is believed to be a source of inequalities in the society. Moreover, the linkages between the two markets are seen as a source of all national and international inequalities.

The common denominators of nationalism, egalitarian nationalism, and Marxism are the beliefs that a free market left unregulated will hurt some important values (a nation, equality, a class), that only a body representing all strata in the society can know best, and that regulation and planning are much older and more dependable than the market.

Although both pluralist and centrally managed (or communist) societies utilize planning, the former are involved in indicative planning

and in "strategic planning"—which, as defined by Charles E. Lindblom, means incremental and piecemeal—whereas the latter are involved in some degree of synoptic planning.[19] Lindblom treated with
skill and insight the differences in incremental and central planning.
But he left out the developing countries, some of which are neither
pluralist nor communist. To fill in this omission I will introduce a second taxonomy, which deals both with what I believe are more suitable
models of state intervention in different kinds of societies and with
their objectives.

For the sake of simplicity I will use two kinds of societies; the first I
call *bargaining societies,* which operate on a consensus built through
threats and promises.[20] A bargaining society thus includes not only
democratic "polyarchies"[21] but also other types of semidemocratic
and not-so-democratic societies, such as corporatist societies with a
bureaucratic and authoritarian state, where bargaining takes place
and where the state, although prominent, responds to the pressures
of different groups. I call the second *compellence societies,* which include not only the centrally managed societies of the Soviet bloc but
also any other society, to the right or left of the political-ideological
spectrum, where a state—military or not—forces its values on the society and where almost no bargaining takes place.

The objectives of a state's intervention are *stabilization, equality, development, security,* and *"postindustrial corrections."* A state can plan to
achieve more than one objective, but occasionally some of these objectives are incompatible, so that the maximization of one may require
abandonment or disregard of another. These five objectives are briefly
described as follows:

> *Stabilization:* economic goals that have to do with disequilibrium in
> an economy, such as inflation, negative balances of payments, and
> so forth, where the aim is not to have more of something or better
> distribution, but to avoid further deterioration in the trouble
> areas.

> *Equality:* intervention aimed at redistributing income equally to all
> members of a society, which previous or present economic and/or
> social mechanisms failed to do. Two basic ways are considered:
> (1) set up a system that would distribute the fruits of growth
> equally; or (2) raise the standard of living of the poor.

> *Development:* economic growth.

Security: self-preservation of the nation-state, both physically within its borders and culturally in terms of its values.

"Postindustrial corrections": the attempt to alleviate and correct problems caused by industrialization and indiscriminate growth. Addressed are ecological problems such as pollution, the search for new and renewable resources, the planning of cities and transportation, guidelines for the use of the sea and space, the prevention of nuclear proliferation and nuclear accidents, and so on.[22]

By juxtaposing the type of society with the planning objectives, we get a taxonomy of planning activities, illustrated in Table 1.

This exercise reveals that what differentiates nations with incremental and central planning is not only the type of political system, but also whether or not there is room for bargaining. This differentiation facilitates the inclusion of the developing countries into the planning taxonomy. In a bargaining society, whether democratic/pluralist or authoritarian, we do find the trend toward control, regulation, and planning, but bargaining imposes a limit on what the state can decide by itself. When various actors in a noncompellence society hold different ideas and ideologies about, for example, equality or postindustrial corrections, the highest decisionmakers of the state have to compromise and only partially decide for the society.

In a compellence society the state has more powerful means to plan than in a bargaining society. For example, it can stabilize an economy or spend disproportionate amounts of money on security by compelling the masses to earn less and by stifling dissent. The price of achieving a goal thus becomes coercion and repression.

Not only are the differences between bargaining and compellence societies in this taxonomy evident, but the number of activities held in common also demonstrates how far some bargaining societies have gone into regulation and planning.

Market and Planning in Science and Technology

There is no such thing as a free market in science and technology. At the national level governments regulate, plan, centralize, or engage in many of the activities concerned with the development of science and its application to production. This includes government- and

Table 1. Planning Activities in Bargaining and Compellence Societies

	Objectives				
	Stabilization	Equality	Development	Security	Postindustrial Corrections
Bargaining societies	Budget Amount of money Fiscal reforms Employment Foreign exchange Forecast Indicative planning	Employment Budget Welfare programs Education Indicative planning	Fiscal and monetary measures Investment Public enterprises Forecast Indicative planning R & D	Budget Forecast Planning R & D Manipulation through the media	Budget Institutions Indicative planning R & D The media Bargaining with industry and labor
Compellence societies	As for bargaining societies, plus: Policies of negative redistribution Economic repression Central planning	Same, plus: Nationalization Confiscation Land reform Central planning	Same, plus: Central planning	Same, plus: Policies of negative redistribution Economic repression	Same, but: No bargaining No need of media Plus: Central planning

public sector–sponsored research and development, government regulation of innovation, government identification of sectors to be developed, and, of course, scientific and technological planning. Some countries, mainly Northern ones like the United States and Germany, have reached a high level of technological development without plans, but they have nevertheless regulated and sometimes planned some activities relating to specific sectors, for instance, energy. Other countries have formal science and technology plans or are involved in economic development planning with some relevant parts dedicated to science and technology.

Nor does a free market operate in the transfer of technology from developed to developing countries. Here, multinationals oligopolistically dominate markets, imposing restrictions on the uses of the technology and its commercialization in the present and the future. This domination has led developing countries to punch more holes in the free market with their attempts to regulate foreign investment and technology transfer.

Interdependence and Self-Reliance: International Choices

The consistent development of knowledge and its applications, the need to increase capabilities for production and domination, have generated a system of relationships based on trade, exchange, communication, movement, and diffusion. The journeys toward progress are thus not only linked, they meet at intersections; and some argue that they have been waiting too long for a green light. The problems and the solutions have transcended the realm of the nation-state: they have become international. Interdependence means many of the benefits the liberal philosophy has promised, but it also means problems and conflicts.

The conflicts have created winners and losers. It took a surprisingly long time for developing countries to realize who the losers were. The 1950s and 1960s were filled with the excitement of new countries joining the "club" and the expectation that all of them were, to a greater or lesser degree, on the road to modernization. But disenchantment led to despair, and the 1973 energy crisis engineered by OPEC heated the blood. Nationalist, egalitarian-nationalist, and Marxist intellectuals of the developing countries had seldom favored interdependence, but not until the 1970s did its negative aspects at both the national and international levels become so clear to them. For some, the journeys toward progress meant regulation and management of these problems and conflicts: "If you can't fight 'em, join 'em." For others it meant autonomy and self-reliance: "If you can't fight 'em, and joining them is no good, then dump 'em."

Interdependence

Definitions of interdependence in the literature vary according to their focus. Robert Keohane and Joseph Nye stress asymmetry and dependence as important analytical variables, and so the simplest definition of interdependence becomes "mutual dependence," which can be symmetrical or asymmetrical.[23] In most cases interdependencies are perceived as asymmetrical, for no two countries in the world have the same amount and kind of capabilities. Others, who focus on outcomes, define interdependence as "the outcome of specified actions of two or more parties (in our case, of governments) when the outcomes of these actions are mutually contingent."[24] Haas, Keohane, and Nye also define interdependence in terms of sensitivity and vulnerability, two elements that relate it to power and make it more intelligible. "'Sensitivity interdependence' connotes actor responsiveness to market forces in shifting their resources from one country or region to another; a highly interdependent actor perceives the need to shift readily. 'Vulnerability interdependence' means actor dependence on the ability of another actor to change the basic rules of the game which apply to the particular group of events; a highly interdependent actor is one who feels completely at the mercy of others."[25]

Following Haas I would like to make the point that interdependence is not something we reach out and grab; rather, it is perceived.[26] This characteristic is what makes it so important to decisionmakers at any level and a paramount factor in the perceptions of complexity, uncertainty, and change both within nations and in international relations. Some, such as Waltz, have tried to demonstrate through the use of trade figures that in fact there is less interdependence now than there was before World War I. He argues that as long as a country has paramount capabilities, its vulnerability will be small, and the more a system tends toward bipolarity, the less vulnerability the major powers will perceive.[27] Waltz's conclusion about diminished interdependence is arguable,[28] but in any case he misses the point, because interdependence does not pertain only to the realm of trade—it pertains to any single issue generated by scientific and technological development and the process of modernization that makes the actors themselves perceive that they are interdependent.

From an approach concerned with change, interdependence means the reciprocal effects of change on the international actors. Thus the type, amount, scope, nature, and direction of change that will occur both within and between nation-states and other actors will differ according to whether these actors are interdependent or autonomous.

The multinational corporations, as one of the most important by-products of interdependence,[29] are linked to the developing countries' journeys toward progress. The involvement of multinationals in the transfer of technology—their goals, methods, bargaining, learning, and, most important, the evolution of the interaction with the technology buyers—will be substantiated in subsequent chapters. This section will be limited to some general remarks about multinationals and interdependence.

Simon Kuznets believes that the foundation of a transnational society lies in the transnational nature of knowledge: "No matter where these technological innovations emerge . . . the economic growth of any given nation depends upon their adoption. In that sense, whatever the national affiliation of resources used, any single nation's economic growth has its base somewhere outside its boundaries—with the single exception of the pioneering nation, and no nation remains the pioneer for long."[30] Whether it was the steam engine in the past or the computer in the present, technological knowledge has always given rise to images of growth and modernization, and the multinationals became the means of its diffusion across nations. They grew because of a great demand in the developing countries for knowledge as a commodity—not only its purchase but its production as well.

Science and technology, modernization and interdependence, therefore, are intrinsically linked to the evolution of the multinationals. "Transnational processes and international interdependence refer, in effect, to overlapping sets of phenomena. . . . By and large . . . as transnational processes increase in number and in scope for a specified set of states, the level of interdependence among them similarly increases."[31] The multinationals in the last quarter of this century have become important in the manufacturing sectors of many developing countries, not to mention the main and almost only source of technology, which through "packaging" techniques also includes the related activities of marketing, distribution, servicing, and financing. These techniques, plus the high prices of technologies and services, the import of foreign values, the demonstration effect problem,[32] and the underdevelopment of a national research and development capacity, have led developing countries to become suspicious and try to steer their "growth machine" in the direction of their own interests instead of those of the multinationals. Thus the relationship between the multinational corporations and the developing countries has become one of love/hate, although the ratio of love to hate varies by country and regime. Some examples will illustrate this fluctuating, schizophrenic relationship.

The multinational corporation is a source of growth in developing countries, but at the same time it is partially responsible for the inequitable distribution of the fruits of growth. A source of much-needed investment capital, the transnational also causes loss of capital through the high prices demanded for imported technology and a loss of control on the part of the host country over its own industrial sectors. It is the source of technology, but in many cases this technology does not fit the values of the host country and may lead to the underdevelopment of a scientific and technological infrastructure. Contracts also usually include restrictions on commercialization and other uses. The multinational may produce jobs, but highly capital-intensive technologies may also cause unemployment. It may bring to the developing countries modernization, rational thinking, and managerial capabilities, but with them may come alienation and pollution.

Self-Reliance

Self-reliance is a reaction of the leaders and intellectuals of some developing countries against what they consider to be the inequality, frustration, and sense of degradation that an interdependent economic and political system has imposed on them. I believe the following words of Mahbub ul-Haq, director of the World Bank's Policy Planning and Program Review Department, express these sentiments and show us that, as with the other phenomena described in this chapter, self-reliance is a perception—a cognition of a past and a present, and a set of expectations about the future.

History has taught us at least one lesson: that the poor and the weak always get exploited unless they get organized. Our present world is no exception. We have often seen our poverty and our weakness exploited in the name of grand-sounding principles.

When we protested against deteriorating terms of trade for our countries, we were quietly informed that this was merely the workings of a free international market mechanism, and we were left to puzzle for ourselves why, by some curious coincidence, market mechanisms always worked to the detriment of the poor nations.

When we embarked on our economic development we were taught to worship the goddess of GNP growth. . . . We were left to discover for ourselves through our own sad experience that the character of growth and distribution of its benefits was as important as growth itself. . . .

At the beginning of a programme of international assistance, we were assured that there would be a true partnership in development and a genuine transfer of real resources. We found out later that there can be no true partnership between unequals. . . .

It is rather an unpleasant truth that poor countries in the Third World have often been swindled out of decent return from their produce in the name of market mechanism, deprived of their economic independence in the name of world interdependence, seduced by imported life styles, foreign value systems, irrelevant research designs—all in these name of freedom of choice. And we have also seen the same grand-sounding principles change with an amazing suddenness when the shoe was on the other foot.

No exploitation can continue for long except with the tacit co-operation of the exploited. . . .

After all, what have we done to change the realm of ideas, to fashion more relevant strategies for our societies, to equip our political negotiations in international conferences with powerful ideas and concrete briefs which could become a rallying cry for the Third World?[33]

It seems that something was done, in some cases in the realm of action, but fundamentally—as ul-Haq remarked—in the realm of ideas, ideas that have to do with self-reliance. Thus, argues Carlos F. Díaz-Alejandro, the frequently lamented hypersensitivity of Southern countries about transnational and direct foreign investment—the symbols of dependence—has placed "the transnationals and direct investment firmly on the agenda of economists, political scientists, and governmental leaders. This is no mean achievement."[34] Self-reliance is eroding some assumptions that before seemed inevitable, such as the necessity of trade. It is seen as the solution to dependency, following the ideology of egalitarian nationalism: allowing for some nuances, self-reliance calls for autonomy of decision or self-determination (nationalism) on the one hand, and the redistribution of wealth (equality) on the other. Concerning equality, many analysts think mainly in terms of the strategy of basic human needs and the provision of minimum services and well-being to the world population that lacks them.[35] In Latin America, although the question of basic human needs is stressed, there is also a concern about direct redistribution of wealth.

We can learn a great deal about self-reliance and its definition and scope from the Pugwash Symposium, which met in Dar es Salaam in June 1975 to deal with "the role of self-reliance in alternative strategies for development." According to the symposium, self-reliance can be dealt with (1) as a concept, (2) as a strategy in national development, and (3) as a possible focus of assistance and multinational cooperation in furthering the self-reliant development of developing countries.

1. As a concept, self-reliance "is to be understood in the light of the harsh realities which prompt such exhortation and may be viewed as an indication of strong motivation on the part of the developing countries to use the concept of 'liberation' to provide at least a part of the emotional force needed to bring about the political, social and economic transformation of their society. It reflects the attitude of wanting to start a 'war' against poverty and oppression."[36]

2. In operational terms, self-reliance "is to be understood at the national level of each developing country as the will to build up the capacity for autonomous decision-making and implementation on all aspects of the development process including science and technology. The character, content, direction and pace of social and economic change—whether in rural areas or in urban areas, whether in industries or in education—has to be defined and executed with reference to national needs and aspirations."[37]

3. "This approach to [self-reliance] is reflected internationally as opposition to all forms of dependence. It calls for changing the mode of incorporation of the developing countries into the international political, economic, and cultural system. . . . This may call, in certain cases, for developing countries to detach themselves

temporarily from the present world system and then recouple with that system on a new basis at a later date."[38] The symposium emphasized that delinking does not mean the end of interdependence and that self-reliance "should not imply autarky. . . . [It] does not imply a lessening of interest in international cooperation but a desire to make the relations, between industrialized and developing countries, reflect genuine interdependence and complete international economic justice."[39]

I must stress several points, however. First, many Third World leaders do not necessarily think along these lines. Although most of them are very nationalistic, they are concerned about gains in their term of office or their lifetime, and so they seek short-term rather than long-term results. A self-reliance strategy involves a change that will bear fruit far in the future. Interdependence, trade, and technology transfer promise short-term solutions, so politicians favor them. Intellectuals do not adhere to these "political rules," and many of them favor self-reliance. But, at times, intellectuals are able to influence politicians to favor selective self-reliance strategies. Second, self-reliance does not necessarily mean stopping growth. On the contrary, it means growth—once pride has been regained and some inequalities are redressed.[40] Third, according to Sagasti, a self-reliant or autonomous development strategy is one that leads to an endogenous scientific and technological base, that is, the utilization of scientific knowledge for production, in contradistinction to the exogenous base that Sagasti sees as typical of Latin America and the rest of the Third World.[41] Sagasti also believes that developing countries should aspire to become not scientifically but only technologically self-reliant.

> The conduct of scientific activities is necessary to maintain an autonomous decision capacity in the majority of technological fields, particularly in those developing at a rapid pace. Without a base of active scientists and professionals it is almost impossible to follow the evolution of technology and to have a clear perception of available alternatives and options. . . . However, it is clear that the concept of self-reliance does not apply to the conduct of scientific inquiry as such. Science, considered as a process for generating knowledge, is an international activity, and, in this sense no country can be self-reliant in science. The reasonable approach is to consider the development of scientific capabilities that provide a basis for technological self-reliance.[42]

What actions would be involved in technological self-reliant development? According to the Pugwash Symposium, such actions would include selecting through an autonomous decision-making process those elements of scientific investigation and technological know-how that could be domestically generated and used; identifying and acquiring, on the best possible terms (financial, institutional, and technical), those elements for foreign supply; blending the two components in such a way that continuous increases in productivity take place; and harmonizing technologies of different "vintages" and capital/resource intensities to make the best possible use of all available domestic resources, including human resources.[43]

This last point is significant, for it underlines the fact that there would not likely be one single approach to scientific and technological development; the variety that will inevitably flow from the diversity of heritages and predicaments of developing countries is itself an aspect of self-reliance, and this variety would also mean walking on many legs—the rural and the urban, the small-scale and the large-scale, labor-intensive and capital-intensive, technologically sophisticated and unsophisticated.[44]

Choice: What Are the Implications of Science and Technology Interdependence?

- Continuation of the flow of scientific knowledge and technology across national borders without barriers.

- Attempts to regulate the transfers so as to compensate the developing countries for the inequalities and other problems of the past that resulted from the unregulated interdependence processes and/or to make it possible for them to learn more about technology and to have more access to information; thus, both their scientific and technological infrastructure and their bargaining position with technology suppliers would be improved.

- Rapid transmission of advanced technologies developed in rich countries to Third World countries; contacts at the educational level, fellowships, instruction abroad, and participation in international forums where knowledge is produced, evaluated, and transmitted.

- The possibility of using knowledge cooperatively to solve domestic and international problems.

- Economic conflicts between the developing countries and the multinational corporations, including protection of markets, obstruction, retaliations, nationalizations.

- The possibility of some technologies needed by developing countries for strategic reasons being withheld in the event of political conflicts, especially if interdependence is strikingly asymmetrical.

- Difficulties experienced by elites in developing countries in allocating social values through the state because their freedom of movement in dealing with problems of development, establishing priorities, pursuing goals through local means, and so forth, is constrained by the mechanisms of technology transfer and the "intrusion" of the multinationals into the domestic economy and political life.

- Loss of maneuverability, creating difficulties for a developing country committed to distributing the fruits of production more equitably among all the population.

Choice: What Are the Implications of Science and Technology Self-Reliance?

- The capacity for autonomous decision making in matters of technology.

- The capacity to generate independently the technical knowledge required for a particular product or process.

- The autonomous potential for domestic production of the goods and services considered essential in the development strategy.

- The possibility of building appropriate technologies scaled to fit the developing countries.

- The possible loss of capital needed for industrialization if the multinationals were to be directly antagonized with the issue of technology and retaliated by drying up capital flows and other non-technology-related investments.

- A decrease in the amount and quality of the technologies available as the developing countries fail to keep pace with the rapid technological change in the rich countries.

- Possible redundancy and loss of efficiency.

- Possible reduction in academic interaction, exchange of students, opportunity to study abroad, etc.

- Possible slowdown of development, industrialization, and exports and a concomitant increase in the amount of manufactured imports, balance of payments problems, and inflation.

As with everything else concerning science and technology, there is no free lunch in the choice between scientific and technological interdependence and self-reliance. The processes of modernization—all the stages of the circular taxonomy developed in this chapter—would take various directions toward progress, depending on the choices made between interdependence and self-reliance. Self-reliance is popular and appeals to nationalist actors who perceive the inequalities both within and among nations, because it means more effective control of the continuous process of change in which they are involved. It is a way to deal not only with inequality but also with complexity. Interdependent scientific and technological development means that even if some of the inequality and nationalist feelings could be settled, the development of new knowledge in one country (which implies change) will induce change in other countries as well. It means, thus, a more complex world. It follows that the choice between interdependence and self-reliance implies a choice between two courses of action and interaction that would lead every nation in particular, and the community of nations in general, in a different direction. For some, the "reinvention of the wheel" for the sake of "calming overreacting nationalist feelings" and the "haste to solve in one blow the problems of inequality" seem ill-advised and wasteful. For others, without autonomy and equality there seems to be nothing to lose.

Three Strategies
for Managing Science
and Technology 3

Goals, Means, Information, and the State

Goals

That the free market should not determine the implantation of new technologies is generally agreed. But what should? "Planning" implies the setting of priorities; but what intellectual magic tells the planner what has precedence? Alternatives can be arranged on a scale of sociopolitical complexity, from the simplest to the most demanding. The principles involved in this scaling imply other goals, at a higher level of policy, where decisionmakers supposedly weigh the alternatives for economic development in general, and scientific and technological development in particular, in terms of growth and economic stabilization or in terms of equality, autonomy, and self-determination. Thus, the principles used in planning are the means of achieving these higher goals; but they are also goals in their own right.

The simplest goal is the creation of rules prohibiting practices that are considered abusive. Foreign investors should abide by the law of the land, observe tax and foreign exchange rules, obey labor and land regulations, and not resort to bribery. These expectations amount to little more than the subjection of foreign firms to rules of conduct customarily observed in industrialized countries but flouted in many underdeveloped states unable or unwilling to enforce their own legislation.

A more ambitious goal calls for the subordination of foreign and domestic investors to the overriding goal of national economic development. Because often there is no national development plan with appropriate guidelines for investment conduct, entrepreneurs are not forced to invest in given sectors or abstain from others. But they are expected to make use of tax incentives, tariff legislation, and other measures designed to attract new industry. Somehow it is expected that incomes and demand will rise as a result, but the amount and kind of change is not worked out in advance.

A still more ambitious goal does involve a national plan and is directed explicitly at the foreign investor. The government has a sense of priority, which may involve stressing capital-goods growth at the expense of consumption, or the reverse, and it thus specifies a growth strategy by individual sectors. The sense of priority may be populist or elitist, but it is clearly articulated, and the foreign investor is expected to act in accordance with the rewards worked out by the state. In this way the national plan governs innovative activity.

Some Latin Americans, however, go beyond this objective in specifying their future goals: they think in terms of technological self-determination. Channeling of investments and innovations into economic development—simple purchase or transfer of plant and intermediate capital equipment—is not enough. Latin Americans may wish to participate not only in choosing but also in designing and adapting the technology for the future they desire to create. This goal implies a judgment about the *kind* of technology to be fostered, not just its quantity; and it implies a critical political objective: the ability to control decisions concerning technology transfers.

Finally, a fifth, even more ambitious, goal can underlie the regulation of technology transfers: the creation of a national or regional cultural capacity to innovate. At stake is not just the fruit of innovation—locally made products and a higher standard of living—but also the knowledge that one is no longer dependent on foreigners for these values. If a nation has its "own technology," it can determine its own future. This goal amounts to a cultural renaissance, with all that implies for building new institutions and habits.

All goals are systematically related. Planning the implantation of new technology according to a principle of global income redistribution requires national planning priorities that take advantage of the global rules. Opting for national development also requires awareness of the existing extranational constraints and demands. Both require that investors adhere to the incentive structure of national rules and guidelines and that they abstain from common abuses of their power. But the reverse is not the case. The goals of eliminating abuses and making the investors adhere to minimal rules can be attained without national or global planning. Technological self-determination thus implies a command of all the techniques and types of information that go with the lesser goals. Cultural autonomy, in turn, presupposes the ability to make all the crucial decisions without having to defer to foreign ideas.

Technology Transfer: Means

The attainment of these goals requires that specific actors be assigned roles. Who is to invest? Who is to train or employ engineers and scientists? Who is to provide backward and forward linkages? The actors include local entrepreneurs, foreign firms, agencies of national government, and international organizations.

Modes of action must also be specified. For example, producing under foreign license or renting patent rights is less complicated than designing a new plant, port, or satellite communications system, and borrowing from private banks or the World Bank and using technical assistance from abroad less demanding than using one's own funds and personnel.

Seven modes of technology can be distinguished.[1] Three stress the transfer of knowledge, three deal with capital, and one is explicitly designed to combine knowledge and capital.

1. The simplest mode of technology transfer is by way of technical assistance. Such assistance may involve a regular program of assigning foreign experts to local enterprises and sending local ex-

perts abroad on fellowships. It can be provided by private firms as part of licensing, direct investment, or participation in joint ventures. Technical assistance is often financed by the U.N. Development Program, the World Bank, and by bilateral and regional aid agencies, either as separate programs or in the context of investments. Finally, private foundations extend technical aid in the implantation of new technologies, especially in agriculture and population control.

2. Knowledge-generating capital equipment purchases constitute a more complex mode of technology transfer. When a developing country imports automobiles or sewing machines, it acquires the products of technology but not the knowledge required for production. This is not quite the case with respect to a large range of capital goods imported whole. The purchase and installation of new telecommunications systems, a port, or a shipyard usually builds technological knowledge even if the equipment is designed, manufactured, and installed entirely by foreign firms. Knowledge is diffused simply because maintenance and operation of the equipment is in local hands, although no immediate capacity to produce is acquired.

3. When local entrepreneurs acquire rights to a foreign patent, they gain the technical knowledge needed to manufacture the patented product, but they use their own capital. Similarly, when a licensing agreement is concluded between a local entrepreneur and a foreign firm, the indigenous enterprise acquires the necessary knowledge and has to pay for this right in a number of ways. Financing for the patent and for licensing operations can be provided by the entrepreneur's own government, a foreign bank, or an international organization.

Transferring technology through the introduction of new capital is primarily the role of multinational corporations. Three types of direct investment can be distinguished.

4. The foreign firm may simply assemble its product locally using imported parts. This practice may create employment and spur demand for new products, but very little technological knowledge is transferred in the process. The Latin American automobile and mining equipment industries originated in this way.

5. Direct investment in mixed operations is more common now, with some of the parts manufactured locally by a subsidiary of the foreign firm and some imported. The transfer of engineering and manufacturing processes thus begins to tie in with local educational and research and development (R & D) activities, although the impetus is still provided by the multinationals' global experience and plans.

6. The multinationals may opt for complete local manufacture of the product, in which case the backward linkage with indigenous institutions and patterns is much more extensive. However, because the capital for doing this remains predominantly foreign and the planning underlying the activity remains rooted in the global strategy of the firm, the operation does not enter the nexus of national development planning.

Latin Americans complain that even direct investments featuring mixed or totally local manufacturing processes—irrespective of the extent to which minority domestic capital participates in the operation—do not trigger "real" technology transfer and the full range of desired backward and forward linkages with the economy and society. They blame the "technological packages" in which multinationals wrap their operations, overreliance on plants and products designed abroad for a global market, inflexible production routines, multinational control over intermediate processes, and imported marketing techniques and possible restrictions on sales to third countries and on patents and trademarks. In other words, although capital is being transferred, knowledge is not.

7. The last mode of transfer is labeled "joint venture," which is an agreement between a local entity—public or private—and a foreign investor to install new productive capacity. Formulas for doing this differ: sometimes the foreign firm must "fade out" after a certain period; there may be licensing of foreign processes; or the capital may be entirely indigenous, but the foreign firm is given a management contract to run the enterprise at first. Voting formulas for assuring management control vary, but all give the local entity a predominant role. Specific means for the transfer of knowledge—of research, manufacturing processes, and marketing—are usually included in such agreements, as are understandings about the market in which the product is to be sold. Joint venture

need not involve foreign multinationals only—organization of Latin American multinationals under the same formula of shared power is now being attempted.

Organizing Knowledge

Modes for domestic scientific and technological innovation and for the organization and storage of knowledge—which are distinct from the modes of technology transfer—must also be specified. These means I call generically the *organization of scientific and technological knowledge*.

In a simpler world of free markets and unplanned scientific inquiry, the very notion of "organizing knowledge" would be incongruous. The fact that most Latin Americans reject the free market, however, highlights their commitment to some form of planning and information management and provides another illustration of their impatience with the older modes of technology transfer. Organizing knowledge refers to all institutions and routines designed to improve domestic innovation, to bring the producers and the consumers of scientific and technological knowledge into systematic contact with one another, and to store information and channel it to the productive system. To the extent that it succeeds, the organization supersedes the "free market" in ideas; innovation is not triggered by random contact among foreign scientists, engineers, and domestic investors but by one or several organizations. As in the case of our goals, the more complex ways of organizing knowledge subsume and require the simpler ones.

The simplest way of organizing knowledge is to channel talent into predetermined directions by training experts. The organization of higher education is therefore crucial: a decision to stress one branch of engineering or medicine over another in university curriculum design may foreclose one group of experts and create a new one. The common complaint in Latin America is that no such direction has been provided and that too much of a free market for young talent prevails—a situation linked to the extraordinary brain drain of university graduates. One way of organizing knowledge, therefore, is to establish policies regarding the branches of higher learning to be encouraged and discouraged.

Another way is to set up and maintain institutions capable of conducting R & D activities in those areas of applied science considered important to economic development. The present dependence on R & D conducted elsewhere and for different purposes, both technological and economic, could thus be mitigated.

Once knowledge has been produced by the university faculties and research institutes, it must be made available to the productive system. Hence, the organization of knowledge has taken the form of retrieval systems and data banks, which exist at the national, regional, and global levels. The purpose of such systems is to provide the users of information with a wide range of choices, thus expanding technology diffusion beyond established contacts between foreign suppliers and domestic users.[2] However, while debate about the characteristics of the "optimal" retrieval system is lively, no final options have been exercised.

Once established, these repositories of information can be used for systematic forecasting. In the past in Latin America, market estimations for new products have been erratic and have sometimes resulted in misinvestments. Forecasting as practiced by large Western firms is a possible corrective. Moreover, commercial forecasting can be improved by including the social and economic benefits and costs associated with various possible patterns of technological innovation.

Finally, new experts, data banks, and models can be used for planning entire economies, or at least significant sectors of economies. The most ambitious way of organizing knowledge is to put these talents and techniques at the service of the state—when and if the state opts for the systematic planning of the nation's future.[3]

Role of the State

There is a general consensus in Latin America that the selection of means of technology transfer and the organization of the appropriate knowledge should involve the state as a core actor. This consensus is symbolized by the wide acceptance of the so-called Sabato Triangle as an almost mystic metaphor informing contemporary discussion. The triangle, conceived by Jorge Sabato, is composed of the Scientific and Technological Infrastructure, the Productive Structure, and Government. Local development of technology requires a developed scientific and technological infrastructure that includes highly skilled scientists, basic and applied research in institutes and universities, and budget allocations adequate to the maintenance and development of this potential. It also requires a modernizing industry willing to take the risk in the creation of demand for local technology (that is, willing to invest in local R & D). The government must also play a central role in the planning and management of scientific-technological policies.[4]

I summarize the possible roles of the state by asking this question: if a particular method of technology implantation is desired, which mode of organizing knowledge should the state sponsor and/or finance? Put slightly differently, which method of public intervention in society and the private economy will the state adopt, given a preference for one or several means of technology transfer? The following spectrum of possibilities then results:

- If the state wishes to aid private entrepreneurs in the purchase of foreign patents and licenses, it must help train national technical personnel to select and use the innovation. This requires appropriate funding of facilities in higher education and the creation of data banks and other information retrieval systems containing systematic information on foreign production processes and their applicability. The state, in this setting, merely plays an *educational/facilitative role*.

- If the state wishes to help private entrepreneurs or public enterprises by contracting for the presence of foreign expert advisers, or to assure training abroad for indigenous technical personnel, through its own funds or with the help of an international technical assistance program, it must use systematic means to evaluate the available options. This calls for dependence on data banks and information retrieval systems, whether operated locally or by international organizations. In this case the state plays a *cooperative facilitative role* together with an international organization.

- If the state wishes to promote industrialization by setting up new manufacturing facilities supported by direct foreign investments but is not concerned with improving or adapting the imported technology, it will promote forecasts of demand for the new products. Typically this activity involves encouragement of forecasts made by the investing firms, often in conjunction with such international organizations as the Economic Commission for Latin

America (ECLA). The state must also design fiscal policies to make the market attractive for the investor. The state here plays a *forecast/promotion role*.

- If the state concludes that the private sector will not or cannot produce goods considered essential for the economy, it will operate plants and services directly, creating a "public sector" to perform the tasks left undone by domestically owned private plants or by foreign-owned facilities. This choice should also depend on forecasts of demand for goods, which can be made by state agencies or by international organizations.

- If the state wishes to promote industrialization and a gradual process of technological self-sufficiency by capturing *new* technology for its private or public sectors, instead of merely assembling components imported from the industrialized countries, it must guide direct foreign investments so that some components or the entire product is manufactured locally. This task requires a more extensive set of knowledge-generating activities. Local technical personnel must be available, hence university training is involved. Indigenous R & D capacity and information retrieval systems must be available to help the foreign investor work out appropriate local adaptations of the technology. Research institutes and centers are publicly funded, though financing may depend on contributions from international organizations. And forecasting of demand for the new product remains critical, although the forecasts must not necessarily be made by public institutions. In this instance, since the state intervenes more widely in the technology transfer and domestic innovation processes, it plays a *guiding role*.

- If the state wishes to implant new technology in the context of an overall program of coherent modernization—which may involve concern for a balance between rural/agricultural and urban/industrial activity, employment, housing, sanitation, or income distribution—it must have a plan. Planning involves forecasting, not merely of demand for a single line of products but also of many macroeconomic variables. Forecasting must therefore become the domain of a public agency: a systematic data base and a retrieval system become imperative, and R & D institutions and the necessary backup in university curricula become matters of priority.

The state must ensure the viability of all these knowledge-generating activities for joint venture agreements with foreign investors to be fully effective. In this case, then, the state actively assumes the *role of planner.*

How does the state "know" which mode of transfer to select? The choice depends on the predominant ideology of development. Before any single option can be chosen, goals must be reasonably explicit, which presumes a national consensus on the kind and direction of development. The scale of goals can be matched with the various means and ways of technology transfer and of knowledge management to clarify the state's possible roles in development.

A state plays an autonomous educational and facilitative role if the national goal seeks, not to restrict freedom of choice in the domestic market, but simply to ensure that foreigners obey the law and to guide investors into development priorities through fiscal and monetary incentives. The role is "autonomous" because the means employed do not require cooperation with or dependence on international organizations or regimes. There is rather reliance on technical assistance in the form of end items and complete systems, and on patented foreign technology to encourage domestic manufacturing. Proper utilization and adaptation of the patents, however, requires the creation of local R & D centers and the reorganization of higher education to stimulate the training of engineers.

A different role is played once the government turns to international organizations for technical and financial support. Its role is still largely educational and facilitative of the private sector, but it is no longer autonomous. Knowledge will continue to be sought and managed as before, but reliance on larger bodies of information encompassing additional economic sectors becomes necessary. Organizations such as ECLA now enter the game as sources of information.

The state becomes a promoter and forecaster once the planned economic development of the nation becomes the official goal, requiring tight control over investments. Foreign firms are no longer wooed indiscriminately but are compelled to adhere to the government's priorities. The earlier means now give way to elaborate rules governing the kind and extent of domestic participation in foreign-owned enterprises, including controls over the kinds of technology to be imported. Economic forecasting, though rudimentary, becomes necessary to establish guidelines that will encourage the private sector.

The state as guide is a still more elaborate role. As a significant portion of the productive sector becomes publicly owned or operated, the remaining private sector must compete with the public sector (state as producer) or work within the rules set by the government (state as regulator). A vital new goal now takes precedence over earlier ones: technological self-determination. The means necessary to play the role include national control over foreign-owned enterprises and exclusive reliance on joint ventures to attract new foreign capital. Joint venture arrangements with other Latin American governments are also promoted. The success of these measures clearly requires sophisticated ways of obtaining, organizing, and retrieving technical information for use in elaborate development forecasts.

The state as planner is an even more ambitious role. The goal of technological and cultural autonomy now joins the more modest objective of determining one's own technological future. The array of means remains the same as before, but the demand for knowledge and the organizations managing it becomes more insistent. Formal planning calls for a large variety of specialized institutions and complex bureaucratic links between ministries, research institutes, and production units.

Goals, means, and ways of managing knowledge, however, are embedded in the ideas people use to conceptualize the future. Technology, in addition to all its other properties, is an essential part of ideologies of development. We cannot understand how Latin Americans choose goals and means without first understanding their ideologies of development.

Strategies and Ideologies

Goals emerge from images in policymakers' minds of what the future should be like. The meshing of goals with means, the management of knowledge, and the definition of the proper role of the state combine to form a "strategy" of technological development. Strategies constitute a patterned rationality for achieving change, embodying a consensus on causes and effects, antecedent conditions, and preferred outcomes.

The least demanding of these strategies, one historically linked to import substitution industrialization (ISI), is that of "technological laissez-faire." It is based on the idea that technology should be produced or acquired according to market, laissez-faire conditions. From this perspective the choice is not between national and foreign technology but between the best, most efficient, and least costly technologies. ISI, or the policy of protecting infant industries by temporarily opting out of a liberal, low-tariff world trading system, accepts the basic premises of economic liberalism but simply suspends them temporarily to enable latecomers to catch up. Made famous by Raul Prebisch and ECLA around 1950, ISI dominated Latin America for a decade and continues to be in vogue in some countries. Liberalism also informs technological laissez-faire science and technology policies when the ISI strategies are challenged and partially replaced by export-led strategies.

However, the underlying liberal premises were increasingly challenged during the 1960s and 1970s by Marxists and egalitarian-nationalists. They argued that overcoming dependency required more elaborate measures for opting out of a liberal world economy, for creating the basis for a technological autonomy that would lead to integral development. I call the two alternative strategies of economic and technological development they proposed "structural" and "pragmatic."

Figure 2 summarizes the ingredients of goals, means, information and knowledge, and the role of the state that are involved in the shaping of science and technology strategies. A rough correspondence of these ingredients to the three science and technology strategies to be described and explained in this chapter is suggested.

At the root of the choice between technological laissez-faire and structural and pragmatic antidependency lies an attitude about international, dependency-generating relationships. We can thus distinguish between "communal" and "coercive" dependence.[5] *Communal dependence* has been described as a relation of support and fulfillment: B depends on A because such dependence fulfills, enables, or protects B in some respect. The relationship of communal dependence is also one of mutual trust. This attitude is inherent in the technological laissez-faire strategy and in fact translates as liberalism in the science and technology field: A may be dependent, but A does not see anything wrong with that as long as the relationship protects, enables, allows for growth, and so on. Communal dependence seems to be

Figure 2. Science and Technology Strategies: Goals, Means, Management of Knowledge, and the State

Science and Technology Strategies[a]	Goals	Means	Organizing Knowledge	Role of the State	Levels of Complexity
Technological laissez-faire	Economic growth: Rules prohibiting abusive practices	Technical Assistance	Train experts / higher education	Autonomous, educational and facilitative	1
Pragmatic antidependency	Subordination of investors to national economic development goals	Knowledge-generating capital equipment purchases Rights to a foreign patent and license	Research and development	Educational and facilitative with reliance on international organizations	2
	Subordination of investors to national plan	Foreign direct investment	Retrieval systems and data banks	Promoter and forecaster	3
	Technological self-determination	Joint ventures / open technological packages	Systematic forecasting	Guide	4
Structural antidependency	"Integral development": national, cultural capacity to innovate[b]	Joint ventures only among Latin American firms	Planning	Planner	5

[a] The correspondence of the strategies to the ingredients is only approximate.
[b] The more complex ingredients subsume and require the simpler ones.

linked to the idea of the well-being of a community as a whole, in which any existing inequalities would be eliminated by a trickle-down process as prosperity increases.

Coercive dependence, however, is based on a power relationship that is "bad," "wrong," and "damaging." It is a zero-sum situation. For B to depend on A means that it is controlled by A and subject to A's coercive power. This coercion can be direct, in the control of key resources and of scientific and technological knowledge. In a mutually coercive-dependent relationship, A and B may coerce each other symmetrically or asymmetrically, with equal or unequal amounts of intensity. In any case, a feeling of distance from others and of lack of community is inherent in a coercive-dependence relationship. Structural and pragmatic antidependency strategies derive from the view that dependency is necessarily coercive.

Technological Laissez-Faire

The liberal underpinnings of the import substitution industrialization (ISI) strategy predict a great deal of its ideological content. Latin America is viewed as a part of the West, a participant—albeit an underprivileged one—in the heritage of material progress associated with free and unguided scientific research and technological innovation. The nationalism associated with the ideology does not seek to set apart any "Latin American culture"; it sees individual Latin American states as societies that have not yet attained the status of developed nations and wishes to push them into the accepted European model. Even though the economic policies preferred by advocates of the strategy stress departures from free-trade liberalism, no parallel guidance for science and technology is provided. Scientific progress is assumed to lead automatically to technological applications and to improvements in welfare. Therefore, the issue of whether the technology is indigenous or imported is beside the point. Planning is not considered desirable because wealth is thought to increase by the "trickling down" into society of the gains from overall economic growth. Emphasis is on an increase in the national product, not on the distribution of the increase among segments of the population. The state is therefore not called on to guide entrepreneurs and investors or to encourage research. Nor does it seem to matter whether the imported technology is in foreign or national hands.

The attitude of ISI advocates toward imported technology is indiscriminate: the more the better, provided a market for the product is likely to develop. The "prestige of local invention" is irrelevant. In short, developing countries need not emphasize national innovation and leave aside foreign technology, because foreign technology is there, waiting to be purchased, waiting to help developing countries progress at much faster rates than the already-developed countries did. "The new countries," writes Nuno Fidelino de Figueiredo, "can obtain licenses for the manufacture of products or for the application of new processes for the development of certain products, all coming with designs, global and detailed, all coming with manufacture instructions, with detailed instructions about process machinery and raw materials."[6]

I cannot improve on this description of ISI in operation:

One of the fundamental characteristics of the import substitution process, which was to have an immediate and longer term effect on the stimulus to science and the development of technological innovation in our countries, was the considerable protectionistic action undertaken by the State, which aided this process through strong tariff and exchange rate barriers to protect industry from foreign competition and to assure profits to national entrepreneurs, even in the cases when they operated at low levels of productivity. This strong protectionism contributed to the formation of an entrepreneur with a mercantile mentality, who, without a clear awareness of his social goals and, at the same time, enjoying an exaggerated "laissez-faire," has not felt the strong need in most cases to look toward science to improve the results of production and to increase his efficiency. On the other hand, most Latin American governments have granted almost unconditional terms for importing technology in the form of patents, equipment, semi-finished products and technical personnel, with the advantages these importations imply for enterprises in terms of rapid decision-making and mounting of operations. If to this we add ignorance and lack of confidence in the internal capacity to supply technology, it is easy to understand how the industrialist's general tendency has been to purchase foreign technology almost indiscriminately by any of the ways mentioned, and that their demand has been almost exclusively in that direction. As the process of import substitution advances reaches the stage of manufacturing intermediate goods and equipment increasing technical complexity, it generates the direct participation of foreign enterprises because of the capital and "know how" needed. This is the scenario for the establishment of subsidiaries of large multinational enterprises who have a natural source of technological supply in their home offices and who orient their decisions exclusively in function of the over-all interests of these companies, interests which in many cases are contrary to the social benefit of the Latin American economies in which they operate. In this way the intervention of these multinational corporations in the productive sectors of Latin America weakens even further the precarious internal demand for technology, giving rise, among other consequences, to an even greater estrangement between the scientific and industrial sectors.[7]

One result of the import substitution process and its corollary, the intervention of multinationals in productive sectors, has been heavy migration of Latin American scientists and engineers to the industrialized countries, where they can hope to find employment in firms committed to R & D. Another is the employment of the remaining technical elite in basic research unrelated to technological innovation. Local entrepreneurs, quite rationally, prefer to license imported patents, thus contributing further to the stagnation of local R & D. State intervention is confined to maintaining a high protective tariff; even state enterprises import technology without a clear strategy. There is no deliberate macroeconomic policy of technological development; innovation results from the discrete decisions of individual enterprises, each motivated by its own short-run economic rationality.

ISI is not the only development strategy linked to technological laissez-faire. In an outward-oriented liberal economy, the technological laissez-faire science and technology strategy remains as strong, if not stronger. Under an export-led strategy, tariffs are slashed, domestic markets are opened to foreign products, production is geared for export, and foreign investment is welcomed. Technological laissez-faire follows as an almost automatic extension of all these actions.

In summary, technological laissez-faire looks like this: its underlying *goals* are economic growth, stability, and efficiency, without triggering a major change in the structure of society and the economy; equality is not usually a stated objective. There may be no explicit science and technology goals, other than the prohibition of abusive practices; therefore, foreign investors are not subordinated to a more rigorous set of controls and guidance. The *means* toward these ends stress the importation of finished technology by way of patents and licenses. Local manufacture of intermediate products generally requires direct investments by foreign firms, and these are likely to be capital-intensive and to shun the establishment of backward linkages with the local economy. No systematic steps are taken for the organization and *management of knowledge and information;* it is up to the importers of the technology to find their own foreign counterparts. Some training may take place, and efforts, although sporadic and decentralized, are made to seek out the appropriate suppliers of technology through small research institutes and government agencies (uncoordinated as they are). The importers must abide by licensing contracts with clauses restricting the use of the technology and its reexportation. The *role of the state* is educational and facilitative. In the ISI case it is confined to maintaining a very high tariff, whose rates, again, are not determined according to a clear set of priorities or aims, and to allocating the amount of foreign exchange to be devoted to imports. But the state may inadvertently stimulate the importation of the "wrong" technology by its administration of import quotas and exchange controls. There is little evidence of the links Sabato sketched as necessary and desirable. Only when import substitution is found to be lagging at the national level is the state called on to consider some international stimulant to lead state policy into a regional free trade strategy. In the export-led case the role of the state is to encourage foreign investment and technology transfers to the leading sectors through incentives.

Structural and Pragmatic Antidependency

The critique of the liberal system in the Latin American context was conceived in terms of the dependency argument. Because of the variety of ideological "parents" and analytical concepts, dependency became an ideology in its own right: a theory of stagnation, a critique of capitalism, a structural condition, and a process. Gabriel Palma, for example, has classified the dependency literature according to three distinct approaches, all of them deterministic.[8]

The first approach takes dependency as a "theory" that describes and explains underdevelopment and stagnation. This approach was first articulated by scholars such as Paul Baran and André Gunter Frank.[9] It continues "the central line of Marxist thought regarding the contradictory character of the needs of imperialism and the process of industrialization and general economic development of the backward nations" and argues that because "economic development in underdeveloped countries is profoundly inimical to the dominant interests in the advanced capitalist countries," these interests will form alliances with precapitalist domestic elites to inhibit such transformations. The approach concludes that the possibilities of economic growth in dependent countries are extremely limited; any surplus is expropriated in large part by foreign capital and otherwise squandered on luxury consumption by traditional elites.[10]

A second approach, based on the Latin American disenchantment with ISI, also arrives at the deterministic conclusion that dependency can only bring about stagnation.[11] Advocates of this approach do not pay attention to growth rates[12] but make an "ethical distinction between 'economic growth' and 'economic development.'" Development does not take place when growth is accompanied by "(i) increased inequality in the distribution of its benefits; (ii) a failure to increase social welfare . . . ; (iii) the failure to create unemployment . . . ; and (iv) a growing loss of national control over economic, political, social, and cultural life."[13] Studies by Maria de Conceição Tavares and Osvaldo Sunkel, for example,[14] are characteristic of this position.

The third approach is a critique of dependency as a theory of stagnation and underdevelopment. It considers that development can take place in a context of dependency but that this development is determined by changes in the world capitalist system. For example, Fernando H. Cardoso believes that *associated dependent development*" is possible because of the

transformations which are occurring and have occurred in the world capitalist system, and in particular the changes which became significant towards the end of the 1950s in the rhythm and the form of capital movement, and in the international division of labour. The emergence of the so-called multinational corporations progressively transformed centre-periphery relationships, and relationships between the countries of the centre. As foreign capital has increasingly been directed towards manufacturing industry in the periphery, the struggle for industrialization, which was previously seen as an anti-imperialist struggle, has become increasingly the goal of foreign capital. Thus dependency and industrialization cease to be contradictory, and a path of "dependent development" becomes possible.[15]

This approach is not a dependency "theory"; rather, it considers dependency an outcome of a peculiar historical situation in which countries are linked to the international capitalist system and thus need to be studied only in concrete terms, in specific countries, and according to the conditions of each country's class relations and class structure.

Palma deals only with approaches based on classical and, especially, neo-Marxist deterministic ideologies. But the intellectual map is more complicated, because dependency ideas are not only arrived at from a Marxist and neo-Marxist set of beliefs; they are also informed by egalitarian-nationalist ideologies. Thus, those who, as in the case of the approaches described by Palma, see the issue as total and global and who find the solution in structural change should be differentiated from others who wish to decouple the components of the issue and who advocate serious less-than-total policies that we can characterize as reformist. I call the first group *structural* and the second *pragmatic dependentistas*. The pragmatists do not wish to wait for broad structural changes; they believe that important changes can be made now. Dependency for them, then, is not totally determined in a structural reality. Moreover, although they also use the term *structural*, they seem to mean something slightly different from their opponents. To a neo-Marxist, "structures" are rigid and long-lasting global property and class relations typical of late capitalism that "deform" the pattern of modernization in Latin America. But the pragmatists, in their search for immediate reforms, wish to identify weaknesses in these relations and hence downplay their rigidity; they refer to typical institutions—such as universities, R & D institutes, legal codes, and types of business and industrial organization—as "structures."

Structural Antidependency

The ideology of the structural antidependency approach to science and technology in Latin America is a restatement of Marxist theories of economic imperialism as applied to the relations between developed and developing countries. However, structural antidependency transcends classical theories of economic imperialism by arguing that the expansion of capitalism places the developing countries in the periphery of world economic relations, thus causing them to be underdeveloped and poor.[16]

The structural antidependency school contends that technological knowledge is the main instrument in the maintenance of relations of domination. Amílcar Herrera argues that "the new dependency does not require that the big powers resort to the direct system of political-military domination that was the most visible characteristic of imperialism in the last century. The new instrument of domination, more subtle but no less effective, is the scientific and technological superiority of the developed countries."[17] Three major propositions follow from this line of analysis. First, ISI distorts developmentally progressive technological innovation by encouraging the wrong technologies. Second, technological dependency results, which in turn aggravates a large number of other conditions that also inhibit socially progressive change. Finally, technological dependency is responsible for the "marginalization of science" in Latin America. This situation must be reversed by making science and technology socially progressive and useful.

The argument is that under import substitution policies the urban middle and upper classes demand the same kinds of goods as consumers in the industrialized countries, so they import technology to manufacture these goods, technology that usually is highly inappropriate. New enterprises are set up through direct foreign investment, or even if enterprises are independent of foreign capital, local firms purchase foreign technology through license agreements or other contractual methods. Thus foreign technology tends to substitute for technologies that might be developed by local R & D. Even where local laboratories come up with some type of local technology, the tendency is to use the foreign version instead, on the grounds that what is foreign is "good." Local science and technology are therefore "alienated" from the productive forces,[18] and "market forces operate so that there are very limited opportunities for 'learning-by-doing,' both in innovative activities (like applied R & D) and in project construction (engineering). . . . The linkages between R & D laboratories, engineering, and machine-building, which have played an important part in relating science to production in the advanced countries, hardly develop—or develop very slowly. This is sometimes referred to as the "'self-perpetuating tendency in technological dependence.'"[19]

Consequently, the structural antidependency approach rejects the theory that scientific underdevelopment in Latin America is a result of cultural and institutional obstacles. The causes are much more profound; they are structural: "scientific underdevelopment is not simply the result of some great lack or of some failure in the system which can be corrected with external aid; *it is a consequence of economic and social structure*" (emphasis added).[20] Thus the problems of development in general and of technological dependency in particular cannot be solved until the "real" causes are removed: "the structural center-periphery dependency and the internal contradictions in those countries that have a deformed capital formation because of the historical process of subordination within the world capitalist system."[21]

Technological dependency, it is argued, ultimately leads to a long series of additional ills that, taken together, inhibit socially progressive change. Where there is technological innovation, change is biased in a capital-using direction. Latin America, however, with its excess of cheap labor and its lack of capital, needs labor-intensive technologies. The importation of capital-intensive technology results in consumption policies biased toward the needs of the middle and upper classes, disadvantaging the lower classes in terms of goods available to them while boosting high-income groups by generating skilled job opportunities. It is further argued that too much foreign exchange, already in critically short supply in Latin America, is spent in the acquisition of foreign technology, especially where tight controls are lacking, and in this way technological dependency only aggravates balance of payments crises. Developing countries then have almost no bargaining power because technology gives a monopolistic advantage to the developed countries. One example is the patent system. According to Constantine Vaitsos, patent licensing does not transfer technology, since the patented products are not produced nor are the patented processes themselves used in productive activities in the patent-importing developing country. If the developing country does not profit from it, the developed one does, because "monopoly privileges granted by patents are basically exercised through the creation of secure import markets for the patented products in developing countries."[22]

One of the pillars of the structural antidependency approach is its theory of the "marginalization of science," which argues that underdevelopment is a historically unique form of economic organization.[23] This form cannot be identified with the early period of development in the advanced countries because it is a product of interaction between "precapitalistic economic forms" and industrial capitalism it-self.[24] This theory—as exemplified by Furtado—calls for the creation of autonomous science to prevent the alienation of local scientific institutions and to end their dependency. Science in underdeveloped countries, according to this view, is largely a "consumption item," whereas in industrialized countries it is an "investment item."

The same structural conditions are again held responsible for this state of affairs. "In contrast to the concept of scientific autonomy stands that of science as a social phenomenon, like something that adapts to its 'environment,' i.e., that vast economic, social, cultural, and political web that we call society. The starting point for this idea of science is found in historical materialism."[25]

Nor is science an absolute good. Science and technology have usually been considered a "magic cure," but according to structural antidependency they are not: science is rather merely a prestigious façade, a tool governments use to solve urgent national problems; it is not a panacea capable of curing the ills of underdevelopment. "There are some fundamental fields of technology," Herrera argues, "in which the research carried out in the industrialized countries is not only not useful to the developing countries, but it is even prejudicial to their economic interest, at least in the short and middle range."[26]

Herrera maintains that Latin America must therefore recapture technology as an integral part of its culture, turning it from an exogenous conditioning element into a legitimate mode of expression of its own values and aspirations.[27] How can this goal be achieved? The principal method must be the creation of an autonomous scientific capability in all fields of knowledge, which "means the capacity to make decisions based on the local necessities and objectives of all the fields of social activity, using both indigenous and foreign scientific research."[28]

Structural dependentistas believe that this cannot be done in a capitalist system, implying that it could be achieved only through a revolutionary structural change. Some wish to learn from the Chinese; others evoke the catalog of "alternative technology" approaches: worker control, demystification of expertise, reform of work rules, low specialization, development under the condition of low capital, local or regional self-sufficiency, balanced economic development, resource conservation, low energy use, reduced technological risks. Although there is little specificity, and no consensus, on what is meant, minimally the argument holds out for labor-intensive technology that is easier for people with a low level of technological capacity to master and that offers cheaper equipment and operates on a low scale of production.[29]

In summary, the structural approach of the critics of dependency is as follows: the *goal* of the strategy is to change the "structure of dependency" so that the pattern of development that relies on liberal economic policies can be redesigned to become "integral development." This change requires the severance of many—but not all—of the chains that link Latin America to the industrialized world, including the ISI-sanctioned pattern of technology transfer. Put positively, the goal of the structural dependentistas is technological self-reliance and cultural autonomy. Integral development substitutes the entire array of social advancement, education, working class participation in decision making, employment, and the economic objective of reducing disparities in income and welfare for the ISI-associated goal of gradual improvement in living standards through reliance on unplanned industrialization. But integral development presupposes a smashing of the existing structures that are held responsible for the sluggish pace of change. Attainment of the goal of technological self-determination would break the hold of the center on the periphery.

The *means* found appropriate for this strategy, of course, stress the role of the state, once it is no longer under the control of the "lumpenbourgeoisie." The economy would be centrally managed, and the important economic activities would be state-owned. The remaining private sector would be permitted to function only within the goals and targets set by the state. Technologies appropriate to integral development would be developed locally, with emphasis on labor-intensive and employment-creating innovations. Consumer demand would cease to determine productive patterns. The importation of technology is not ruled out, but it would have to conform to the socio-economic goals of the state, particularly the priority of producing goods needed by the poorer population. All private investment is suspect, but joint ventures with other Latin American firms are desirable.

The organization and *management of knowledge and information* would also be centralized, with all scientific and technological institutions linked to the state, either through central coordination of the existing institutions or through their direct subordination to an appropriate ministry. Local R & D institutions would abandon their reliance on foreign models of research and foreign industrial routines. A central council of science and technology would coordinate the government with the scientific and technological institutes, developing technological alternatives and presenting them to the government in matters where scientific and technological knowledge is required to facilitate decision making. Once the decisions were made, the council would transmit them to the appropriate institutions. It would also have some functions of oversight and control in the implementation of these decisions.

The *role of the state* is the core of the strategy, not just an aspect, especially as the exerciser of rigorous control. Centralized control, through a science and technology council as well as through laws, must assure that private industry acts within the framework imposed by the government; that the scientific and technological infrastructure innovates technologies that are in accordance with the government's socioeconomic and cultural goals; and most important, that the importation of foreign technology fits the state's goals of integral development. Direct foreign investment, especially by multinational corporations, and other means by which technology is introduced would be tightly controlled as to quantity and kind.

Pragmatic Antidependency

Basically, pragmatic dependentistas say that dependency is a real problem: while Latin American nations must learn to produce their own technology, they should not cut the links with the external world[30]—to question the validity of science and the importance of technology under rigid ideological canons is not only impractical but also self-defeating. Thus Luis Carbonell argues that

> many social scientists, especially economists and sociologists of the structural school, believe that real scientific and technological change in a developing society can only be achieved after the prior transformation of its basic structures. . . . We believe, however, that currently it is feasible and necessary to design and initiate the progressive execution of programs aimed at effectively incorporating technology into the productive process.[31]

Francisco Sagasti, one of the main advocates of this approach, has conceptualized the principles on which a technology policy of pragmatic antidependency in Latin America should be based. I will summarize his argument before analyzing the approach further.[32]

Technological progress is the "continuous and cumulative process of creation, diffusion, and utilization of knowledge." Because the process is now dominated by foreign interests in Latin America, an indigenous technological capacity must be created. This goal calls for a plan linking the educational, industrial, labor, and R & D sectors. The more complex the actual interrelationships between these sectors become, the more sophisticated the plan must be, inspired by economic, rather than purely technological-industrial, objectives. "'Market forces' are not enough to promote by themselves technological development and insure its correspondence with socio-economic objectives." The state must help compensate for the disadvantage at which Latin American importers of technology find themselves vis-à-vis their foreign suppliers. Latin America's international bargaining position must be dramatically improved, so as to obtain the *most appropriate* new technology, not stop the inflow of technology. "The development of an indigenous technological capacity must be guided by a strategy of *selective interdependence,* choosing research fields according to the possibility and feasibility of importing technology, local comparative advantage, the specific needs of the country and the possibility of exporting the technology." The demand for technology is as important as the supply; hence policies aimed essentially at improving the supply through education and promotion are inadequate.

How does this formulation differ from the analysis of the structural antidependency school? After all, many of the specifics sound much the same. What is missing is the structuralists' universal solvent: the internal contradictions produced by class interests and class struggle. The center is still pitted against the periphery, but the sharpness of that contradiction is weakened—class conflict within and between nations is no longer an all-or-nothing game. Consequently, the pragmatists see possibilities for accommodations that the structuralists dismiss as utopian.

A number of practical conclusions follow from this position. Technological dependency on the industrialized world can be reduced without a global revolution, and a new symbiosis with the North can be worked out. Because technology is *not* conceived of as embedded in the structural givens (Herrera's "implicit" setting) of Latin American society, technology can be considered a marketable merchandise whose utility, given certain economic priorities, can be evaluated in its own right. Technology, in short, is relatively autonomous. Therefore, who *owns* the technology is no longer the sole criterion underlying policy, though it remains central. Since "structures" do not determine "functions," foreign ownership can be tolerated under appropriate controls.

Because in this approach no zero-sum conflict situation exists, a simple reassessment of the relations of the developed and the developing countries will allow Latin Americans to achieve more control and independence in their development decisions, making them masters of their own destiny.[33] In the very short run, pragmatic anti-dependency may imply some conflict between Latin American governments and foreign sources of technology, but after a period of readjustment "ways of doing business" beneficial to both sides will be found, especially in the form of joint ventures. If foreign enterprises could gain the confidence of the Latin American governments and the latter could have confidence in themselves, it is believed, a mutually supportive relationship could develop.

Pragmatists believe that science and technology are important to economic and social development, and the aim is to create the proper conditions for them to fulfill their role. At the same time, the "negative" effects—that is, dependency on technology—must be minimized, and now, because it is impractical to wait until drastic socio-economic structural change occurs.

Pragmatists do not deny that technology is embedded in culture. But they take for granted that Latin American culture requires technology and should assimilate it. They are relatively indifferent to the cultural critiques of technology of some structuralists; hence questions of ownership are less important than considerations of design, production, and inclusion in policies of overall modernization. What matters is functional control over technological innovation.

The attainment of this objective requires a well-constructed scientific and technological policy properly integrated into the nation's development plan. As Wionczek puts it, "In operative terms this means the creation of an assemblage of direct science and technology instruments and the readjustment of many instruments of other policies that, indirectly, have an influence on the functioning and the development of national science and technology."[34]

The foreign presence in a local economy thus fulfills an indispensable function for the achievement of national development policies. Pragmatists agree with structuralists in preferring indigenous to imported technology, but they admit that foreign firms, if properly guided, can furnish the necessary knowledge and capital much more rapidly than could a policy of complete self-reliance. Foreign technology is not to be rejected; it must be mastered and subjected to a national development plan. Pragmatists take for granted the continuation of a considerable technology gap between the industrialized and the developing countries. However, they believe that a policy of guided importation would nevertheless give Latin America the ability eventually to disaggregate the multinationals' "packages" and adapt the components to locally devised priorities. "There is, thus, a generative relationship between the import and the generation of technology; through an intelligent handling of the former, we can supply 'work' to the system of generating technology, providing for its support and for its expansion."[35]

In contrast with the haphazard practices of ISI, the marriage of imported with indigenously developed technology demands an active Latin American R & D policy of deliberate innovation, geared to meeting the economic and social priorities shared by the pragmatists and the structuralists. But the pragmatists, unlike the structuralists, believe that this can be done now, even in the absence of social revolution, by strengthening the scientific and technological infrastructure, thereby weakening dependence on foreign research. This calls for systematic coordination of the state, the private sector, and the educational-research system, as urged so eloquently by Sabato. But the pragmatists differ from the structuralists here again by arguing for a full-fledged national technology "regime," perhaps even an international one: a system for coordinating the separate parts rather than for centralizing all power in the state.

The state thus plays a central role in this approach, but industry enjoys considerable independence. Under the broad framework of control and restrictions imposed by the government, industry is free to buy foreign technology. Government and industry together stimulate technological innovation and the diffusion of both imported and nationally developed technology. Scientists must act within the framework of the scientific and technological plan, which reflects the development plan of the country but which permits much independence—more than under the structural antidependency approach. The state intervenes, but not too much; industry acts within governmental guidelines, but not exclusively; a free market is retained, but only to a certain extent; the scientific establishment remains independent, but only under the framework imposed by the state's science and technology plan.[36]

An insightful comment by Charles Anderson summarizes the beliefs of the pragmatic antidependency school:

Rather than true believers, Latin America has been more apt to produce maddeningly eclectic political ideologists. Although this ideological heterodoxy is the despair of sterner folk who like to keep their doctrines in good order, it may be one of the greatest assets of Latin American politics. The Latin American intelligentsia has, by and large, remained skeptical, open to experience. Seldom has an alien philosophy been accepted nationally as defining the "right way" to change. The dominant theme in much of Latin American social thought seems to be that it is a bit peculiar and impractical to anathematize either the Marxists or the British utilitarians, to consider either as the prime enemy of the social order.[37]

Although pragmatic dependentistas sometimes refer to technological autonomy as their ultimate goal, their immediate *goal* is technological self-determination. This goal is instrumental: its basic purpose is the achievement of socioeconomic development, defined as growth and programmed domestic income redistribution. It differs from the goal of the technological laissez-faire strategists in stressing the nonautomatic nature of welfare improvement; and it differs from that of the structuralists in accepting as necessary some intermediary steps that would lead to technological self-reliance and not considering dependency derivative of the international class system. Because the pragmatists postulate no single cause of dependency, they see no need to sever technological links with the North. Instead, technological self-determination means simply the creation of an autonomous capacity to innovate and adapt, implying the ability to channel, control, and subordinate the foreign investor to overriding national economic development goals and plans. The pragmatists do not find it necessary to define their objective as the creation of a Latin American cultural identity that equates imported technology with Western cultural exploitation.

These goals call for the creation of technology *regimes:* "the assembly of dispositions that regulate the production, trade, distribution, and utilization of the technology necessary for the achievement of the goals of industrial policies."[38] Such a regime includes both a technology plan, which serves as the "knowledge component" of the overall economic plan, and an ambitious informational infrastructure, to be discussed below. In gearing technology to economic planning, Sagasti distinguishes between horizontal and vertical policies. Horizontal policies formulate technological rules applicable to all economic sectors, whereas vertical policies are careful to distinguish between the productive entities to be influenced. Contrary to the structuralists, Sagasti advocates policies that allow for differences in size and number of factories and in patterns of ownership. No single technology policy is then possible, and planning becomes much less monolithic—though perhaps not less demanding. Technology is seen to possess some attributes related to economics and engineering, not simply ones derivable from historical materialism. The notion of regimes thus acts as a conceptual roof over more detailed considerations of means, the management of knowledge, and the role of the state.

The aims of coordinating the activities of domestic and foreign investors and of guiding technological innovation and diffusion make the *means* for guiding and controlling investments the core of the pragmatic approach. Pragmatists want to motivate domestic entrepreneurs to innovate; therefore means must be found to subsidize risk taking. Because they wish to encourage learning-by-doing, direct foreign investments featuring "technology packages" must be discouraged. Multinational operations that depend on foreign-controlled subsidiaries are suspect. Uncoordinated licensing of foreign patents is undesirable. Joint ventures defined and controlled by Latin American authorities are the preferred means of encouraging transfers.

Learning-by-doing requires that enterprises perform the key technological activities themselves, including product-design, market research, and marketing. Pragmatists therefore call for disaggregation of the technological package. This disaggregation—the lack of which is most vividly typified by the use of "turn-key" plants (*llave en mano*)— allows the national enterprise to differentiate between the necessary and the superfluous and to identify the components of technical knowledge that will enable it to master imported technology. Otherwise, buying a "black box" package might lead to the unnecessary purchase of many components that could easily be produced locally. The disaggregated technology is called a "white box." Because it is not easy to jump from a black box to a white box, a "gray-box" stage has been suggested in which one turn-key project at a time is disaggregated. When the white-box stage is reached, technology and capital goods can even be exported.[39]

Obviously, none of these means can be deployed effectively without vast improvement in the *management of knowledge and information.* The pragmatists' list of things to be better understood is long and exacting. Information and coordination systems must be created to effectively link demand for technology—redefined according to forecasts, not the free market—with the supply to be produced—again, in non-market terms. This demand for technical knowledge involves many new actors, including, for example, the ministries of industry, economics, agriculture, and mining; industrial credit organizations; chambers of commerce; professional associations; and trade unions.

A new network of institutional interconnections must be created, one clearly different in organizational structure from traditional science, which was supported mainly by small research centers and educational entities. This network serves an informational function of "systematic forecasting." It requires the reorganization of higher education to train technical personnel in foreign technology and its adaptation to local needs, as well as the enlargement and further centralization of existing R & D institutions. Patents cannot be selected wisely without information retrieval systems, and technology cannot be selected unless the machinery is understood as part of a productive process directed at the social and economic objectives of income redistribution. But without systematic forecasting of trends and possibilities throughout the economy, there would be no way of estimating the task of the R & D centers, the need for technical manpower, the credits and fiscal incentives needed to motivate the private sector, or the kind and type of joint ventures to be contracted for with foreign firms.

The *role of the state* and its agencies is most important in the pragmatic approach. The main forecasting role is assumed by the national councils of technology, which give orientation in selecting scientific and technological alternatives, coordinate the components of the national technological regime, and stimulate the government, private institutions, and universities to be active members of the regime. These councils straddle the public, private, and educational sectors. In addition, interministerial councils coordinate the scientific and technological activities of governmental agencies proper, and there is room also for private independent councils, either geographical or functional in form.

The main task of the councils is the preparation of an indicative national plan for science and technology. The plan should explicitly enunciate the guidelines for science and technology's contribution to the broader goals of national development. It should define the principles, goals, means, institutional framework, budgets, and regional and international arrangements by which national science and technology policies could be pursued.

The state must also issue the regulations that determine what kinds of foreign enterprises may establish themselves, the percentage of foreign capital allowed in foreign-owned plants and banks, the conditions governing the repatriation or reinvestment of profits, and the rate at which multinationals must "fade out" from their subsidiaries. International collaboration may become desirable if such regulation has the effect of merely encouraging multinationals to move from one country to another. In short, part of the state's role is to elaborate a comprehensive code of conduct for technology transfer. So is participation in regional investment allocation schemes and in the exploration of possible joint ventures among Latin American enterprises, leading to the formation of Latin American multinationals.[40]

Moreover, the state is itself an actor in the process of industrialization. In Latin America some manufacturing enterprises and many public utilities are state-owned and -operated, and state selection and purchase of new technical processes and equipment for extractive enterprises that have been nationalized in past years also has a major impact on overall technological development. The state is also able to insist on the modification of patent agreements thought to be overly restrictive for their local licenses. There is arising

> a gradual consciousness in different Latin American countries that state purchases can be used as a powerful instrument for technological development. . . . The state is the most important customer of the production of goods and services in the majority of the developing countries, and it undertakes . . . large investments that incorporate complex technologies. Thus, it can broaden the market for the national products through orderly, yet increasingly complex, programs for the purchase of goods and services, which allows it to impose technical requirements to enhance quality, to reduce costs, and to establish strict terms of delivery, thus promoting the technological progress of the productive sector.[41]

These constitute the universe of ideologies and strategies that have informed scientific and technological policy in Latin America. No single one is expressed perfectly in actual government policies. Each ideology, however, acts as a cognitive simplifier in policy making. Each suggests what nations can be expected to achieve and what behavior vis-à-vis foreign actors should be.

The Policy-Making Process and the "Subversive Elites"   4

Science and Technology Policy Making

Policy-Making Level of Analysis

Many of the studies on science and technology in Latin America overlook policy-making questions. Most of them are written by economists
who rarely pay attention to political processes. The study of public administration and policy making in developing countries, and in Latin
America in particular, is a very small field, and few attempts have
been made to link it with political economy in that setting.[1]

The policy-making process in science and technology can be seen
in the context of Graham Allison's seminal study of decision making
during the 1962 Cuban missile crisis. Allison distinguishes three conceptual models of decision making and its implementation—rational,
organizational process, and bureaucratic politics—which he then ex-
emplifies with particular events of the crisis from the U.S. point of
view.[2] Each model represents a different cognitive map on which
knowledge is arranged and explained. My concern is with the conceptual approach of each model, rather than with the way Allison applied them.

The rational model is the one most commonly applied to the subject of science and technology. Thus the major portion of the literature focuses on what the state has done or should do with regard to science and technology policy, dependency, and foreign investment and the multinational corporations. This tradition is rooted in both economics and comparative politics.

But viewing the state as an actor when we *do* deal with policy, especially with policy change, has merit only if we recognize the other two models. The reason is simple. People within the government, state organizations and bureaus, interest groups, and private associations may have ideological differences over an issue. Thus only through the organizational-process and bureaucratic-politics models, which disaggregate the actors in a policy-making process, can we become aware of the various ideologies and how their differences affect policy. That is why when we study what the state has done, is doing, or should do we are in fact "covering up" for all that happens among individuals and organizations, and although we may well be able to describe a set of policies and actions, it is unlikely that we could explain them or the reasons why they change. The organizational-process and bureaucratic-politics approaches in particular allow us to look for piecemeal, incremental changes—"old-boy" networks, special ad hoc relations, informal relationships—that otherwise remain out of sight.

Because policy is made by people with differing perceptions, when a change in political regime occurs, not only do the individuals change but so do the criteria for evaluation, the orientation toward policy and bureaucracy, and the relative weight given to industrial development and science and technology. Similar changes occur within state institutions, where some may share the ideas and ideology of the new regime while others may disagree but will nevertheless stay in their jobs.

Public Policy Making in Latin America

We can best characterize Latin American bureaucracies by what Fred Riggs calls overlapping: "The new formal apparatus, like the administrative bureau, gives an illusory impression of autonomousness, whereas in fact it is deeply enmeshed in, and cross-influenced by, remnants of older traditional social, economic, religious, and political systems."[3]

Overlapping is linked to what Riggs terms formalism, which can be explained through the metaphor of a map that is supposed to represent reality but, because it is poorly drawn, does not. "A law which is formalistic sets forth a policy or goal which is not, administratively, put into practice. Social behavior does not conform to the prescribed norm. Thus legalistic administration is a particular kind of formalistic system."[4] Formalism leads to ambiguity, which allows for a variety of personalized choices by the enforcement officials.

In Latin America there is frequently no separation between the politics of an issue and its public administration. Where the distinction is made, it is blurred in the policy-making and policy-implementation processes.[5] This has its consequences. For example, entrepreneurs must rely on political "tricks" to circumvent rules and regulations to get what they need or even to get information about current science and technology regulation to really understand it. Because these regulations change frequently, uncertainty is high and only the political tie or influence confers some stability on the process. As one Argentine industrialist said: "The government is very inefficient. Once something has been introduced and approved, in our favor and at a favorable time, we need to take advantage of it immediately. Otherwise one becomes a victim, for the legislation has no ideological consensus and no unifying plans."

In contrast to the tight operating procedures of bureaucracies in developed countries, oral communication, family ties, and friendship often form the basis of Latin American decision making. It may also happen that actions are taken even in the absence of policies, laws, or regulations—they may be carried out in the spirit of the ideology of the political regime in power. Several Argentinians who participated in the Peronist government between 1973 and 1976 said that restrictions on foreign capital and technology transfer were applied before the new regime had time to consider a new law and enact it. They described such actual changes instituted prior to any formal legislative changes as "the regime's spirit."

An additional feature of Latin American policy making is that governmental decisionmakers become "bureaucratic entrepreneurs"[6] themselves when the government sets up state enterprises to promote some sectors or to compete with foreign enterprises. And because

> development administration implies change and major change
> implies the need for political support, it is not surprising that
> development administrators and bureaucratic entrepreneurs
> tend to diagnose the possibilities close to the seats of power: the
> office of the president, an important minister, or the head of an
> agency. . . . It provides the reason, of course, for proposals to at-

tach national planning agencies to the office of the president or to the cabinet, or to relate them, at the very least, to a very powerful ministry, such as Finance. In order for a new service or organization, or a transforming old one, to be effective, however, there must be the awareness of need, at political levels, and continuing commitments of support by political elements.[7]

Latin American bureaucracies deal with science and technology issues in a compartmentalized way. Each ministry, institution, or group is staffed by people of differing ideologies and therefore acquires its own organization ideology, which leads to conflict and confrontations with other organizations involved in the process. Sometimes these conflicts are solved in the give-and-take fashion characteristic of bureaucratic politics. Often, though, agreement cannot be reached, and ministries and institutions continue to work at cross-purposes, each using all the means at its disposal to further its own position and ideology and even to obstruct action on the part of its colleagues.

Military governments sometimes introduce expediency and some discipline to the bureaucracy. Although under strict military rule nonstate institutions, and even state institutions, might have only very restricted leverage, there are often good reasons for technocrats and the military to work together (see n. 15, below). Because the military is interested not only in authority but also in legitimation, in many cases the desire for the latter dilutes the impact of the former.

The high level of centralization in Latin American governments makes influencing the decisionmaking process difficult. Thus the arena for change, either by influence or by obstruction, is the implementation process itself, especially when the state does not possess the necessary penalty and reward mechanisms to elicit compliance with laws and regulations. As Merilee Grindle remarked:

> Political elites, by choice or by inability to change the situation, may regard the implementation process as a political one in which a considerable amount of adjustment must occur. Flexibility in policy execution may even be part of a polity-wide accommodation and conflict resolution system used by political elites to maintain the often tenuous cohesion of the political community itself. . . . Given the concentration of political activity on the implementation process, it is likely that policies and programs will be even more difficult to manage and predict and even more subject to alteration in the Third World than elsewhere.[8]

Laws and regulations are thus only the tip of the iceberg.

Planning and Policy Making

In mixed economies such as those of Argentina and Brazil, planning *is* policy making. Once a plan has been approved by the political actors it becomes a command for action—action that sometimes does not occur. Robert Daland, in a study on Brazilian planning, has found that the preparation of a national development plan tends to have positive values for the maintenance and survival of the regime, while the implementation of such a plan tends to have the opposite effect.[9] On the one hand a plan, as a declaration that a regime knows what it wants, serves to legitimate the regime and build consensus; but on the other hand the plan becomes a divisive factor owing to the conflict and lack of consensus in the implementation process. Thus planning remains the very personal product of its immediate sponsors,[10] be they the ministry of planning, the president, or a bureau. Given that the planner is usually an economist who might react to short-range considerations, and in view of the vagueness of and conflict over goals, the implementation of science and technology plans (in contrast to their enactment) tends to depend on the degree of ideological consensus within the bureaucracy. Where competing interests, low consensus, conflicts, and stalemates characterize the bureaucracy, we could expect enacted plans to belong to the formalistic gallery of laws and regulations.

To illustrate this point we can use the bargaining and compellence types of societies discussed in Chapter 2. The results of their juxtaposition with a low and a high ideological consensus are shown in Table 2.

The "Weathermakers":
Intellectuals and Political Action

In his old age [Hermann] Hesse wrote a lengthy novel, *The Game of the Glass Beads*, which deals very much with the subject being discussed here. He described a utopian state, Castalia, ruled by philosophers whose pastime is a difficult and sophisticated game played with glass beads. Hesse has woven into the novel a story about a weather maker of olden days. After learning the trade from the preceding weather maker of the tribe, this man had acquired, following a lifetime of observation and experience, the

Table 2.   Planning, Type of Society, and Ideology

	Ideology	
	Low Consensus	High Consensus
Bargaining societies	Planning interferes with bureaucracy and leads to bureaucratic conflict. Plans remain formal, and only bits and parts are implemented. Very little resemblance between plan and action.	Bureaucracy helps to implement plan. Input of different groups is added to preparation and implementation of the plan. Plan has greater chance of influencing policy.
Compellence societies	Either the plan fails to influence action because the compelling actors have no mechanisms for producing and implementing it, or the plan is imposed on bureaucracy through repression and purge, without bright prospects of success.	The political leaders demand, the bureaucracy goes along, and planning is turned into action.

ability to sense the course of the weather, to predict it, in some cases even to change it. He was the medicine man, the *shaman* of his tribe, not its chief. But, in his way, he was just as important to the tribe, for the chief ruled within the climate the weather maker predicted and shaped.[11]

In their role as producers and diffusers of ideas, intellectuals can be seen as weathermakers who shape political events and change.[12] The weathermakers predict and shape the political climate in the following ways: Intellectuals create the assumptions, preconceptions, and presuppositions that people "absorb almost by osmosis from their mental environment, of which they are frequently not fully aware, or in any case seldom mention since they take them for granted."[13] Intellectuals are bearers of ideology and creators of utopias, and by their role, usually indirect, in the policy-making process they are also partly responsible for turning utopias into reality.

Intellectuals create and mediate new values. Because they usually go abroad to study, intellectuals are exposed not only to information but also to new values, and these they transmit to their own societies and see integrated into the traditional system to form new hierarchies of values. Intellectuals are also instrumental in the creation of developing countries' self-images, both about themselves and about themselves vis-à-vis other peoples. Images that incorporate nationalist and indigenous cultural notions, such as "autonomous technology" and "culturally relevant technology," are representative.

Intellectuals are involved in and have an effect on the policy-making process in their role as advisers to policymakers and as policymakers themselves, and by creating and contributing to ideas that "stay around" in books, articles, and other communication media, they influence other actors. But one of the most common, and most significant, ways the intellectual affects the policy-making process is by becoming a *technocrat*.

As Aloysio Biondi remarks:

> But, who are these technocrats, really? And who are the humanists, these beings with a long political vision, really? It is simple. The economist who teaches in the university, and takes a position critical of the process, is a humanist. The same economist, when he accepts a position as a state official, automatically should be called a technocrat—even if in his position he applies his humanist vision to the administration. A sociologist who spends his life studying the world of the shanty-town dwellers (*favelados*) is a humanist. The same sociologist, when he accepts work with Housing Planning (*Plano Habitacional*), automatically is downgraded to a technocrat, even if, in his position, he will apply the experience he gained in the "humanist" phase.[14]

Whereas technocracy exists under all kinds of political regimes, the military-authoritarian type of government tends to upgrade its importance.[15] For example, in Latin America the *técnico economista,* or economist-technician, has "become an integral element in the decision-making process."[16] Economists themselves agree that "the days are over when politicians made arbitrary decisions independently of technical considerations. Today, if the President says do this or do that, he usually says it in view of some studies or advice he has received from technicians. It is a changing situation, a gradual involvement of technicians into the political discourse, it is a going process."[17] Many Mexican economists feel they have insufficient access to power, and this may be partly true; but as Daniel Cosío Villegas has pointed out, although these professionals come to their positions as technicians, and never in a political role, "as long as they enjoy the confidence of the president, *their secondary or derived power is considerable*" (emphasis added).[18]

Subversive Elites

We can find in different societies individuals whose aggregation I call a *subversive elite*.[19] Their ideology about the nature of politics and economics differs from that of the established elite, nevertheless, because they have some access to the decision-making channels and structures, they succeed in getting their ideas through to the decisionmakers. This ability to influence policy and action, either overtly or indirectly, is what makes them an "elite."

Who are the individuals who qualify for membership in this subversive elite, this group capable of the subtle penetration of ideas, ideologies, beliefs, and expectations? Only those who deal with such things professionally, those in charge of producing and propagating ideas: intellectuals—professionals in the arts and the humanities and social sciences as well as scientists and technicians. They all share a very valuable resource: knowledge, which can be used both for production and for domination. The subversive elite is thus composed of experts: economists, sociologists, political scientists, and historians, physicists, engineers, and chemists.

If intellectuals agree with the main precepts of the established elite's ideological positions and actions and are involved in the policy-making process, then we can say that they are part of the elite—the cultural elite, as some would call them.[20] If they disagree with the ruling elite ideologically but still influence, at least somewhat, the evolution of ideas and the decision-making process, they are part of the subversive elite. As the state's strength increases many technocrat-intellectuals respond to the call to "serve the state," producing a clear channel for the flow of subversive-elite ideas.

Egalitarian-Nationalist Weathermakers in Latin America

Intellectual Nationalism and the Philosophical Roots of Dependency and Self-Reliance

At any given time there are philosophical conceptions that characterize a nation or a group of nations. They influence intellectuals and their aspirations; therefore they also influence planning, policy, and change. Adrien Taymans asked:

Is it not remarkable that both Marx and Schumpeter developed a theory of economic change, precisely at a time when scholars in almost all sciences were interested in evolution and progress? And is it not more noteworthy that Marx worked along a deterministic line, at a time when determinism was favored, while Schumpeter exalted personal initiative and responsibility? . . . This is only one instance to show how far economic theory is engaged in the ideas of the time.[21]

The ideology of egalitarian nationalism as expressed in Latin America and the ideas of dependency and self-reliance it generates are central to this study, being representative of the ideas "pushed up" to the circles of power by subversive elites. I will therefore attempt to provide some clues as to why egalitarian nationalism is so strong among intellectuals in Latin America and why Latin Americans have been pioneers in conceiving the ideas of dependency and self-reliance and then diffusing them to the Third World, to the point where today they figure in the agenda of international deliberations between North and South. The following paragraphs do not pretend to be a philosophical study of the Latin American intelligentsia; it has too rich a history to be dealt with in such a simplistic way.

These ideas are most deeply rooted in the concern of many Latin American intellectuals over questions such as, "Who are we and in what way are we different from others, especially North Americans?" This is, in the words of Francisco Miro Quesada, a very real "crisis of identity."[22] On the one hand, egalitarian-nationalist intellectuals, educated in the European or American tradition, embrace the goal of modernization. Thus they take from other cultures scientific and technological concepts, devices, and knowledge that they believe will help in the process of modernization and economic development. On the other hand, they are imbued with traditional values and a belief in the culture of their people. They want their nation to be recognized and they resent those who consider it inferior.

The egalitarian-nationalist intellectuals also suffer from a crisis of confidence that reflects present and, even more, past differences in administrative, managerial, educative, and scientific approaches. These intellectuals resent the fact that the tools of modernization cannot be acquired within their own nation-state, and they experience the "nationalistic pain" of not being able to "catch up."

Because of these crises of identity and confidence, the egalitarian-nationalist intellectuals perceive dependence as coercive. They do not experience inferiority complexes, at least not always, but they do feel that as long as something native is not created to compete with the foreign, the native society is not vindicated. The type of nationalism involved is in part an "effort to find self-respect, and to overcome the inferiority of the self in the face of the . . . power of the foreign metropolis." [23]

The egalitarian part of the ideology has its roots in Marxist theory, on the humanist rather than the historical-determinist or "scientific" side. These Marxist-humanist ideas have a symbiotic relation with nationalism. The politics of egalitarian-nationalist intellectuals contain both socialist and populist elements, but these are "secondary to and derivative from their nationalistic preoccupations and aspirations. Economic policies have their legitimation in their capacity to raise the country on the scale of the nations of the world. The populace is transfigured in order to demonstrate the uniqueness of its 'collective personality.' The ancient culture is exhumed and renewed in order to demonstrate, especially to those who once denied it, the high value of the nation." [24] This accounts for the intellectuals' advocacy of state intervention in economic matters.

The humanist-idealist orientation does not conflict with the Latin American intellectuals' modern educational background and ideas of progress; it is merely added to it. The combination produces the eclectic mixture of philosophies that I described in the last chapter as "pragmatic antidependency." Thus, when these intellectuals make "choices" about desired goals and means for national progress, they apply the modernist, rational, and utilitarian tradition (so much resented in José Enrique Rodó's *Ariel*) [25] of their Northern counterparts as well as a measure of European humanism and idealism. In this way they are able to look at the future without losing sight of the past.

The egalitarian-nationalist intellectuals cannot separate the two traditions: they cannot simply look back to humanism and tradition and cut the links with modern symbols and sources of knowledge, nor can they remain indifferent to their perception of nations "working for other nations" and "human beings remaining outside the benefits of modernization." They are thus pragmatic; they do not want to get rid of the foreign investor and foreign technology, but they are emphatic about the terms they want and they look forward to the day when their country will no longer need the foreigner.

For example, egalitarian-nationalist intellectuals who study in foreign universities bring back new methods, values, and ideas about how to connect bits of information and about physical causes and effects. Systems analysis is a case in point. But when applied to local problems, these tools often prove inadequate to the task. While they may or may not acquire a new meaning to fit the new circumstances, these imported techniques certainly validate the nation's desire and need to create something, different if possible, by itself.

Dependence is seen as coercive because it hurts the intellectuals' national as well as humanistic, idealistic, and egalitarian conscience. The solution for all these conflicts thus becomes self-reliance. If Rodó were living today, that is what he would advise: fulfillment and self-creation. The reduction of coercive dependence and the achievement of national or regional self-reliance is thus the "affirmation of man through his own creation,"[26] through his own science and technology.

Egalitarian-nationalist ideas in Latin America are not confined to the last part of this century—the Mexican Revolution may have been the first large event influenced by them. But it is possible that too much post–World War II optimism about the prospects for rapid development hindered the maturation of ideas and theories of dependency and self-reliance. "Magic" solutions such as ISI were thought to be the trick, and egalitarian-nationalist aims were thus "externalized" or put on hold. But when solutions proved to be not so magic, frustration touched the most vulnerable areas of the egalitarian-nationalist sensibility; the result was a cry for reform embedded in a theory of dependency: for "self-determination" and "self-reliance."

The Pragmatic Antidependency Guerrillas

"Immediately after the end of the war," writes Edward Shils, "I met Leo Szilard through some old Weimar friends. For a short time we formed a two-member alliance; Szilard conducted and tried to control a vast network of such two-member alliances. He was a wonderful man but he was a conspirator, a benevolent conspirator, . . . benign and warmhearted . . . , but a conspirator nonetheless. He regarded politicians as instruments, not as collaborators."[27] Conspirators, as Shils calls Szilard, the famous nuclear physicist, are "subversive": they use guerrilla tactics, with ideas as their weapons.

In the first country I visited during one of my research trips I talked with an energetic intellectual who specializes in science and technology questions for his government and for regional and international organizations. Today he is considered a major Latin American authority. As I asked about the story of science and technology in his country and how policies were born and evolved, he began to talk about the role he and others played, which in the 1970s was quite considerable. "You see," he said, "we were all the time trying to influence these people to accept our ideas. In fact, I am going tomorrow to meet with government officials to discuss what our strategy should be and how we can preserve our achievements. We cannot let them rest, *we work like a guerrilla, creating a space for maneuver.* . . . We need to create a new ideology, to provide a reinterpretation of the role of science and technology in the condition of underdevelopment." Another bureaucrat-intellectual, working for the Andean Pact, added: "Once we are in the administration we make heroic decisions."[28]

These two gentlemen, like many others I came to know, are part of what I call the Latin American *pragmatic antidependency guerrillas.*[29] They use ideas as weapons in their studies, in their work for and within the government, in their formal and informal linkages with politicians and other high administration officials, always trying to influence policy with varied degrees of success. What makes the pragmatic antidependency guerrillas a subversive elite to the full extent of the term, and what makes them powerful, is the authority these intellectuals can acquire in their own countries and in regional and international forums where they represent their countries. "Because of this authority," writes Hirschman, "the process that in the realm of science and technology is known as the protracted sequence from invention to innovation often takes remarkably little time in Latin America with respect to economic, social, and political ideas."[30] In Latin America, then, cognitive change occurs rapidly and creates the conditions under which the pragmatic antidependency guerrilla operates.

The pragmatic antidependency guerrillas are not a monolithic group. Each individual has his or her point of view and gives different advice to policymakers. Policymakers, as will be shown in the empirical section, tend to look to intellectuals for help that fits not only their ideological perspectives but also their political interests. And in many cases policymakers are good at setting one intellectual against another. Nevertheless, from a broad ideological perspective, the pragmatic antidependency guerrillas stand as a distinctive dependency-conscious group that has driven and probably will continue to drive the policymaking process toward scientific and technological autonomy.

These intellectuals were not always understood or accepted by their governments, but through their publications and their work as technocrats they often succeeded in influencing the policy-making process, both directly and indirectly. They helped in the creation of new institutions and became part of the planning process and decision making. Although their impact has been haphazard, their influence on the consciousness of other actors and on the process of cognitive change, that important prerequisite for political change, has in some cases been extensive.

An organization that has actively and passively influenced policy through the transmission of ideas and ideology and from which pragmatic antidependency has drawn abundant inspiration is the Economic Commission for Latin America (ECLA; Spanish, CEPAL). In all the subject countries there was general agreement that ECLA was very often the catalyst in the process of cognitive change. "In a certain way," said a former Argentine minister of economy, speaking of the change in economic policies in Latin America, "they [ECLA] were the baby's parents." This influence was twofold: ECLA influenced the Latin American intellectuals, who later influenced politicians; and it also influenced politicians directly. For example, José Pelúcio of Brazil, an economist who is regarded as the major promoter of science and technology in Brazil during the 1970s and who strongly influenced João Paulo dos Reis Velloso, former Brazilian minister of planning for ten years, admitted that ECLA's effect on him had been considerable. At the same time, Velloso said that he too had felt an impact from *ideas cepalinas*. In effect, ECLA succeeded in turning itself into a "focus of Latin American identification . . . of autonomous, creative, and new thinking," because it had the capacity to "recruit the best people . . . and . . . provide them with adequate financial support for their studies, their geographical mobility, and their standard of living."[31]

Ramon H. Schulczewski, deputy director of the science and technology division of the ministry of planning in Bolivia, said: "Nobody knew what development was until the first CEPAL. They conquered Chile, and Chile with its ideas conquered Bolivia. Development became fashionable. These ideas were taken up by a group of engineers and economists with influence in the government. What they did was conceive a ministry of planning."

Figure 3 graphically portrays the ways intellectuals affect the policy-making process with the aim of bringing about economic change. One important means by which a subversive elite such as the pragmatic antidependency guerrillas can influence policy making, given that in

Figure 3.   Intellectuals and the Policy-Making Process

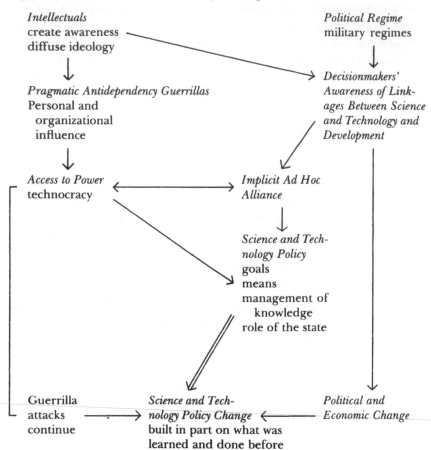

most cases the politicians have other ideological views, do not understand many of the problems involved, or even care little about science
and technology questions, is what we may call the *implicit ad hoc alliance,*
by which the king rules and the intellectual "makes the weather." [32]
Such alliances are based on mutual needs, mutual interests, and the
short-term objectives of the decisionmakers. [33] I will review some of
the elements that describe the creation and workings of such an alliance and that allow the pragmatic antidependency guerrillas to place
their ideas at the center of power and to implement them. (The following points should be viewed as hypotheses, not as facts; there may
also be significant differences between countries.)

1. When dealing with economic development, with its complex political-economic issues, both national and international, the decisionmakers rely on the *técnicos* and their expertise to explain the available alternatives. The "strength of the technicians," wrote Raymond Vernon, describing the Mexican *técnicos*, "lies not so much in their powers to shape policy directly as in their capacity to choose the technical alternatives which are presented to their political masters. But this is a very potent force in itself. And when the instructions from above are ambiguous or when the situation calls for technical action in the absence of instructions, the power of the technician is enhanced even further."[34]

2. Decisionmakers often lack knowledge and understanding of science and technology and of their implications for economic development and, even more, for society. Intellectuals may then not only serve as technical advisers, but they also may be the first to confront the decisionmakers with the problems involved in development; they become the "eye openers" and only source of knowledge for the political elites.

3. There is often a sense of urgency in the actions of decisionmakers, who are keenly concerned with the question of legitimation. This is especially true when leaders have come to power through undemocratic means and must demonstrate their superiority over the vanquished regime. Thus they turn to intellectuals for quick solutions. But this urgency often does not allow them time to choose among plans or even among intellectuals: they grab what is available.

4. Many top decisionmakers, whether civilian or military, were long ago sold on state planning as a strategy to achieve economic goals and to allocate values and resources to society. Other administrators gained an appreciation of planning through education at home and abroad, and still others may have been receptive to planning because they saw it as a trend.[35] As planning becomes more fashionable—and we have seen this happen in Latin America—more planners enter the ranks of technocracy to staff planning ministries and other ministries and agencies. The affinity of pragmatic antidependency ideas with planning also increases their infiltration into the governmental processes.

5. Decisionmakers may feel comfortable with pragmatic antidependency policies for their own ad hoc reasons:

   a. The nationalist side of these policies is very appealing to many
      Latin American (especially military) leaders.
   b. Decisionmakers, particularly if they belong to the military, per-
      ceive a close link between science and technology and military
      power. Military leaders, concerned primarily with national
      security, are therefore often quite receptive to the ideas of
      bureaucrat-intellectuals, who may desire to implement a nation-
      alist policy for very different reasons of economic development
      and equality.
   c. Political leaders are also motivated by national prestige. They
      may want to go along with a self-reliant project—such as setting
      up an autonomous nuclear program, capital goods sector, or
      electronics industry—because of its potential to boost their
      country's prestige.
   d. Some pragmatic antidependency policies may address short-
      range economic problems; the most typical and most frequent
      example is a balance of payments deficit. A decisionmaker who
      places a high priority on this problem may decide to favor in-
      digenous technological projects and control the remittances
      that the subsidiaries of multinationals send abroad.
   e. Finally, decisionmakers may be unaware of the ideology of the
      pragmatic antidependency intellectuals. There might occasion-
      ally be strictly technical cases where these intellectuals would
      separate their personal views and their jobs, but very often
      when intellectuals choose to work for a regime that they do not
      accept and may even resent, it is because they desire to try to
      influence it from within.

   Two anecdotes from Brazil provide good examples of this last
point. In a state technological institution I interviewed a high state
official responsible for domestic technological production. His dependency outlook became clear to me during our conversation, and at the
end of the session I asked whether he had published his views. He
replied, "Do you think if I had published anything I would have this
position?" The second event occurred during a visit with a technocrat-
intellectual working for a state institution. I had learned that he was
part of a group responsible for the creation of a certain sector's self-
reliant capability, and I was almost positive he would be a pragmatic

dependentista. He received me in his office, where another man was working. We spoke for ten minutes or so, and I was very surprised at the tone and content of his answers to my questions. Suddenly the other man left the room, and my interviewee said, "Now we can talk; I could not talk while that man was around because I do not want him to know my position." He then explained that many of his colleagues had left the government because of their disappointment with the policies but that he preferred to stay and influence policy from within.

Is the objective of influencing policy from within the only reason these intellectuals become technocrats? It is certainly a significant one. But military governments may also create an incentive by what Di Tella called "technocratic Nasserism." That is, military regimes make it possible for intellectuals to influence military officers or groups and serve as their advisers for as long as these military are in power or in favor.[36] Another motivation is a negative one: lack of opportunity and poor remuneration for technical jobs, with the only immediate alternative being to leave the country. Some intellectuals prefer to stay and fight the system from a technocratic position in the hope of getting a more intellectually rewarding position if things change. Furthermore, as part of their job they may even create an institution, which might later recruit them as directors, high officials, or scientists. And of course, personal motivations such as financial security cannot be discounted.

How are the intellectuals who make up the pragmatic antidependency guerrillas recruited? From what we have seen so far, it appears that recognition of their work—their technical expertise—is the primary factor. They may be prominent in a certain field at a national university or have publications on the subject, or they may have become internationally known through their work in regional organizations such as the Organization of American States (OAS) and the Andean Pact or an international organization such as the United Nations. The guerrilla gains access to the center of power as a result of the compartmentalization and conflict among bureaus and ministries of the government. A minister can bring his intellectual "pal" into the system, whereupon this intellectual becomes a fighter against policies that are at odds with the views of his recruiter—from whom he gets his power and to whom he reports directly. Usually his bureaucratic life will last only as long as that of his recruiter.

The guerrilla also gains access to the system through friendship or camaraderie, an important aspect of public administration in Latin America. Some positions may be offered to repay a favor, with the recruiter unaware of the intellectual's ideology. But one of the most, if not *the* most, important means of access is what Rafael Kohanoff, Argentina's chief adviser for science and technology in the second Peronist government, calls the "confidence channel": one or more experts get into the power circle, get some political clout, and open the doors to colleagues in their primary ideological group, seeking advisers and planners among people they know and trust. As Kohanoff stated, "Without the 'confidence channel' the experts' influence would be but secondary." When Kohanoff became assistant to José Gelbard, minister of economy in the Peronist government, he called on experts whose ideas followed his own, such as pragmatic dependentistas Alberto Aráoz and Carlos Martínez Vidal. When Isaias Flít was named director of the Institute of Industrial Technology and Technical Norms in Peru (the institution in charge of technological development and technology transfer in the early 1970s), he brought along Gustavo Flores, Francisco Sagasti, and others. Once inside the government, bureaucrat-intellectuals have at their disposal the existing bureaucratic organizational apparatus. But they can also devise new institutions, new linkages among institutions, and their own "innovations" to facilitate transmission of their ideas.

II

Argentina's Science and Technology Policy, 1966–1982

5

Two Ways to Travel

One way societies differ, Hirschman has argued, is in the kind of problems their policymakers decide to take seriously. Furthermore, he says, in Latin America, where attention to the ideologies behind the problem is a common feature of problem solving, elaborate theories are sometimes needed to forge a causal link between a privileged problem, such as economic development, and a stepchild problem, such as technological development. Such theories therefore frequently have a strong ideological component.[1]

Both Argentina and Brazil identified the same problem, but they chose contrasting roads to its solution. An ideological consensus about modernization allowed the Brazilians to overcome feelings of failure, form positive expectations about the future, and develop a science and technology policy that accorded with that image of the future. The Argentinians, however, were continually hounded by visions of failure and by the lack of a national consensus with regard to the future. Hirschman has appropriately called this behavior *fracasomania*.

Fracasomania is "the habit of interpreting as utter failure policy experiences that actually contain elements of both failure and success.

The very tendency toward systematic fracasomania is, of course, an important ingredient of the subsequent real fracasos."[2] Fracasomania leads to the perception that everything has to start from scratch over and over again.

Fracasomania also means "to shut oneself off from newly emerging cues and insights as well as from the increased confidence in one's capabilities which should otherwise arise."[3] This partly explains the great puzzle of how a country with large physical and human resources could have failed so dramatically in its "journey toward progress" over the past thirty years. Hirschman, writing in general, described the Argentine situation precisely: "When there are special difficulties in perceiving ongoing change, many opportunities for accelerating that change and for taking advantage of newly arising opportunities for change will surely be missed."[4]

I call Brazil's behavior *generative,* where the present is regarded as being influenced as much by the past as by images of a brighter future. In the case of Brazil these images have been nurtured by nationalism, by visions of power and equality at the international level, and by the expectation of becoming a modernized and technologically advanced country.

Generative behavior differs from fracasomania mainly in that development is not undertaken as a reaction against the past or to parts of the society, and it does not start from scratch every time. Rather, a present stage of development represents continuity and purpose, having evolved on a highway broad enough to hold many differing positions, but with definite, agreed-upon borders.

Argentina's Science and Technology Policy

Argentina's journey toward progress is a peculiar case of stalemate and "reverse development" in contemporary political economic history. The reasons are ideological, for Argentina has been a split nation, harboring two conflicting ideas of progress. One has been represented by a loose coalition of conservative, laissez-faire–oriented landowners, industrialists, and financiers, supported by a significant segment of the armed forces. This coalition finds inspiration in the (pleasant) past; it sees Argentina's role in the international economic system to be that of provider for a hungry world and believes that industrialization is economically "bad" because it goes against Argentina's comparative advantages in cereals and meat and politically "bad" because it nurtures nationalist and populist movements, mainly Peronism. The other idea of progress has been represented by entrepreneurs of middle and small enterprises, labor leaders, some members of the urban middle class, the nationalists among the military, and the bulk of the Peronist movement, all of whom believe that Argentina's future lies only in industrialization and autonomy.

Argentina's scientific and technological development, mainly as applied to production, should be understood, at least in part, in light of the above. Whereas one group with its particular ideas of progress has taken science and technology to be part of a cultural and educational endeavor linked to the prestige of the nation and the health of its citizens, the other group has tried to develop a technological system to serve the needs of economic development, industrialization, and economic independence.

Historical Summary

From the end of the nineteenth century, economic development was led by the expansion of cereals and meat production. Between 1900 and 1930 the economy grew at a yearly average of 5 percent, and exports and imports remained in balance.[5] Foreign investment in the growth areas was large, and machinery and equipment and other technology, mainly for transportation and refrigeration, were imported. At the beginning of the 1930s the world economic collapse forced Argentina to abandon this bonanza and turn to industrialization, more specifically to import substitution industrialization (ISI), to generate growth and employment. Owing to strong state support, by 1946 there were approximately 85,000 industrial establishments with almost 900,000 workers.[6] In 1958 industry represented approximately one-third of the gross domestic product (GDP).[7] By then the manufacturing industry had become the main purchaser of technology. Easy manufacturing needs could be met by domestic technology, but most capital goods and core technologies had to be imported.

Whereas the export stage of Argentine economic history established vested interests in the export sectors and left memories of a model that worked, industrialization created an economic and social reality that encouraged nationalist feelings, especially in the domestic capitalist sector, as well as egalitarian demands for redistribution. As the economic bases for growth changed, so did ideologies, political actors, and the industrial capability of the country. One of the unexpected results was the creation of a working class that later embraced Perón and his movement of *descamisados* ("shirtless ones"), which eventually led to the creation of a national private entrepreneur.

These two stages of export- and industry-based growth, arising from contrasting ideas and ideologies, could not but affect institutional development in science and technology or the focus of scientific activities. In the first thirty years of the twentieth century not much happened beyond the previously mentioned developments in transportation and refrigeration. "At the time, no one spoke of scientific policies or of industrial research. State action encouraged and sheltered scientific work as a function of culture, following the old European tradition of the Academies of the eighteenth century."[8] This situation changed considerably in the 1930s and 1940s with the creation of several scientific and technological institutes such as the Argentine Association for the Advancement of Science in 1935, the Institute of Biology and Experimental Medicine and the Technological Institute of the Department of Industry and Commerce in 1945, the Institute of Biological Research in 1947 and the National Atomic Energy Commission (CNEA) in 1950. This institutional building process continued and was even enhanced during the 1950s. In 1951 the Argentine Antarctic Institute was created, followed by the Armed Forces Center for Scientific and Technical Research three years later.

The decade that followed can be called the "golden age" of Argentina's scientific institutional development. Public institutions were established to promote scientific research, and scientists, on personal initiative, created several private scientific institutes. The National Institute of Agricultural Technology (INTA) was set up in 1956, the National Institute of Industrial Technology (INTI) in 1957, and the National Council of Scientific and Technical Research (CONICET) in 1958.

CONICET was created by an act of the president to coordinate, orient, and promote scientific and technological research and to advise the national government on such matters. It fulfilled only the role of scientific promoter. It failed as an advisory body because the government at the high decision-making level had not established objectives as to the role of science and technology in social and economic enterprises. In the end, research policy was based on the needs and objectives of the scientists themselves, the interest of the armed forces in security-oriented research, and goals of the ministry of education.

CONICET's most important activities were the awarding of fellowships for training in Argentina and abroad; the institution in 1961 of a new career, scientific researcher (*investigador científico*); grant subsidization; participation in cooperation agreements, mainly with the universities; and creation of research institutes. By 1967 CONICET had already supported 349 scientific researchers.[9]

INTA was created to promote and coordinate research in the agricultural sector through a complex network of research stations. It was financed by a tax of 1.5 percent on agricultural exports. INTI provided technological services to industry through laboratories and a network of research centers (in cooperation with industry and universities). Financed by a .25 percent tax applicable to all bank credits to industry, INTI was also supposed to provide industry with financial aid. It did not, however, have any planning power for the industrial sector.

Although there was a large increase in the number of researchers (with a growth rate of about 60 percent between 1961 and 1966, well beyond the overall population growth rate) and university students (from approximately 142,000 in 1958 to 235,000 in 1965), the developments in science and technology (besides the CNEA) reflected a clear lack of concern with domestic research and development. Thus between 1961 and 1966 the share of R & D in the GDP fluctuated between "extremes" of .31 and .33 percent. Only small fractions of public expenditures for R & D were devoted to the industrial sector, and these decreased with time.[10]

Even in the late 1950s, when the state sought to fuel the substitution process by actively promoting capital goods, domestic technology was not an issue. Industries in the dynamic consumer goods sectors, such as electrical machinery, automobiles, and petrochemicals, grew much faster than the national technological capacity to supply industry's demands. These developments coincided with the period of internationalization of big American companies. So Argentina called on the multinationals, and they responded—at least for some years.

Importation of technology was considered the best and even the "natural" option, given the lack of trained personnel, R & D incentives, and technology transfer controls. Technological laissez-faire was embraced by decisionmakers at high levels of government and industry, not for ideological reasons, but rather because the linkages between technological and economic development simply were not recognized. In short, there seemed to be no choice between foreign and national technology or between technological laissez-faire and self-reliance. This situation had direct bureaucratic implications.

Public science and technology institutions, facilitative rather than normative and guiding, were the outcome of personal initiative, as were the private institutions that proliferated in those years. Funding came from the overall science and technology budget, but the institutions remained relatively autonomous, deciding on programs and their implementation based on their own prerogatives. Thus the lack of coordination was almost complete. For example, because CONICET was founded on the personal initiative of Bernardo Houssay, a Nobel Prize scientist, it stressed the natural sciences, his field of research.

In 1966 military repression of universities and scientists had a devastating impact on the national capacity to develop an effective science and technology policy. From this event onward (mainly after 1969), the ideological, political, and institutional struggle was set. Those who wanted to define a science and technology policy, establish technology transfer controls, and develop the Argentine capability to innovate pitted themselves against those who opposed all this for ideological reasons or even sometimes because of indifference. The result has been a science and technology system characterized by inadequate university training, which has prevented graduates from grasping the importance and peculiarities of science and technology; insufficient training and information for managers, both in industry and in the universities; lack of classification regimes adequate to each type of activity for technological researchers; scarce communication between sectors; little interest on the part of researchers in the necessities and problems of industry; no state industrial policy to guide the development of science and technology; very little research by productive enterprises, which ignore national capabilities and fail to have an open mind toward science and technology; anomalies in the field of industrial norms; and lack of economic incentives for national technological innovation.[11]

The Evolution of Science and Technology Policy, 1966–1982

We can distinguish four periods in the erratic evolution of science and technology policy in Argentina between 1966 and 1982. The first, from the 1966 military coup that overthrew the government of Arturo

Illia to the revolt by workers and students in Córdoba in May 1969 (known as the "Cordobazo"), was characterized by a *technological laissez-faire* strategy.

The second period covers 1969 through 1972, during which a strategy of *pragmatic antidependency* was attempted. Goals and policies for domestic scientific and technological development were explicitly formulated, and normative, institutional, and administrative measures were set to provide some foundation for its achievement.

But because of bureaucratic rigidities, the fact that the process involved other ideological groups opposed to pragmatic antidependency, and previous policies and measures that were slow to radically change, the pragmatic antidependency approach was applied only in part. Between 1969 and 1973, then, we see an overlap of a pragmatic antidependency approach with technological laissez-faire.

The third period, characterized by a mixture of *pragmatic and structural antidependency,* ranged between March 1973, when the Peronists regained power, and the military coup of March 1976, although from a practical point of view it may well have ended by late 1974, when Juan Perón died (1975 was so politically and economically chaotic that science and technology policy-making was almost nonexistent). During this period strong measures were taken against foreign investment and technology transfer to keep the multinationals, who were seen as a coercive and threatening element, in check. There were no significant changes regarding domestic technological innovation, however. The strategy was more radically egalitarian-nationalist than that of 1969–1973 and was set to counter practices perceived as leading to "a system that distorts the economy of developing countries and increases social inequalities; the transfer of technology is thus one of the main instruments of the so-called neocolonialism."[12] But in spite of a high level of rhetorical attack against multinationals, foreigners, monopolies, and foreign technologies, many key decisionmakers still believed that foreign technology could be important and necessary for national economic development.

The last period (and swing) encompasses March 1976 to 1982 and was characterized by a strong attempt to counter nationalist antidependency policies and to restore technological laissez-faire, with only partial success. Once again the result was neither the former nor the latter—there was an overlap. The persistence of some technology transfer controls and the maintenance of a science and technology law and a policy of contract registry reflected both bureaucratic inertia toward the adoption of the new strategy and, most fundamentally, the impossibility of retreating to an unequivocal technological laissez-faire strategy like that of the 1950s and 1960s. Nevertheless, the commitment to liquidate any vestiges of the former government's antidependency policy was enough to set the fracasomania pattern of behavior in motion. By 1981 a new technology law and de facto elimination of registry led the science and technology strategy as close to technological laissez-faire as it had been since the 1960s.

Technology Transfer

1970   National purchase law (18.875) subjects state enterprises and other entities to the utilization of goods and services that can be produced by national industry.

——   Industrial promotion law (18.587) encourages technology assistance for national industry.

1971   First technology transfer law (19.231) creates the National Registry of Contracts, Licenses, and Technology Transfer, under INTI's jurisdiction, to control foreign technology and conditions of purchase in accordance with the development goals of Argentina.

——   Foreign investment law (19.151) replaces the foreign investment regime in place since the end of the 1950s. Authorization for investments is tied to the incorporation of modern technology and participation of Argentine professionals.

——   Law 19.135 passed to control foreign technology transfer in the automotive sector.

1973   Law 20.545 restricts the importation of foreign technology in the machinery and intermediary goods sectors.

——   New foreign investment law (20.577) severely restricts foreign investment and defines enterprises of foreign capital as those in which national capital participation is less than 51 percent.

1974   Second technology transfer law (20.794), even more stringent than the previous one, fixes ceilings for payments and considers technology payments of multinational corporations as payments for regular investments.

1976   Still another foreign investment law (21.382) takes a 180-degree turn, revising the foreign enterprises definition and eliminating limits on dividends and profits remittances.

1977   Third technology transfer law (21.617) eliminates many of the previous restrictions, while maintaining effective control of licensing agreements and payments. The Registry is placed under a subsecretary of the ministry of industry.

1981   Fourth technology transfer law (22.426), described as "among the most liberal in Latin America," does away de facto with most restrictions and the Registry, which is transferred back to INTI as a powerless institution.

Science and Technology Innovation

1966   Establishment of national planning system and creation of National Development Council (CONADE) and National Security Council (CONASE).

1967   American chiefs of state, meeting in Punta del Este, Uruguay, decide to promote science and technology for development.

1968   Creation of National Council of Science and Technology (CONACYT) as the third leg of the national planning system to formulate, promote, and coordinate state science and technology policy.

1969   CONACYT produces the first science and technology census in Argentina.
——   Creation of National Commission for Geoheliophysics Studies (CNEGH).

1970   Law 18.587 encourages the creation of R & D and technical assistance organizations.
——   National policies decree (46) encourages state promotion of R & D and development of new technologies.
——   National purchase law (18.875) further promotes national industry and domestic technological development.

1971   INTI promotes domestic research by linking domestic entrepreneurs with the scientific and technological infrastructure (its institutions and centers).
——   National Plan for Development and Security 1971–1975 calls for autonomous science and technology development.
——   Formulation of science and technology national plan, "Objectives, Goals, and Action Guidelines 1971–1975."
——   Creation of Finalidad 8, a special item for science and technology in the national budget.

—— Law 19.276 merges CONADE, CONASE, and CONACYT into a planning and government action secretariat, under which CONACYT becomes a subsecretariat known as SUBCYT.

—— Creation of Institute of Biophysics Research.

1972   Establishment of Argentina's legal metric system, with INTI responsible for its scientific aspects.

—— Creation of laboratories for sensory research and for embryological research.

1973   SUBCYT becomes Secretariat of Science and Technology (SECYT), under the ministry of culture and education. Transfer of CONICET from the presidency to the same ministry.

—— Three-Year Plan for Reconstruction and National Liberation.

—— "National Science and Technology Policy: Operative Plan 1973" establishes 22 priorities, research objectives, and national research programs in the priority R & D sectors.

—— Formulation of the first four national science and technology programs (Food, Electronics, Endemic Diseases, and Housing).

—— Industrial promotion law (20.560) creates incentives for 100-percent-national-capital enterprises.

1974   Creation of a national system of scientific research institutes and centers; it doesn't work. Same fate for the National Scientific and Technological Advisory Center and the National System of Scientific and Technological Information.

1976   Creation of planning ministry.

—— Creation of Center of Science and Technology Information (CAICYT) under the aegis of CONICET.

—— CONICET, following indications of the "Initial Measures" plan, establishes a system of regional centers to help develop science and technology.

1977   Formulation of a fifth science and technology national program: Nonconventional Energy.

1978   Planning ministry is scrapped.

—— INTI creates the Technological Development Program regime (a risk-sharing venture between INTI and enterprises) to promote technological purchase.

1979 The Interamerican Development Bank (IDB) lends $66 million to Argentina for science and technology. $42 million goes to CONICET.

1980 Elimination of the .25 percent tax out of bank credits to industry, which constituted the bulk of INTI's financial support.

—— Resolution SECYT 341/80 establishes the CONICET Development Program 1982/1985.

—— Formulation of three additional national programs: Radio Transmission, Renewable National Resources, and Petrochemicals.

1981 INTI, instructed by Law 22.426, creates means to provide industry with computerized science and technology information.

—— Reestablishment of SECYT as Subsecretariat of Science and Technology (SUBCYT).

1982 Transfer of SUBCYT to the planning secretariat; creation of science and technology consulting commission (CACYT) to help SUBCYT.

Goals

Until 1968 the Argentinian state had not formulated explicit goals for scientific and technological development. While the overall goal of the regime headed by Gen. Juan Carlos Onganía (1966–1970) was to achieve *economic growth* through backward linkages of industrialization and export of manufactured goods, technological development was not perceived as explicitly linked to that goal. In fact, foreign investment was wooed intensively, which may have contributed to the unrestricted technology transfer practices of the Onganía administration.

The creation of CONACYT heralded the beginning of the government's preoccupation with science and technology for development. The new goal of enhancing state control over technological innovation and adaptation was established, but this goal was still not connected to industrial policy, or to specific industrial sectors, and the origin of the technology to be used for development was of no concern.

Between 1969 and March 1973 science and technology policy goals reflected a concern of the new leaders with domestic innovation for the benefit of national enterprises. These new goals were *subordination of foreign and domestic investors* to the objectives of national economic development and achievement of *self-determination* or freedom of decision in scientific and technological matters. Subordinate objectives included: providing a plausible alternative technological supply for national industry, reinforcing the country's science and technology infrastructure, raising the bargaining power of the state and of private entrepreneurs with respect to foreign capital, working toward a solution to the balance of payments problems, and helping to create an industrial structure capable of sustaining healthy GNP growth rates. According to Aldo Ferrer, the objectives were "not intended to close the door to private foreign direct investment, but to place it in a new framework in which the development leadership would be in the hands of national interests. These instruments constitute, in the final analysis, a new bargaining position vis-à-vis foreign interests. They accept the latter's participation but on different terms than in the past." Thus "the treatment of foreign capital and technology transfer was based on pragmatic and bargaining criteria." [13]

In March 1973 the goals of *self-determination* in science and technology and development of the *national cultural capacity to innovate* were promoted. These goals were attuned to Peronism's objectives of achieving economic independence, enhancing national sovereignty, strengthening national industry in relation to foreign enterprises, and redistributing welfare.

Three years later when the military took power the goals of the Peronist government became heresy, and once again the goals were changed. This time they focused on *preventing abuses* and *mildly subordinating foreign and domestic investors* to the objectives of national economic development, especially modernization and increased industrial efficiency. Modernization was preferred over expansion and growth; and because modernization was equated with foreign technology, foreign technology was preferred over domestic. On this point the government refused to compromise. In addition, a tightly restrictive science and technology policy such as existed before 1976 was seen as a roadblock for foreign investment that had to be removed, and so science and technology objectives became subordinated to and dependent on foreign investment goals. José Alfredo Martínez de Hoz, then minister of economy, summarized the policy of objectives as "having the smallest possible participation of the State in the contracts and, instead of restricting the transfer of technology

with *what may be called an inferiority complex,* the State will use all its resources to aid the business sector to update their activities also with
regard to technology" (emphasis added).[14]

Means

Knowledge-Generating Capital Equipment

Capital goods purchases during the first period were governed by a
special regime created at the end of the 1950s by President Arturo
Frondizi. Among other things, the regime promoted free importation
of entire production-line processes and of assembled pieces of machinery and equipment, such that by 1969 an estimated two-thirds of
imported machinery was duty free.[15]

In 1971 an import licensing system went into effect by which capital
goods imports could be denied if equivalent goods were available domestically. In late 1973, during the third science and technology policy period, capital goods imports were further restricted by the im-
position of quotas and the elimination of all concessions of reduced
tax rates or tax exemptions. Approximately $2.25 billion worth of
capital goods were imported between 1971 and 1975.

During the first part of the fourth period, capital goods still required
prior authorization and an import license, but financing and payment
conditions were eased. By mid-1978 most of the restrictions on the
import of capital goods were eliminated to ease the way to modernization. Imports of machinery and equipment totaled $1.020 billion in
1978, and in 1979 an estimated $1.737 billion. Between 1976 and 1980
some $6.584 billion in capital goods were imported, almost three times
as much as between 1971 and 1975.[16]

Technology Assistance, Foreign Patents, and Know-How

Argentina's first technology transfer law (19.231) was enacted on September 10, 1971. It created the National Registry of Contracts, Licenses, and Technology Transfer—under the jurisdiction of INTI—
which required the registration of contracts governing transfers of
trademarks, patents, industrial designs and models, detailed engi-

neering, and technical know-how and assistance. Registration could be denied when the contract did not involve technology transfer, as is the case with many trademark contracts, or:

- When the contract's objective would include the import of technology that can be proved to be domestically available.

- When the price or compensation would not be related to the contracted license or the transferred technology.

- When the rights to be awarded would allow direct or indirect regulation or influencing of national production, distribution, commercialization, investment, research, or technological development.

- When an obligation to purchase equipment and raw materials from a specific and foreign source would be established.

- When a prohibition to export, or to sell with the aim of exporting, national products would be established, as well as when the selling rights would be subordinated to foreign authorization or when exports would be limited or regulated, by whatever means.

- When the requirement to give up patents, trademarks, innovations, or improvements that could have been achieved through the contracted license or the transferred technology would be established on onerous or gratuitous pretense.

- When sale or resale prices for national production would be imposed.

- When the hearing and resolution of cases would be deferred to foreign courts rather than submitted to a national court having jurisdiction in the particular matter.[17]

A few months earlier, Law 19.135, on the reconversion of the automobile industry, had set similar regulations: "It made mandatory the presentation of all contracts related to technology imports and permitted only the registration of those which did not impose license restrictions. . . . The law established that royalties could be calculated only as a percentage of net profits generated by the corresponding licensees. It also set a ceiling of 2 percent of net sales on such royalties."[18]

A new technology transfer law (20.794), enacted on September 27, 1974, required prior official approval of any agreement having transfers of foreign technology as its principal or accessory object. Agreements were not approved if the technology was "opposed to national policies," had a negative consumption or distribution effect, or did not promote progress; if the technology was locally available or could hamper national technological development; if there were not enough guarantees that licensees' capabilities to use the technology would be maintained; if the technology's price exceeded its benefits; if the product was not disaggregated into its component parts; if the price was undetermined; if there were mandatory joint licenses; or if the contracts imposed foreign courts in the case of conflicts.

In addition, the Registry could fix ceilings—generally related to net sales—on amounts and terms of technology transfer payments for any sector, activity, or specific goods. Until ceilings were fixed, contracts either with prices higher than 5 percent of net sales or with terms longer than five years required special approval. For multinational firms and their local subsidiaries, technology payments were considered investments; approval was required for remittances abroad.[19]

On August 12, 1977, the pendulum swung again with the enactment of a new technology transfer law (21.617) that eliminated many of the previous restrictions while maintaining effective control of licensing agreements and payments. The new legislation was generally more flexible, recognizing Argentina's need for foreign technology. It gave officials more latitude in exercising control, and it specified which agreements required registration and approval and which did not.[20] For example, the entry of foreign technicians to install or start up a plant, repair equipment, or engage in basic engineering work or gratuitous transfers of licensing and trademarks would require no approval. The law was also stipulated that approval could be denied if the contract carried restrictive clauses; but this did not happen frequently.[21] The legislation allowed technology payments between a subsidiary and a parent company, but it did not permit prepayment of royalties to parent firms, lump-sum payments, or trademark royalty-payments between parent and subsidiary companies. Royalties were limited to 5 percent of net sales, but in case of high technology the limit could be exceeded.[22]

On paper, the law was more restrictive than the policymakers were claiming. But on March 12, 1981, the government of Gen. Jorge Rafael Videla, before turning its mandate over to Gen. Roberto Viola, enacted a new technology transfer law (22.426). Described as "among the most liberal in Latin America," this law eliminated percentage ceilings on royalty fees, and it no longer required state approval for contracts between unrelated firms, while those between related firms were judged acceptable if the fees fell within the going market rates for equivalent technology.[23] It also stipulated that registration of technology contracts (besides those between affiliates and domestic companies) was for information purposes only and that technology contracts were legally binding whether or not they had been registered or approved by the state, although companies that failed to register their agreements were liable to certain penalties. The new law omitted clauses to prevent abuses and avoided imposing restrictions on the terms allowable in a licensing contract. Thus, for example, it did not prohibit the licensor from setting resale prices or limiting exports.[24]

Table 3.   Argentina. Technical Assistance, Foreign Patents, and Know-How, 1968–1981

	Number of Authorized Technology Transfer Agreements	Authorized Technology Transfer Payments ($ million)	Remittances Abroad ($ million)	Technology Transfer Law
1968		64.3[a]		
1969		127.7		
1970			70.5	
1971			79.8	19.231 (September)
1972	1,706	81.6	54.3	
1973	129		82.0	
1974	125		100.9	20.794 (November)
1975	111	54.4	66.7	
1976	116	32.0	38.7	
1977	120	34.0	53.7	21.617 (August)
1978	323	157.0	169.6	
1979	510	321.0	129.6	
1980	495	582.0		
1981	528	579.0		22.426 (March)

Sources: SECONACYT, *El sistema científico-técnico y su relación en el sistema socio-económico* (Buenos Aires: National Planning Directorate, Dec. 1970), 32; internal memorandum of the Technical Subsecretariat of the Secretariat of Industrial Development; draft documents supplied by INTI; UNCTAD, *Legislation and Regulations on Technology Transfer: Empirical Analysis of Their Effects in Selected Countries* TD/B/C.6/55 (Geneva: United Nations, Aug. 1980), 9, 24.

[a]Licenses and patents only.

Technology transfer payments grew at an average annual rate of 26.9 percent between 1965 and 1970, but only at 9.6 percent between 1970 and 1978.[25] These rates reflect the restrictive measures of the second and third periods. Table 3 shows an overview of authorized technology transfer agreements, payments, and remittances abroad for accrued royalties and other fees. This information is far from complete, but it is clear that there was some reduction in these activities starting in 1973 and then an increase after passage of Law 21.617.

It is likely that either the antidependency strategy of technology transfer was not adequately implemented or the time available to each set of regulations was insufficient to bring about a significant change. Thus, for example, the ratio between technology transfer payments and R & D expenditures climbed from 2 : 1 in 1969 to more than 3 : 1 in 1972. And according to an INTI study on 1,408 contracts registered between March 1 and December 31, 1972, only one-fourth of the contracts placed no export restrictions.[26]

Control was similarly lacking in the area of patents. Even during the years of the most stringent legislation, the number of patents presented by foreigners and granted by the patents registry was far higher than for nationals (at an annual average of approximately two-thirds of the total number of patents applied for and granted).[27]

Direct Foreign Investment

The foreign investment and industrial promotion laws (14.780 and 14.781, respectively) enacted by the Frondizi government in 1958 and 1959 remained in place throughout the 1960s, indirectly affecting science and technology. The foreign investment law aimed very liberally at attracting foreign capital through various incentives, such as tax exemptions. It provided unlimited official bank guarantees for credits obtained abroad as well as free importation of entire production-line processes and many individual pieces of machinery and equipment.

In 1970 and 1971 new industrial promotion (18.587) and foreign investment (19.151) laws were passed, expressing a preference for developing national technology and for improving the conditions for reception of foreign technology. Incentives were created for foreign capital investment concerns to associate with nationals, with authorization dependent not only on the potential contribution to national economic development, balance of payments, and export considerations but also on the extent to which modern technology and the participation of Argentine technicians and professionals would be furthered.[28]

When the Peronists came to power they expressed their antipathy for foreign investment through a new and very restrictive law. It defined enterprises of national capital as those in which national capital represented 80 percent or more of the total. Article 2 of the law stipulated that in the case of associations between foreign and national capital, the technical, administrative, financial, and commercial direction of the enterprises would be in national hands. Other important clauses stated that all foreign investment contracts required approval of the executive power; no ventures would be approved that restricted export possibilities or that could not be submitted to Argentine courts in case of conflicts. The law mentioned some "strategic" sectors in which foreign investment was banned: public services, insurance, financial activities, public information means, and agricultural production and commercialization. Also, foreign enterprises were forbidden to repatriate earnings in the first five years of operation. The lid on profit remittances was placed at 12.5 percent, or the interest rate in the currency in which the investment was registered plus 4 percent.[29]

The tide changed again in August 1976 with the enactment of a new, strikingly different foreign investment law (21.382). It reduced the amount of capital that an enterprise had to own to be considered "national" from 80 percent to a simple majority, and it eliminated the limits on dividends and profits remittance but imposed an excess profits tax (over 12 percent of the registered capital). Capital could be repatriated after three years. The government retained some control of approval, particularly with regard to contract registration. No areas were off limits, but investments in defense, public services, mass media, energy, education, and banks, insurance, and other financial institutions required previous approval.[30] Foreign firms and national firms were now considered equals where incentives and attempts to invest in certain geographical areas were concerned.[31]

The liberal laws toward foreign investment during the late 1950s and the 1960s allowed multinational subsidiaries to establish themselves firmly in the chemical and petrochemical, automotive, agricultural machinery, and mining sectors as well as in pharmaceuticals, food, and tobacco.[32] "By the late 1960s foreign-controlled enterprises are estimated to have controlled from 25 to 30 percent of the Argentine market for domestically produced manufactured goods; and out of the ten largest firms in Argentina, which together account for about 10 percent of total industrial output, seven are subsidiaries of foreign companies. Foreign firms also tend to be concentrated in technology-intensive industries."[33]

A look at the technology transfer contracts registered in 1972 tells us that 100-percent-national-capital enterprises represented less than half of contracts and only 19 percent of the technology transfer payments. Of the remaining 81 percent of the payments, roughly 19 percent were attributed to 100-percent-foreign capital enterprises and 62 percent to joint ventures between foreign and national enterprises. Firms with 50 percent or more of foreign capital amounted to 43 percent of the contracts and 64 percent of the payments. A very high percentage of sales were under license: almost 100 percent for wholly foreign-owned enterprises and more than 60 percent for enterprises of 50 percent or more foreign capital.[34]

Table 4 illustrates the ups and downs of foreign investment in Argentina between 1959 and 1980. It is clear that foreign investment reacted faithfully to enacted legislation and thus to the goals of the policymakers in power. When the Onganía government created incentives

Table 4.   Argentina. Foreign Direct
Investment, 1959–1980

Year	Investment (in $ million)
1959	209.3
1966	2.5
1967	13.1
1968	31.5
1969	59.1
1970	9.8
1971	20.3
1972	8.9
1973–1975	a
Mar. 1977–Sept. 30, 1980	2,200.0

Sources: BIC, *Investing, Licensing and Trading
Conditions Abroad: Argentina* (New York), Oct. 1975, 3,
and June 1981, 2; Guillermo O'Donnell, *1966–1973:
El estado burocrático-autoritario* (Buenos Aires:
Belgrano, 1982), 424; Juan V. Sourrouille, "La
presencia y el comportamiento de las empresas
extranjeras en el sector industrial argentino,"
Estudios CEDES 1, no. 2 (1978): 59.

aVirtually nonexistent.

to bring multinationals to Argentina, significant increases in foreign investment resulted between 1967 and 1969 (although well under the amount of investment registered at the height of the Frondizi developmentalist actions in 1959).

After 1970 foreign investment once again decreased, and it almost stopped during the Peronist period. But still, in contrast to the almost complete lack of foreign investment in the four years preceding the March 1976 military coup, $2.2 billion poured in to Argentina from 1977 to the third quarter of 1980. This amount is especially impressive if we consider that the total direct investment to Argentina up to 1974 was approximately $2.3 billion.[35]

Management of Knowledge and Information

Education, training of experts, and R & D have been the most significant activities in knowledge and information management in Argentina, mainly during the last three periods under scrutiny. There were a few attempts to set up information retrieval systems during the last two periods, but with no significant results. No substantive forecasting was ever undertaken and there was only minimal planning during the second and third periods.

Science and Technology Training and Research

Civilian research activities were centralized in the National Council of Scientific and Technical Research (CONICET) until the creation of the National Council of Science and Technology (CONACYT) in 1968, when the task of coordinating research became shared. Most scientific investigation was undertaken at the other large institutions, such as CNEA, INTA, INTI, and the universities. Very little contract research was performed by industry—industrial firms were seldom willing to pay for technological services provided by local organizations, and most research activities were not oriented toward industry.

Higher education.   Argentina has had a highly developed, relatively efficient educational system with a large number of higher education institutes in the basic sciences and engineering. But the total number of university students per 100,000 inhabitants declined from 2,352 in 1975 to 1,811 in 1978 and 1,867 in 1979. Total expenditures for higher education, 2.6 percent of the GNP in 1975, decreased in 1977 to 2.4 percent but then rose to 3.2 percent in 1979.[36]

A few institutional measures were taken to improve the level of technical expertise at the universities, such as the creation of the secretariats or departments of Science and Technology and Scientific Research within the universities (for the design and implementation of policies to promote research in specific fields of study) and the setting up of university extension courses, often organized via interuniversity agreements.

Higher education suffered greatly from the politicization of the universities. The military crackdown on the universities in 1966 caused irreparable damage, leading to a massive brain drain and the waste of much potential technical skill. The repression reached its height during the third period, especially after 1974. Even more harmful to higher education, though, was the reaction against Peronism after the 1976 coup, when the universities became the stage for violent struggle with the students suffering high casualties. By 1979 the number of university students had decreased 27 percent from 1976 levels.[37]

R & D.   R & D indicators for the periods under study show that although there has been some improvement, levels are still well behind Argentina's possibilities. Table 5 gives the total personnel and the

Table 5.   Argentina. Personnel Involved in Science and Technology Activities, 1969–1980

	Personnel (per 10,000 population)				R&D Personnel as Percentage of Total
	Scientists and Engineers		Technicians		
	Total	R&D	Total	R&D	
1969	100.1	1.9	625.5	2.5	.6
1972	139.2	3.0	763.0	4.1	.8
1976	165.2	3.1	894.2	4.3	.7
1980	—	3.5	—	4.9	—

Sources: UNESCO Statistical Yearbooks: 1971, 650–51; 1974, 642–43; 1978–1979, 844–45; 1982, V-126–27; *Economic Information on Argentina,* no. 124 (Sept./Oct. 1982): 41.

Table 6.   Argentina. Percentage of Total Personnel Engaged
in Research and Development, by Sector, 1976–1980

	Productive Sector			Higher Education Sector	General Services Sector
	Integrated R & D	Non-integrated	Total		
1976	4.0	1.4	5.4	37.4	57.2
1978	4.6	1.8	6.4	38.2	55.4
1980	3.7	17.8	21.5	48.5	30.0

Sources: UNESCO Statistical Yearbooks: 1978–1979, 751; 1980, 787; 1982, V-41.

R & D personnel involved in science and technology activities. The total number of scientists and engineers, and the number of scientists and engineers involved in R & D, increased incrementally during the 1970s, with the largest increases occurring during the second and third periods. At the same time, it is significant that, as shown in Table 5, less than 1 percent of the total scientific and technological personnel was involved in R & D during the 1970s. Furthermore, the number of scientists and engineers working in engineering and technology, as a share of total R & D scientists and engineers, was historically low; in 1978 the figure was 14.3 percent.[38] As shown in Table 6, only a small percentage of R & D personnel worked in productive enterprises and integrated R & D; more than half were in the general services sector and over a third in higher education, except in 1980 when the latter category gained appreciably over general services.

The manufacturing industry employed only 245 scientists and engineers in 1976. (Comparable 1980 figures for Japan, Korea, and Canada were 163,867, 4,644, and 7,746 respectively.) Of the R & D scientists and engineers in productive enterprise in Argentina, 54.4 percent were in manufacturing, and they constituted 3 percent of the total number of R & D scientists and engineers.[39]

R & D expenditure indicators do not show a large improvement either. The percentage of GDP devoted to R & D averaged .31 between 1961 and 1966, went down during the late 1960s to .28, and increased during the second and third periods to reach .40 in 1975 and .45 percent by 1980 (see Figure 4).

Most R & D expenditures were current rather than capital expenditures. The latter reflect improvements in the science and technology

infrastructure; these amounted to only 15 percent of the total in
1968, 21 percent in 1972, 11.5 percent in 1974, and 28 percent in
1980.[40] There was no change in R & D expenditures according to
type of research. In 1976, experimental research remained almost as
low as it had been in 1968 (23.6 percent), and basic and applied research accounted for the rest (26.4 percent and 50 percent, respectively). The share of R & D expenditures for industrial purposes also
remained at a low 11.5 percent.[41]

Information Retrieval Systems

No significant efforts were made to create information retrieval systems during the first period, and most attempts during the second
and third periods were unsuccessful, although data systems were instituted within some national programs, such as the National System
of Information and Agricultural Sciences within the National Program of Food and Natural Resources, and the Program of Endemic
Diseases.

INTI's new role (as of 1971) of registering contracts and looking

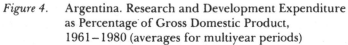

Figure 4.   Argentina. Research and Development Expenditure
as Percentage of Gross Domestic Product,
1961–1980 (averages for multiyear periods)

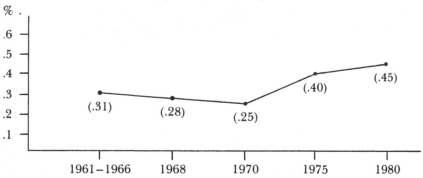

Sources: CONICET, *Programa de centros regionales de investigación científica y tecnológica*
(Buenos Aires, Sept. 1977), vol. 2, chap. 2, 44; Alberto Aráoz and Carlos Martínez
Vidal, "Ciencia e industria: Un caso argentino," *Estudios sobre el desarrollo científico y
tecnológico*, OAS, Regional Program of Scientific and Technological Development,
no. 19 (Washington, D.C.: OAS, 1974), 21; *Economic Information on Argentina*, no. 124
(Sept./Oct. 1982), 42.

for available national technologies created the potential for its laboratories and centers to be linked to national industry, thus making it a useful source of information for domestic entrepreneurs. That happened only sporadically, however. After 1976, INTI emphasized foreign information sources, acting as adviser (through procedures laid down by Law 22.426) to companies wishing to incorporate new imported technologies into their production processes. The means used were to develop and maintain permanent contacts with organizations abroad that could provide information on technology, such as UNIDO (United Nations Industrial Development Organization), foreign chambers of industry, and public offices of the developed countries concerned with technological developments; to encourage subscription to publications on technology; to increase Argentina's participation in fairs, exhibitions, and conferences held in developed countries; to gain access to data banks in the United States and Europe; and to maintain contacts with leading foreign firms and brokers of the private technology markets.

Planning

The 1973 science and technology plan identified 22 research priorities and proposed that national programs be instituted for their consummation.[42] Four national programs were instituted in 1973: Food, Endemic Diseases, Electronics, and Housing. Nonconventional Energy and Radio Transmission programs were set up in 1977, and Renewable Natural Resources and Petrochemicals in 1980.

Role of the State

The state's role in science and technology between 1966 and 1982 was mainly educational and facilitative, and mostly autonomous, although with some reliance on international organizations. From 1969 to 1976 it attempted to play an additional promotional role without engaging in systematic forecasting, but the plans from those years, with the possible exception of the 1973 science and technology plan, had no long-lasting effects.

The role of the state changed with successive regimes, and hence as ideologies changed, but at no time did the state coordinate, let alone guide or plan, scientific and technological development. As Sagasti argued, because the government's administrative decisions lacked unifying guidelines, they served merely to accentuate the ambiguous and contradictory character of the policy instruments, all of which were passive, thus leaving the initiative to the private sector. Not a single policy was coercive, and the state had no means to treat enterprises differentially, to define priorities in technological development, or to affect the key technological decisions of industry.[43]

State Financing of Science and Technology

Finalidad 8 (the science and technology item of the national budget) has been only a very small fraction of the total budget and an even smaller part of the GDP. Its share of the total budget decreased significantly during the third period, increased slightly in 1977 and 1978, and from 1979 to 1981 maintained a level close to that of 1972 and to the ten-year average (1972–1981) of 1.83 percent (see Figure 5). Its share of the GDP grew between 1972 and 1975, went down slightly in 1976, then increased at somewhat higher rates, reaching .37 percent by 1980—still a far cry from a solid investment in science and technology.[44] The average for 1972–1980 was .28 percent.

The relative amounts allocated to state institutions in Finalidad 8 indicate quite clearly the state's implicit priorities in science and technology (see Table 7). For example, the share of Finalidad 8 funds to CNEA and CONICET increased significantly, but funding for INTA and the ministry of education declined. Most striking is the almost nonexistent allocation to industry.

The universities fared poorly, especially after 1975. Their relative share of Finalidad 8 dropped from a 1972–1975 average of 24.1 percent to an average of 7.8 percent between 1976 and 1981.[45] What the universities lost over the years, CONICET gained. Nevertheless, much of CONICET's support goes to the universities: over 70 percent of its fellowship recipients and 74 percent of the 900 CONICET-trained scientific researchers work at universities.[46]

CONICET's relative share of resources dropped during the third period (CONICET did not count among the Peronists' priorities), but increased strongly after 1976: the average relative share for 1976–1981 was almost double that of 1972. This increase is due in part to the establishment of research institutes and regional centers. By mid-1979 CONICET had already established 75 institutes, of which

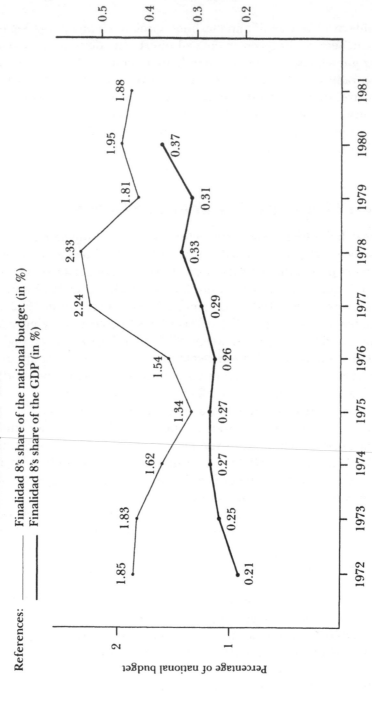

Figure 5. Argentina. Finalidad 8, Relative to National Budget and to Gross Domestic Product, 1972–1981

References: ———— Finalidad 8's share of the national budget (in %)
 ———— Finalidad 8's share of the GDP (in %)

Source: National Administration General Budget and Ministry of Economy.

Table 7. Argentina. Allocation of Finalidad 8, 1972–1981 (in %)

	1972	1973	1974	1975	1976	1977	1978	1979	1980	1981
Presidency (CNEA)	9.9%	4.3%	5.3%	7.2%	13.3%	8.9%	9.5%	15.0%	21.0%	23.1%
Defense	5.5	11.9	6.3	8.3	12.5	19.0	19.9	19.6	11.2	10.2
INTI	3.6	0.9	0.6	0.7	0.3	0.5	0.3	0.3	0.2	0.1
INTA	37.0	35.8	43.9	38.1	29.6	29.2	34.6	26.7	26.3	24.8
CONICET	12.5	10.2	10.2	12.8	25.4	15.9	19.1	23.5	31.4	31.5
Universities	27.9	19.6	23.1	26.0	8.0	5.8	8.0	9.5	8.2	8.3
Others	3.6	17.3	10.6	6.9	10.9	20.7	8.6	5.4	1.7	2.0
Total	100.0	100.0	100.0	100.0	100.0	100.0	100.0	100.0	100.0	100.0

Sources: CONICET, *Programa de centros regionales de investigación científica y tecnológica* (Buenos Aires, Sept. 1977), vol. 2, chap. 2, 48; and Lic. María Luján Marcon (CONICET).

16 are in the technological areas and 38 in the natural and medical sciences.[47] A system of regional centers to promote science and technology in various areas of the country was endorsed in August 1976, to be organized by CONICET and partially financed by an IDB loan.

INTI's share of Finalidad 8 was 3.6 percent in 1972, very low for a country whose manufacturing sector was more than a third of the GNP, even taking into account INTI's independent revenues, the .25 percent tax on bank loans to industry. The portion decreased to an average of .5 percent during the third period and to .25 percent during the fourth. In 1980 this already critical situation was exacerbated when the bank loan tax was eliminated; INTI's 1981 share of Finalidad 8 sunk to .1 percent. In spite of this tight financial situation, however, INTI managed to increase the number of professionals and technicians working in its laboratories and centers.[48]

INTA received the greatest share of Finalidad 8 through 1979—an average of 32.6 percent between 1972 and 1981—and also enjoyed considerable independent financial resources. During the periods under consideration, agriculture accounted for a large proportion of both Argentina's national product and its exports. The elites therefore had a substantial interest in agriculture, which translated to strong state backing for agricultural technology. In addition, since developed countries undertook R & D and training in the agricultural areas most necessary to Argentina at only minimal levels, the Argentinians were forced to do it themselves.

Educational and Facilitative Role

During the four periods the state has mainly played an autonomous educational and facilitative role. Thus it has supported universities, research institutes and centers, and individual scientists. In the fourth period it also helped enterprises (private and public) to have easier and wider access to foreign technology sources—it saw this as support for the "young blood that would revive the system." And through INTI it tried to help private industry make its own technological decisions, based on the belief that the national entrepreneur needs help to abandon the system of protection and adjust itself to a competitive free market.

Regarding the reliance on international organizations, Argentina followed an OAS suggestion that international cooperation must complement a national effort oriented by the country's own priorities and necessities. Among the programs in which the state took an active part were the following:

1. The U.N. Development Program (UNDP). During the second period one UNDP contract financed experts (66.2 percent), fellowships (13.5 percent), equipment (12 percent), and subcontracting (9.3 percent); 17 percent of the total was planned to go to industry.

2. UNESCO. Argentina is a member of the Programs of the Biosphere, International Hydrology, and International Oceanography, the International Program of Geological Correlation, and the General Information Program.

3. OAS. The Argentine state contributed to the OAS's Regional Program of Scientific and Technological Development, and it is also a recipient of various programs. These programs are divided into ordinary and special projects and are carried out within the context of the Interamerican Council for Education, Science, and Culture. Several institutions such as CNEA, INTI, INTA, several universities, and CONACYT became part of the projects.[49]

4. IDB. In 1979 the government secured a bulky loan of $66 million from the IDB, of which $42 million went to CONICET and $24 million to the National Institute of Hydric Science and Technology.[50]

5. Technical Cooperation between Developing Countries. With the purpose of complying with the Buenos Aires Plan of Action adopted in 1978, Argentina sought to strengthen its science and technology links with developing countries outside the American continent.

Promotional Role

The state has been the largest source of R & D funds, accounting for 87 percent during the 1970s, with the rest coming from productive enterprises (7.2 percent), foreign sources (3 percent), and other sources (2.8 percent).[51] During the second period the state promotional role became prominent, as exemplified by the creation of Finalidad 8. Legislative documents encouraged the development of the domestic science and technology infrastructure, promoted research "in the national interest," provided tax incentives for public industrial enterprises engaged in R & D,[52] and instituted a tariff reduction for materials imported for use in scientific and technological research. The beneficiaries of these actions were official national, provincial, and municipal institutions and civil associations. CONACYT also instituted a program to promote R & D in foundations, civil associations, and private entities for the public good, including special support for R & D programs attaining financial self-sufficiency of at least 50 percent.

The state also required its own enterprises to use goods and services that could be provided or developed by national industries (Law 18.875 of December 1970, known as National Purchase Law). One of its most important dispositions was that "engineering and consulting services will be contracted with local firms and professionals," and it prohibited the receipt of credits from foreign consulting firms.[53] Another regulation (Decree 46 of June 1970, called National Policies) encouraged state involvement in and promotion of R & D, the development of new technologies, especially those not being developed elsewhere, and adaptation of complex technologies. It also urged that R & D focus on the adaptation of imported processes and equipment to national conditions and on the nuclear area. The sectors to be promoted "with the objective of increasing national autonomy" were nonferrous metals, capital goods, chemical industrial products, cellulose and paper, and the activities linked to the needs of the armed forces.[54]

The establishment of specific national R & D priorities and guiding programs in 1973, during the third period, provided the state—mainly SECYT—with additional means to promote the sectors and technologies it considered important for the national interest. Also during this period, the National Development Bank opened a special line of credits to promote the manufacture of capital goods with local technology, and executive power was used to give credits or other incentives for the acquisition, study, and perfecting of manufactured product prototypes.

A new regime for industrial promotion was approved in November 1973 (Law 20.560), giving direct benefits in the form of tax relief, credits and guarantees for credits abroad, and tax deferments to 100-percent-national-capital enterprises only. Law 20.852 of September 1974 stipulated further that national industry would be given preference and tax benefits in international bidding on IDB-financed projects.

Among the main industries promoted during this period were mining, steel, and petrochemicals. A national petrochemical program was instituted in December 1973. The state also promoted technological exports through reimbursement benefits of 40 percent for the export of industrial production and transformation plants that fell under the modes of complete plant or turn-key plant.[55]

In the fourth period this reimbursement was reduced to 25 percent, and a special third-period credit to Cuba for industrial technological export promotion was canceled. Between 1973 and 1980 Argentina exported ten infrastructure projects for a total of $247 million and 61 industrial plant projects, some of them turn-key, valued at $160 million.[56] The active promotional role of the state in this last period was focused mainly on the purchase of foreign technology. However, the 22 priorities set in 1973 were maintained, four new national programs were created after 1977, and an important additional promotional role was undertaken: INTI's Regime for the Promotion of Technological Development, by which INTI participates in an enterprise's venture into technological innovation, sharing the risks.[57]

Planning

Although several economic and science and technology development plans were issued, especially between 1970 and 1976, they were only indicative and had little effect on policy. Political and economic instability forced policy making to be based on very short-term considerations that plans could not take into account. Thus, planning and economic and industrial policies coexisted almost without engaging each other.

CONADE prepared the National Development Plan 1970–1974, which had only a formalistic role, and the National Plan for Development and Security 1971–1975, which called for the development of autonomy in scientific and technological development to reduce what it saw as a "strong cultural and economic dependency." It set out to achieve a growth rate for private investment in R & D higher than that of the national product; to achieve that by 1975, not less than 50 percent of the investment in science and technology had to support research relevant to the various priority economic sectors. Legislation, the Registry, and improvement of industrial research were highlighted as possible channels to achieving these objectives.[58]

The 1973 science and technology plan's main goals were to assure the exercise of national sovereignty by raising the self-reliant decision-making capacity based on a solid national science and technology system; to raise the system's efficiency by way of better use of human and material resources; and to promote efficient interconnection between the science and technology system and the productive system to achieve "a more dignified and integral human development and a regional structuring of scientific and technological development that would be more just, harmonic, and equilibrated."[59] The plan was indicative and did not set allocations, but it did initiate three types of programs: priority, oriented, and free.

The three-year economic development plan issued in 1973[60] attempted to *guide* science and technology activities by setting objectives for the various research institutions; however, these objectives were very general and gave only rhetorical support to science and technology.

The military government that came to power in 1976 created a ministry of planning, but no development or science and technology plan was issued. The Initial Measures Plan of 1976 endorsed the development of regional research centers. In December 1979 the military junta released a document entitled "Armed Forces Political Bases for the National Reorganization Process," which called for an active role by the state in science and technology and no bias regarding the source of the technology or expertise; it added that the productive sectors could be treated in terms of "efficiency and comparative advantage." As a sequel to the document, SECYT in its Resolution 341/80 established the CONICET Development Program for 1982–1985. The document called for decentralization, equilibrated development of different science and technology disciplines, improvement of higher education and human resources, and the establishment of information retrieval systems.[61]

Ideology and Policy Making: Fracasomania

6

Technological Laissez-Faire

When the military came to power in Argentina in 1966, they set up a government under the presidency of Gen. Juan Carlos Onganía. Soon they moved against the universities, which they regarded as the focus of subversive activity, and purged faculty members and students, thus inducing hundreds of teachers and scientists to leave the country. This onslaught, known since as the "night of the long clubs," dealt the universities and scientific research a blow from which they have not yet recovered.

The coup was supposedly not the result of a loss of faith in liberal democracy but rather a measure taken to solve an economic crisis and foster rapid economic growth.[1] To cure some of the country's economic ills, such as irregular growth in real production, significant annual price increases, low international reserves, and an appreciable and recurring fiscal deficit, Adalbert Krieger Vasena, who became minister of economy in January 1967, instituted a comprehensive program to coordinate price, exchange, wage, agricultural, and industrial objectives. According to Robert Kaufman, the immediate purpose of these measures was to contain inflation, but they were also devised as the groundwork for three larger developmental goals: expansion and diversification of exports; development of the economic infrastructure; and promotion of massive private-sector investment in heavy industry, primarily with the aid of multinational capital.[2] Export diversification was to emphasize manufactured products (which by 1966 accounted for a third of the GDP) and be fueled by the investment in heavy industry. The policy to expand exports of manufactured products, initiated by Arturo Frondizi's government in 1959, was particularly stressed in this period and, indeed, was regarded as the main engine of growth up to 1976. It is significant for science and technology that inflation was in fact partially controlled, allowing the state to consider more long-range objectives; but at the same time the balance of payments problem was aggravated, a condition the next political regime inherited.

Industrial production grew in 1967 and 1968 by over 5 percent and in 1969 by approximately 11 percent. Manufacturing exports expanded by an average of 169 percent annually between 1966 and 1970, as compared to an average 10 percent annually between 1960 and 1966. In 1972 manufacturing represented 23 percent of all exports; in 1965, 6 percent.[3] This last figure indicates that the policy to stimulate manufacturing exports was working. Also, because Argentina was exporting more sophisticated products, its industries' technological needs were growing. Foreign firms were already heavily involved in export-oriented manufacturing, and many national entrepreneurs supported this participation and called for its continuation.

This economic situation created the propensity for change in science and technology, but it was not enough; explicit choices needed to be made—choices informed by ideologies not to the taste of Onganía and Vasena. The main actors in the policy-making process were Onganía; his assistants; Vasena, who fulfilled the role of economic team leader, supervising the activities of the other "economic" ministries; a group of Argentine technocrats, most of them economists; and the National Development Council (CONADE) and the National Security Council (CONASE). (CONADE, however, had only a marginal role in science and technology planning, concentrating on human resources.)

Also involved in the policy-making process were the Argentine Industrial Union (UIA), representing the largest industries and, usually, the liberal-minded industrialists; and the General Economic Confederation (CGE), "a middle-class organization of smaller, newer, and provincial firms established in 1951 by entrepreneurs from the interior and others excluded from or opposed to the older industrial and commercial groups. . . . Generally, the CGE has favored a statist-nationalist approach including limitations on foreign firms and foreign investments, strict regulation of the transfer of technology, greater tariff protection and creation of a self-contained heavy industrial base."[4] Scientific and technological institutes such as CONICET, CNEA, INTA, and INTI also played a role in the process.

The initiation and formulation of economic and industrial policy were in the hands of Vasena and his economic team. Consultation was limited to the small circle of liberal economists; policy was sanctioned by the president and executed by a mix of bureaucratic controls and market incentives. Vasena's policies received the enthusiastic endorsement of the UIA but were opposed by the CGE, which saw in them a great threat to national industry. Unable to influence policy, the CGE "devoted itself instead to a campaign directed at embarrassing the government by drawing public attention to what it claimed was the 'denationalization' of Argentine industry and finance through increasing foreign takeovers of weakened industrial firms."[5]

The institutes determined science policy, and there was no technology policy as yet. Several science and technology institutions, such as CONICET and INTI, acted more or less independently in response to their own needs—but of course, within the resource allocation framework set by Vasena's policy to control inflation.

The creation of CONACYT in 1968 was largely divorced from the economic policy-making process described above; it occurred rather on the initiative of one of Onganía's advisers, who proposed that CONICET's legally defined scientific and political tasks be separated. This adviser wanted CONICET to continue with its basic research activities, which its director, Bernardo Houssay, strongly supported, but preferred that the policy-making function be placed in different hands. And so CONACYT came into being. But a clash soon developed between this new policy-oriented organization and the scientists, headed by Houssay. The disagreement was ideological, involving the question of whether science should be managed by the state, and it was also understandably political, a question of power. The 1970s saw no easing of the conflict.

CONACYT's main functions were the drafting of a national science and technology policy "to be based fundamentally on the objectives formulated in the General Plan of Development and Security"; integration of the domestic and international security policies on all matters that concerned science and technology; coordination of its activities with CONADE and CONASE "so as to achieve the joint objectives of development and security"; coordination and stimulation of research; formulation and promotion of the science and technology plans assigning special resources and distributing them in accordance with national objectives and the demands of balanced development; assignment of resources for science and technology according to development and security needs; evaluation of the scientific and technological activities undertaken by the public sector; adaptation or creation of new organizations to reinforce and complement CONACYT; and intervention in all other matters concerning science and technology.[6]

The Quest for Technological Self-Determination

The May 1969 revolt of students and workers in Córdoba shook the stability of Onganía's regime and made the top decisionmakers aware that important groups of the population could not cope with the orientation and results of the regime's economic policies. The "Cordobazo" led to Vasena's resignation and his replacement by a slightly more nationalist intellectual, Dr. José María Dagnino Pastore. For all practical purposes, then, the "Cordobazo" became the breaking point between the technological laissez-faire policy and that which replaced it. But although some mild pronationalist legislation was enacted at the end of 1969 and in February 1970, the major changes occurred after June 8, 1970, when the Onganía regime came to an end. Among the factors that contributed to the fall of the Onganía government was the reaction of nationalist groups such as the CGE and the nationalists within the armed forces to its pro—foreign investment attitudes.

The new president was Gen. Roberto M. Levingston, but he was replaced within a year by Gen. Alejandro Lanusse, who held the presidency until the elections of 1973. Policymakers under both regimes firmly believed in a national development strategy, and although this had happened before in Argentina, in 1970 it made a large difference. Ideas raised by intellectuals about a strong national scientific and technological infrastructure and about controlling technology transfer were embraced by segments of the military and by other political actors. And because they were the "right ideas" for the "right time," they were taken to the center of power and implemented.

The strength of these ideas at the time stemmed in part from the April 1967 decision of the American chiefs of state, gathered in Punta del Este, Uruguay, to "put science and technology at the service of our people. Science and technology must be given an unprecedented impetus at this time . . . their organization and implementation in each country cannot be effected without a properly planned scientific and technological policy within the general framework of development."[7]

Prior to the Punta del Este conference Latin American intellectuals met to discuss the science and technology project that the United States was to propose at the conference. As it stood originally, the project envisioned the establishment of two or three large technological institutes in Brazil, Mexico, and/or Peru. This proposal did not satisfy the intellectuals, however; their aim was a comprehensive program. Thus they decided to lobby the Argentine, Chilean, and Uruguayan delegations to submit a program that had been formulated by the OAS with the help of a group of intellectuals, including Emilio Rosenblatt, J. Arias, Máximo Halti Carrére, and Jorge Sabato. The Punta del Este resolution and the subsequent OAS program were the culmination of a decade of efforts by intellectuals who in national and international forums had been creating an awareness of the linkages between scientific and technological development and the ills of technological dependency.

During the 1960s seminars and discussions kept these ideas alive and nurtured them, and economists and technologists began to actively interact. Then, according to Sabato, came the idea to link foreign investment and technology transfer. The result was Decision 24, the Andean Common Market's foreign investment code, which subjected direct foreign investment and technology transfer from abroad to the harshest control measures yet enacted by a regional forum.

Aldo Ferrer, an Argentine economist, was heavily influenced by Sabato's ideas. An outspoken advocate of national industry, Ferrer called for control of foreign investment and technology transfer and advocated an aggressive policy aimed at adapting foreign technology and producing domestic technology. This would probably have been politically irrelevant had he not become minister of public works and later minister of economy during the science and technology policy's second period. Years later, Ferrer acknowledged the effect of the principles and legislation of the Andean Pact on Argentina's actions: "The inspiring principles of Decision 24 and of Argentina's policy in this period are convergent. . . . In the Argentine case a series of dispositions were adopted along the same line as Decision 24."[8] Ferrer's economic program consisted basically of an approach opposite to that of his predecessors: economic expansion and stimulation of domestic industry. However, one important element of the previous economic strategy, the commitment to manufactured exports, remained unchanged.

The GDP did fairly well during these years. The manufacturing sector grew to almost 38 percent in 1972, and there was significant growth in manufactured goods exports.[9] Sophisticated industrial exports (with a high-technology component), in turn, grew from 9.7 percent of total exports in 1969 to 16 percent in 1973.[10] Capital goods, which in 1969 were 28.7 percent of manufactures production, increased to 34.2 percent in 1971, and investment in machinery and equipment as a percentage of gross domestic investment also showed some growth. Machinery's share of total manufacturing, which during 1965–1969 averaged 33.9 percent, grew to 35.9 percent in 1970, 37.3 percent in 1971, and 41.9 percent in 1973.[11]

A strong trend toward heavy and more sophisticated industrialization is thus apparent, with an increase in export manufactures. However, although the emphasis on industrialization created a greater need for sophisticated technology, it did not necessarily induce a policy of technological autonomy. Nor was the development of a nationalist science and technology policy linked solely to economic disequilibrium, even though the negative balance of payments may have intensified the drive toward control of foreign investment and foreign enterprises. Indeed, the nationalist technological policy of the second period can only be understood if we take into account the importance of technological self-determination ideas, so popular at the time. Aided by favorable circumstances such as Argentina's two mild nationalist presidents; a reaction, shared by some senior members of the military, against Vasena's pro–foreign investment policies; and a balance of payments crisis at a time when heavy industrialization processes were generating the need for more and new technologies, these ideas were placed on the political agenda and found—mainly, though not exclusively, via Ferrer—an avenue to political power.

The major actors in science and technology policy making were the two presidents; Ferrer; the CGE, headed by José Gelbard, who in 1973 became the minister of economy; the Registry; CONACYT; and the scientific and technological institutions, such as CONICET, INTA, INTI, and CNEA. Levingston favored national development, a fact that Ferrer considers important in the overall shift in policy during these years. Lanusse, although more liberal-minded than Levingston, continued the nationalist approach. A former secretary of industry said that Lanusse was one of the most "civilian" among the military, that he was in favor of national industry from the start, and that he used to consult with Gelbard on industrial and science and technology matters.

Other interviewees pointed out Ferrer's relative importance in this process. An industrialist said it was not by chance that Ferrer was called on to make economic and industrial policy, because the idea was to react against the liberal policies of Vasena. Another said that Ferrer was able to change science and technology policies in the nationalist direction because there were conditions that from an economic point of view justified the actions: the productive system was functioning with good results; it was the right moment. Sabato, in turn, said that what happened in 1970–1971 was "purely ideological." Asked why in his opinion Ferrer and others with the same ideological position were able to impose their views, Sabato identified two basic reasons: Ferrer's intellectual power and confusion within the military on the subject, with many officers finally taking the nationalist stand.

Ferrer represents a typical pragmatic antidependency position. He accepted dependency as a fact but believed that "there are options to change the course of events and break dependency linkages." The choice, he argued, "is not imposed from outside. It is immanent in each country. The options between dependency and national development lie, then, on domestic political decisions." He added that "in the transnational context, the claim of each society, its aspirations, and its political system's maturity decisively affect the course of events."[12]

Ferrer gives the subversive elite credit for the political change: "The role played by such researchers as Vaitsos, Sabato, Wionczec [sic], and others in this process constitutes one of the most notable contemporary examples of the application of imagination and talent to the management of our countries' economic policy."[13] Marcelo Diamand, industrialist and economist, agrees that the ECLA literature and scholars such as Katz and Wionczek were influential. Juan Valeiras, former director of the Registry, argued moreover that nationalist policies, laws, and regulations derived from a self-reinforcing process by which past successes (for example, CNEA and Sabato's work in

SEGBA [Electrical Services of Greater Buenos Aires]) encouraged efforts to achieve more *argentinización* and independence.

However, Argentina's industrialists, as a group, did not support the new technological policy, mainly because they, too, were divided along ideological lines. Whereas in the main the CGE approved the nationalist measures, private entrepreneurs not affiliated with this institution continued to back foreign capital and laissez-faire policies.[14]

The creation of the science and technology registry and its placement under INTI jurisdiction consolidated for the first time the tasks of domestic technological development and those of technology transfer control, revealing to all concerned the nature of the contracts, their flaws, and the opportunities available to national industry to substitute for foreign technology. The final decision of the reviewing process rested nominally with the secretariat of industry, on which INTI institutionally depended, but in practice, said Valeiras, the Registry made the decisions. Financial questions related directly to balance of payments were the concern of the central bank, which created some bureaucratic links with the other organizations to coordinate policy.

In 1971 CONADE, CONASE, and CONACYT were combined into a national planning system, and CONACYT became the subsecretariat SUBCYT. Conflict between SUBCYT and CONICET and its scientists continued; the latter objected to the moves toward nationalization and planning in science and technology and resented the fact that SUBCYT took control of Finalidad 8. The disagreement led to increasing implementation problems and Registry delays; troubled relations with the central bank, whose policies did not always coincide with those of other state organizations; and worker unrest, which led to delays in technological decision making by national private entrepreneurs. SUBCYT could wield little influence through economic sanctions because some institutions, such as INTI, had functional budget autonomy thanks to the .25 percent tax applicable to all bank credits to industry.

From Pragmatic
Antidependency to Chaos

Free elections held in March 1973 saw the Peronists reemerge with the upper hand. Not only were they more nationalist than Levingston and Lanusse, they were egalitarian as well. The regime's first president,

Héctor Cámpora, held power temporarily until Juan Perón assumed leadership at the end of 1973. But even Perón could not solve the movement's toughest problem—internal conflict.

There were clearly two regimes within this period. The first, lasting until October 1974, was egalitarian-nationalist, represented by moderate egalitarian and extreme nationalist forces such as CGE and CGT (General Confederation of Workers). The mainstream ideology of many Argentine intellectuals merged with that of the government on science and technology matters. There was no need for a pragmatic ad hoc alliance: what in other regimes was a subversive elite became in this regime part of the elite. The government sought its advice, and its members gave it willingly.

This situation came to an end in October 1974 when Perón died and his wife, Isabel, replaced him. She was faced with a choice between two courses of action: she could continue to support the uneasy alliance between the CGE and the CGT, or she could shift to a position favoring José López Rega's conservative faction by breaking with the CGE and repressing CGT dissidents and antigovernment guerrillas.[15] She chose the latter. Gelbard, the former CGE president and first minister of economy, was dismissed, and the "Social Contract," created to seal the bonds of the government with industry and labor, was broken. From then on, under the guidance of López Rega, the president repressed Peronists and non-Peronists alike.

Economic and political chaos followed. In the year and a half before the Peronist regime fell in March 1976, there was practically no economic or science and technology policy making, only political intrigues and emergency economic measures to save the country from complete bankruptcy. Three ministers of economy walked out on the government. Nevertheless, the events of this period were of utmost significance because they influenced the policies and actions of the military, who when they regained power in 1976 set out to deliver what was to be the final blow to what *they* perceived as a "bunch of terrorists that ruined the country."

The basis of economic policy between March 1973 and October 1974 had been the Social Contract, a wide range of strategies intended to achieve price stability, progressive income redistribution, and national economic independence.[16] When this contract was broken, economic indicators changed dramatically. GDP growth was over 6 percent in 1973 and 1974, but the minus signs for 1975 and 1976 (-1.3 percent and -3 percent, respectively) tell an opposite story. With regard to the productive system, the manufacturing share of the GDP remained approximately constant over the period. In 1973–1974 industrial production showed some growth, but then it declined (−2.8 percent and −4.5 percent). High technology sectors experienced negative growth in 1975–1976: machinery and equipment (−7.9 percent and −3.8 percent), basic metal industries (−1.9 percent and −24.1 percent), and automobile production (−16.1 percent and −19 percent).[17]

Science and technology policy between March 1973 and October 1974 was similar to the nationalist economic policy, with a trend toward heavy industrialization, industrial exports, and reduced foreign investment. (The productive sector crisis of 1975–1976 had very little to do with industrial changes, however, and much to do with political changes and the consequent economic mismanagement.) The main political actors in the period were the two presidents, Cámpora and Perón, and the "economic tsar," Gelbard. Key players in the technology area were Gelbard's cabinet assistant, Kohanoff, who was president of the commission to draft a new technology transfer law, and Secretary of Industry Alberto Davie.

When the ministry of economy was reorganized in October 1973, nine secretariats were created, among them two directly connected to the science and technology issue: coordination and industry and economic planning. Through this reorganization Gelbard gained direct control over economic and industrial policy and acquired jurisdiction over the promotion, coordination, orientation, and evaluation of scientific and technological research in the industrial, mining, forestry, and fisheries sectors and over foreign technology controls. Centers for the study of technology transfer and administration were created in most of the ministry's secretariats, each with the task of proposing a technology policy adjusted to the needs of its particular sector.[18]

Technology transfer policy was initiated by Gelbard and his advisers, who were responsible for the formulation of the new foreign investment and technology transfer laws. Also involved were INTI, headed by Jorge L. Albertoni, and its dependency, the Registry, headed by Valeiras. With the technical advice of the INTI centers and laboratories, the Registry suggested what to approve or reject, although INTI had the last word. (The formal final decision was that of the secretariat of industry, but in most cases this was merely a formality.)

Concurrently, a commission was established by the central bank to make sure that technology transfer policy met the requirements of the balance of payments policy. With the participation of several bureaus involved in the making of the aforementioned policies, this commission oversaw the Registry's activity. In practice, though, the commission, following the policy criteria set by Gelbard and his CGE associates, generally simply supported the Registry's decisions.

The CGE during these years was the main source of ideology for the formulation of technological policy. The small and middle-sized (and nationalist) entrepreneurs organized by the CGE found in the Peronist movement an open avenue to political power. The Peronists, at least while Cámpora and Perón were in office, did not object to this—they shared a basic ideological affinity with the entrepreneurs, and besides, they needed CGE support as much as they needed the CGT.

> That the CGE had moved so quickly and convincingly prior to the election is not surprising. Since its creation . . . it had enjoyed access to the Peronist leadership, and during the 1960s had developed the kind of technical expertise which the Peronists themselves were not prepared to mobilize immediately before the 1973 elections. In the short run then, the Peronists needed the expertise of the CGE, and the CGE the political power of the Peronists if the CGE was finally to have an impact on economic policy. Yet, in accepting the CGE proposal, the CGT and Perón were also conceding a great deal, for it tied them to moderate short-term measures to combat inflation and postponed the adoption of the more radical, redistributive measures they had long promised their constituents.[19]

An outgrowth of the CGE was COPIME, the Small and Middle-Sized Enterprises Corporation, dominated by the same personalities as the CGE and sharing its ideas and ideology.

There was a general consensus among the CGE, the ministry of economy and its secretariats, and many Argentine intellectuals on technology, and this proved instrumental in the making of technological policy. Where egalitarian-nationalist intellectuals used both the CGE and the Peronist movement as vehicles to turn ideas into reality, the national entrepreneurs and the CGE used the Peronist movement alone—it was a situation of double access to power. During these eighteen months, then, a close cooperation developed between the decisionmakers and intellectuals such as Aráoz, Chudnovsky, Mario Kamenetzky, Katz, Sabato, and Vidal. As Kohanoff remarked, Gelbard called Aráoz, Kamenetzky, and Vidal and asked for opinions. He was their friend and trusted them. Their help was important—they pointed the way, which saved the politicians some embarrassment.

Meanwhile, in the realm of general scientific and technological research policy an innovation took place. The subsecretariat of science and technology (SUBCYT) was taken out of the planning secretariat's orbit, turned into a secretariat (SECYT), and, together with the National Council of Scientific and Technical Research (CONICET), placed under the aegis of the ministry of education. Although one of the aims of this move was to end past institutional and personal conflicts and thereby rationalize science and technology policy making, the education ministry's newly acquired power and jurisdiction arose largely from strong political pressures within Peronism. One of the consequences of this change was to emphasize the divorce between scientific and technological research, on the one hand, and Gelbard's industrial technology and technology transfer policies, on the other. Gelbard met on several occasions with senior officials of the ministry of education to discuss some administrative way to coordinate political action, but to no avail. Nor did the feud between CONICET and SECYT stop, for both agencies continued to dispute jurisdictions and the type of research to be promoted.

While CONICET continued with its traditional role of supporting research in its institutes and at the universities, SECYT's mandate was to statistically analyze science and technology activities; to establish budget needs and then distribute the budget among the various institutions by means of Finalidad 8; with the foreign ministry, to coordinate scientific and technological cooperation with international organizations; with the ministry of economy, to coordinate the use of international financing for science and technology; to coordinate and promote research; to organize the training of scientific researchers and technical personnel; to draw up fellowship plans and postgraduate programs; and to organize services such as information and data centers. SECYT had to make the decisions about scientific and technological capabilities, which were then to be sanctioned by the cabinet. SECYT was also entrusted with the national programs as well as a special budget allocation, called "Reinforcement Actions," of which up to 50 percent went to the national programs.

In January 1974 a national system of scientific research institutes and centers was created to coordinate scientific research, but the system never really worked.[20] An attempt to establish an advisory center to identify scientific and technological developments as part of the national productive activity met likewise with little success. The larger and older science and technology organizations were too independent, the conflicts between institutions such as SECYT and CONICET too pronounced, and one year—the duration of this experiment—too short a time to change the entire infrastructure. So at the end of 1974 the experiment was over.

Erasing the Peronists' Legacy

By late 1975 the only open question about a military coup was when it would occur. It happened in March 1976, and a new military regime was set up under the presidency of Gen. Jorge Videla. The political change could not have been more drastic, representing the shift once again from a civilian democracy to a military dictatorship. But it also reflected a radical ideological change away from populist and national- ist ideas and toward policies favoring laissez-faire, landowners, financ- ing interests, foreign investment, and imported technology, brought on in part by the revulsion the military and its supporters felt for Per- onism and the economic and political chaos. However, the Peronists and some leftist groups refused to retreat without a fight, and an armed conflict followed the coup (which itself was bloodless) that left thousands of Argentinians dead or "missing." Most of what happened in the realm of economic and science and technology policy must be appraised in the light of these events. As an Argentine economist put it, the 1976 economic reforms were

> more suited to the condition of Argentina in 1960, before the oc- currence of substantial growth both in industrial productivity and in agricultural output and before the surge of manufactured ex- ports, etc., rather than to the conditions of the 1970s. The revival of this diagnosis, and of all the associated prescriptions derived from economic liberalism, must be attributed to the political crisis of the first half of the decade, a crisis which both the Armed Forces and the private sector perceived as threatening to destroy the prevailing social system.[21]

The new economic outlook represented a dramatic cognitive shift from an ideology that stressed equality and economic development through industrialization, with the state playing an active role in the development process, to a laissez-faire ideology mixed with a politi- cal commitment to halting and even reversing the industrialization process.

Events after 1976 emphasized the financing side of the economy; it was believed that capital markets should be free to function without interference and that efficiency and the allocation of resources should become the main economic goal.[22] The fight against inflation became the most important short-range objective.

The Videla government had claimed to have a clear industrial policy based on opening up the economy and instituting laissez-faire measures to bring a high degree of efficiency to the "ailing," "overprotected" industrial sector. The post-1976 economic strategists thought it would be better for one-third of existing industry to survive and become efficient, productive, and internationally competitive than to continue with the old system of protection. With this goal in mind, modernization was imperative, and free technology transfer became the means to achieve it.

This period was therefore particularly bad for industry. With the liberal policy on imports and tariffs, national industries lost large markets to imports, resulting in an increasing number of bankruptcies among industrial firms. Thus, while in 1976 the percentage of the GNP represented by manufacturing was 36.7, in 1980 it was only 25.4.[23] Other signs of distress included the dismantling of R & D and export departments, migration of skilled labor and even of managers and firm owners, reduction of the domestic market, and stalemate of investment.[24]

The main actors in the policy-making system were President Videla and the minister of economy, José Alfredo Martínez de Hoz. Inasmuch as all the issues concerning industrial policy and technology transfer depended on the ministry of economy and its secretariat of industrial development, Martínez de Hoz was the real architect of the new policies. His liberal economic policy with regard to industry was formulated and implemented by the secretariat of industrial development, which was under the direction of Pablo Benedit. In 1978 a technical subsecretariat was put in charge of science and technology policy; it became the authority for the application of the liberal new technology transfer law of 1977 and was also responsible for the drafting of the 1981 technology transfer law in accordance with the liberal economic policy.

INTI's policy role was much reduced from what it had been prior to 1976. It no longer had charge of the Registry, played no policy-making role with regard to technology transfer legislation, and was not a factor in industrial policy. The 1981 technology transfer law eliminated the Registry and invested in INTI the authority to register contracts, which from then on would be done primarily for statistical reasons.

SECYT was alienated, turned into a passive element in the science and technology policy-making system and relegated to a secondary role. It still had the power to perform budget appropriations, but this was not very significant politically, given the independence of many of the scientific and technological institutions. Because emphasis was now on short-range economic goals, the education ministry's concerns were perceived as low-key and routine compared to the "relevant" concerns of inflation and economic efficiency. In April 1981, SECYT, continuing its historical instability, was once again made a subsecretariat (SUBCYT), within the ministry of education.

As if this were not enough, Argentina "lost" hundreds of scientists and teachers who were persecuted, had to leave the country, or were killed for their ideas or simply for being part of an educational system considered to be part of the "subversive" network. The ideas of a strong science and technology policy, of self-determination, of autonomous development were placed on hold, and the science and technology institutions, with the exception of the CNEA, languished into oblivion, at least until the realization in 1982 that Martínez de Hoz's policies had brought a curse on Argentina. The feelings of desperation and the "rediscovery" of science and technology were also spurred by the Malvinas (Falkland Islands) debacle. (Sabato summarized the feelings of those committed to a strong science and technology system, saying that in 1976 Argentina entered an ahistorical period.) In January 1982 SUBCYT was transferred to the planning secretariat of the presidency, and once again science and technology planning and domestic technology became part of the decisionmakers' rhetoric. The pendulum was swinging again.

Summary and Conclusions

Science and technology policy between 1966 and 1982 developed in zigzags, faithful to fracasomania. The fast pace at which the educational and research activities had been progressing up to 1966 was halted by the "night of the long clubs." Technological laissez-faire had meant small progress toward the achievement of some measure of scientific and technological autonomy until 1969–1970. The mild pragmatic antidependency policies of 1970–1973 attempted to bridge interdependence and national development by creating both controls and incentives for domestic production. But this policy was succeeded too quickly by a strong rhetorical, almost structural, antidependency policy of the Peronists, which in turn gave way after March 1976 to an attempt to bring back technological laissez-faire.

Throughout all this, institutional instability was a clear reflection of ideological instability. For example, CONACYT was turned into SUBCYT, then into SECYT, and then back again into SUBCYT. And since the roles of CONACYT and CONICET were never clearly differentiated, political quarrels among them have been the rule. INTI was given political control power in 1971 but lost it in 1977. Technology transfer and foreign investment laws changed every couple of years, adding to the uncertainty and often actually preventing real technology transfer. Whatever steps were taken in one period were almost certain to be undone in the next. Dependency, perceived as coercive by policymakers of the first half of the 1970s, was explicitly seen as communal in 1976.

Argentina's ideological, political, and institutional zigzags provided no opportunity for the intellectual pragmatic antidependency guerrillas to have a lasting effect on political processes and policy implementation, as in Brazil. The clearest case of ideologically motivated intellectual action in science and technology started at the beginning of the 1970s and reached its peak when the CGE was dictating economic policy. But the intellectuals and the policy fell victim to the ideological polarization and struggles of the period and had too little time to achieve a more lasting effect before the ideological tide turned in the opposite direction.

Argentina's science and technology development process has been hampered not so much by lack of ability or resources as by economic and political instability, on the one hand, and lack of consistency, awareness, and broad support for domestic technological development, on the other. And while Argentina was locked into fracasomania, the world would not stand still . . . and neither would Brazil.

Science and Technology in Brazil, 1962–1982   7

The development of a science and technology policy in Brazil is the history of an idea; it started with an image of the future, which filtered through the political power structure and emerged as alternatives for action. Since the early 1960s, but especially after 1968, the Brazilians have been involved in a strong and unprecedented effort to create an institutional structure and an explicit policy for science and technology, to direct investment toward specific goals, to promote domestic innovation, and to improve the management of knowledge and information. Most science and technology policymakers have defined the aim as being technological autonomy. This objective will probably not be achieved. But in the search for it Brazil's technological development has changed significantly, and both the government and entrepreneurs—foreign and national—have adapted to the new situation.

Many factors account for what has happened. A new political regime came to power in 1964 and proposed new economic development and industrialization policies. But there were also ideological and institutional factors—ideas that "burned" in the minds of some people and influenced institutional and political events.

The aspirations that fueled this progress were rooted in nationalism and, increasingly, in egalitarian nationalism with its ideas of autonomous development and greater equality at the international level. The two ideologies shared a growing, and mutually reinforcing, awareness of the linkage between scientific and technological development and technological dependency. Structural dependentistas, constrained by their own ideology, had little influence on policy. But pragmatists were offering a different "merchandise," a different set of ideas, beliefs, and values that had the potential of fulfilling the aspirations of both the political elites and the egalitarian nationalists. These ideas had enough political, institutional, bureaucratic, scientific, and technological impact that they ultimately defined a state policy that has become, at least partially, irreversible.

Brazil's Pragmatic Antidependency Science and Technology Strategy

Industrialization: The Main Event

The main event in Brazil's twentieth-century history has been the growth of industry. In contrast to Argentina, the determination to industrialize was not hampered by alternative models of progress based on the export of primary commodities. Even matters related to national security were then, as they are even more today, linked to industrialization, giving rise to the set of beliefs that became so important after the 1964 revolution—*segurança e desenvolvimento* (security and development)—and that in turn fueled more and "deeper" industrialization.

Although the industrialization process started early in the twentieth century, Brazil was clearly a coffee export economy until the 1920s. When the Great Depression brought the collapse of the coffee economy, however, industrialization began to grow—not as a result of design or planning, but from a need to adapt to a situation beyond Brazil's control.[1] Events after 1929 were related to the foreign crisis and also to the Estado Novo, an authoritarian, corporatist state instituted by Getúlio Vargas in 1937 and lasting until 1945.

The Estado Novo had enduring effects, not only on economic development policy but also on ideological and political attitudes about the roles of the state, the military, and the bureaucracy. Regardless of the factors that led to the Estado Novo, I want to emphasize its legacy: the centralized political machinery, the political power of the military, the personal power of the president, and, most important of all from my point of view, the image it presented of the future—Brazil as a military and economic power.

To promote industrialization the state relied mainly on import substitution industrialization (ISI). Among the measures taken to control imports were the Law of Similars, which restricted the importation of products similar to those produced in Brazil, and a multiple exchange system. Effective tariffs averaged over 250 percent for manufactured goods.[2]

The results of this strategy in the 1940s and 1950s were impressive from the point of view of industrial growth.[3] This success reinforced the idea embraced by intellectuals and politicians from the beginning of the century, that Brazil could develop not simply into a self-sufficient economic entity but, even more, into an economic power. This idea, in turn, further fueled the industrial growth process.

As a result of the developmentalist drive to industrialization of President Juscelino Kubitschek's administration in the late 1950s, industry became more mature, with a diversified capital goods sector. These transformations again had a self-reinforcing effect. In "the course of the industrialization process, as the forces which stood behind industrial dynamism gained in importance and influence, a growing insight into the nation's problems was acquired. . . . This reflected a political urge, a nationwide endorsement of the idea that the country must develop, which provided an explicit frame of reference for Brazil's economic policy."[4]

Shipbuilding was born of these efforts, and industries such as paper and pulp, heavy mechanical and electrical equipment, steel, and tractor production were expanded. Another major development was the birth and growth of an automobile industry in which by 1962 ten firms were already producing 190,000 automobiles and trucks with about 90 percent of the parts domestically manufactured.[5] But as in the course of the industrialization drive the functions of the state broadened,[6] so did the role of foreign investment. It is estimated that between 1949 and 1962 the total growth produced by foreign firms amounted to 33.5 percent of the expansion in manufacturing and 42 percent of the import substitution industry growth.[7]

Industrialization and
Technological Laissez-Faire

Advancement of the ISI process required technological expertise and training in order to operate industrial plants and to produce technology on the basis of foreign know-how and engineering. The government did not interfere with capital goods and technology purchases or with the contracting of national and foreign firms for the acquisition of licenses, patents, and trademarks. Thus although up to the beginning of the 1950s the Brazilian state was involved in mild control of foreign investment, and from the 1950s with efforts to encourage local capital goods production, these measures were not concerned at all with science and technology. Before 1962 almost no attention was paid to the commercialization of technology, with the possible exception of Income Law 3470, which will be reviewed later.

Brazil was a newcomer to both science and higher education. Its first professional schools in law and medicine date from the nineteenth century, its first university was established in 1920, and science as an organized activity barely existed before the 1930s.[8]

The Brazilian Scientific Society was created in 1916, and in 1922 it was transformed into the Brazilian Academy of Science, which, like most Latin American academies, had a cultural and intellectual rather than a scientific role. São Paulo University, years later to play a role in training scientists in biology, physics, and chemistry, and the Technology Research Institute (IPT), to become one of Brazil's most successful technological institutions, were created in 1934.[9] Between 1930 and 1949, 160 institutes of higher education were created, among them the Federal University of Rio de Janeiro. In 1949 the Brazilian Center for Physics Research (CBPF) was set up under the direction of César Lattes and José Leite Lopes; it later became a base for physicists who wanted to master the new nuclear genie. One year later the Brazilian Society for the Progress of Science (SBPC), similar to the American Association for the Advancement of Science, was created, which is now a forum for scientific exchange and for political discussion and ideological development and confrontation.[10]

CBPF and SBPC were created on the initiative of scientists. There was no formal scientific policy, and research nuclei were the fruit of isolated individual efforts. The higher education system was limited to a few schools with training in careers indispensable for the society, such as engineering and medicine.[11]

Whereas many scientists in the early 1950s still continued to regard technology as "science—patrimony of all mankind"—an increasing number of scientists began to link scientific and technological development with the mastery of nuclear energy. A growing awareness on the part of policymakers about this linkage and a desire to be second to none in Latin America's nuclear power development led the Brazilian state in 1951 to create the National Research Council (CNPq) and the Company for the Improvement of Higher Education Personnel (CAPES).

CNPq was to become a leading institution for science and technology development in Brazil. In its first years, its concerns were (1) maintaining contacts with researchers and research institutes, (2) granting aid for research and (3) fellowships for professional training, (4) support for national and international scientific meetings, (5) for scientific exchange both within the country and with foreign and international institutions, and (6) for atomic energy research, and (7) other activities in support of scientific and technological research.[12]

The early 1960s saw new institutional developments. At the end of 1960 the ministry of education created the Supervisory Commission of the Institutes' Plan (COSUPI) for the support of technological education, and in 1963 the Program for the Expansion of Technological Education (PROTEC) was set up.[13] Both were incorporated into CAPES in 1964. The São Paulo Foundation for the Support of Research (FAPESP) and the University of Brasília were created in 1960 and 1961, respectively.

The Evolution of Brazil's Science and Technology Policy

The initiation of an explicit science and technology policy in Brazil cannot be dated, per se; it concerns not formal guidelines and bold actions, but rather a *perception* on the part of decisionmakers and analysts that certain actions in the aggregate constituted a science and technology policy. The Programa Estratégico de Desenvolvimento 1968–1970 (Strategic Development Program), issued in 1968, is generally regarded as the "breaking point." I share this view, but only insofar as arguing that what occurred in 1968 and after was more accurately the culmination of a process of events that took place between 1962 and 1968. Thus it seems reasonable to locate the policy origins in the mid-1960s, even if the first explicit document reflecting the new policy was issued in 1968.

There were four stages in the development of the science and technology policy. The first (1962–1968) I call the *policy formation* stage. Here the ideas regarding policy evolved into science and technology ideologies, which penetrated the political system. The key Strategic Development Program was issued, containing an explicit statement about science and technology autonomy.

The *creative* stage (1969–1975) involved a proliferation of measures aimed at the formulation and implementation of a science and technology policy. An institutional "system" was created that included partially self-sufficient financial institutions in charge of promoting autonomous scientific and technological development, and norms to regulate technology transfer were issued. This "creativity" was not unrelated to the economic miracle Brazil experienced during those years, but it should be pointed out that the two most creative years were 1974 and 1975—after the miracle was over.

During the third stage of *stabilization* (1976–1978), new normative, planning, and institutional measures gave more coherence and teeth to the previous policy developments.

I call the period from 1979 to 1982 the *drought* stage: because of recession and a financial crisis, resources became scarce and the build-up process slowed down. Nevertheless, the science and technology system continued to grow, albeit at a slower pace, and it retained its importance and financial backing.

The following is a chronological description of the highlights of science and technology policy from the early 1960s to 1982. This description, divided according to the policy's basic elements, the regulation of technology transfer and the creation and support of a domestic capacity to innovate, also serves as an introductory summary of the rest of this chapter.

Technology Transfer

1962   Law 4131, basically concerned with foreign investment, also controls technology transfer contracts for the first time, restricting the remittance of earnings abroad, fostering the absorption of technology, and prohibiting the remittance of earnings between multinationals and their affiliates for technology transfer.

—— Law 4137 creates the normative means to control abuses in technology transfer contracts.

1964 Law 4390 and Decree 55762 (to regulate the law) repeal some of the restrictions instituted by Law 4131 but maintain the registration of contracts, the control of real transfer, and the prohibition of remittances between multinationals and their affiliates.

1965 Creation of SUMOC (Superintendency of Money and Credit) to regulate the operations of machinery and equipment imports.

1970 Creation of National Institute of Industrial Property (INPI) to administer the technology transfer system, the patents and trademarks systems, and the linkages between domestic industrial technology supply and demand.

1971 Industrial Property Code institutes some modifications to systematic regulations of technology transfer aimed at speeding up the "real" transfer and increasing the bargaining power of the government vis-à-vis technology suppliers. Contract approval by the Central Bank requires INPI's endorsement.

1973 Creation of Industrial Technology Commission to analyze technology transfer contracts.

1975 Normative Act 15 divides technical know-how into five categories: patents, trademarks, supply of industrial technology agreements, technical-industrial cooperation agreements, and specialized technical services agreements; and requires separate contracts or agreements on each so as to open the "technological package."
—— Brazil becomes a member of the World Intellectual Patents Organization.

1978 Normative Act 30 bars payments by automotive subsidiaries to their parent companies for R & D on new models.
—— Normative Act 32 establishes a consulting stage in which companies must pass along to INPI all details of a negotiation before INPI will decide on the contract.

1979   The Law of Similars is placed under temporary review and then repealed.

1981   Reinstatement of the Law of Similars.

Science and Technology Innovation

1964   Creation of Scientific and Technical Development Fund (FUNTEC) within the National Economic Development Bank (BNDE) to support research geared to increasing both domestic supply of technology and demand by national enterprises, for which it finances the purchase of national equipment.

——   Creation of Special Agency for Industrial Investment (FINAME) within BNDE.

——   Law 4506 regulates the capitalization of research costs.

——   CNPq broadens its role to include the programming of science and technology activities.

1965   Creation of Studies and Projects Financing Agency (FINEP), which would become prominent only in the 1970s.

——   Creation of COPPE (Coordination of Graduate Programs in Engineering).

1967   Establishment of Fund for the Support of Technology (FUNAT), to train industrial technicians, under the ministry of industry and commerce (MIC).

1968   Formulation of Strategic Development Program for 1968–1970, the first document of an explicit science and technology policy. The main ideas of pragmatic anti-dependency appear in this document: incorporation of science and technology into the productive system through "real" technology transfer and through development of the Brazilian capacity to innovate.

——   CNPq issues a Five-Year Plan, mostly unknown outside CNPq, as a result of which an Experimental Graduate Program of Research and Education is set up.

1969   Creation of National Science and Technology Development Fund (FNDCT), which became the principal instrument for the support of science and technology in Brazil.

1971   Publication of I PND (first National Development Plan) iden-
tifying science and technology as one of the grand national
objectives, outlining a policy of autonomy, and instituting the
Basic Plan of Scientific and Technological Development
(PBDCT), which became the main policy instrument.
——   Establishment of Minas Gerais Technological Center
(CTMG).

1972   Decree 70553 creates the National System for the Development of Science and Technology (SNDCT) to formulate and
execute policy, with CNPq in charge of scientific and technological matters and the ministry of planning and BNDE in
charge of financial matters. These actions formalize the relations that already existed between CNPq and the planning
system.
——   Creation of Industrial Technology Secretariat (STI) as an
agency of MIC to develop and implement industrial technology policy. INPI and the National Institute of Technology
(INT) are integrated under the MIC/STI hierarchy.
——   Creation of National System of Science and Technology Information (SNICT), a network of data banks covering economic and social sectors.

1973   I PBDCT issued for 1973–1974, describing science and technology as the most dynamic element of Brazil's development.
Its stated aims: to accelerate technology transfer and to
strengthen the national capacity to innovate.
——   Program for the Support of Technological Development of
the National Enterprise (ADTEN) begins to provide funds
for equipment purchases.

1974   II PND issued, setting in motion a major reorganization of
the science and technology apparatus. It continues to embrace the goals of the past plans, placing strong emphasis on
the energy sector and industrial technology, especially basic
industries with a high technological content such as computers, capital goods, petrochemicals, steel, and aeronautics.
——   Ministry of planning becomes the planning secretariat
(SEPLAN), with broader powers and attached to the
presidency.

——   CNPq becomes National Council of Scientific and Technological Development and the central organ for planning, coordinating, and implementing science and technology policy; placed under SEPLAN. FINEP is subordinated to CNPq.

——   CAPES is restructured and its powers enlarged.

——   National Council for Metrology, Norms, and Industrial Quality (CONMETRO) is restructured and placed under STI's jurisdiction with the name National Institute for Metrology, Standardization, and Industrial Quality (INMETRO).

——   Creation of subprograms for the Demand and Utilization of Technology and the Creation of Technology Supply under BNDE/FUNTEC to provide incentives for R & D projects.

——   Creation of three additional BNDE institutions with the aim of strengthening the domestic machinery and equipment sector: Brazilian Investments, Inc. (IBRASA), to invest in Brazilian enterprises; Brazilian Mechanics, Inc. (EMBRAMEC), to provide capital for national capital ventures in the capital goods sector; and Basic Goods, Inc. (FIBASE), to provide capital for basic goods.

1975   Creation of Scientific and Technological Council (CCT) under CNPq, for which it becomes the central policy-making body.

——   Establishment of Scientific Consulting Group (CCI) under CNPq. A consulting body of scientists, many of them egalitarian nationalists responsible in one way or another for the ideas that have been turned into policy, it institutionalizes the role of the pragmatic antidependency guerrillas.

——   Creation of Superintendency for International Cooperation (SCI) under CNPq to coordinate the Brazilian international activities related to science and technology.

——   Restructuring of SNDCT. Sectoral organisms called technology secretariats are placed in every ministry, all integrated to the SNDCT.

——   Creation of Nuclei of Articulation with Industry (NAI) for the benefit of public enterprises; in charge of promoting the purchase and development of national manufactured equipment, they serve as a "buy national capital goods" instrument.

—— Creation of Coordinating Commission of Articulation with Industry (CCNAI) under FINEP's jurisdiction.

—— Enactment of National Plan of Graduate Studies.

—— Creation of National Alcohol Program, in which IPT and INT assume major research tasks.

—— Signing of formal agreement between CNPq and SBPC, enlarging support for the latter institution.

1976 II PBDCT issued for 1975–1977. Like the previous PBDCT, it emphasizes the role of science and technology as a force of dynamism and transformation, upholding the autonomy objectives and budgeting almost $2.5 billion for the plan's implementation.

—— Establishment of Scientific and Technological Research Incentives Fund (FIPEC) within the Bank of Brazil to promote scientific and technological research.

—— Creation of Program of Research Administrators (PROTAP) within FINEP with the aim of strengthening human resources for the development of science and technology in national enterprise.

—— Creation of Brazilian Institute for Scientific and Technological Information (IBICT) under CNPq to manage science and technology information.

1977 Establishment of Industrial Technology Foundation (FTI).

1978 Industrial Development Council (CDI) issues incentives for the modernization and technological development of the chemical-pharmaceutical sector.

—— Creation of Center for the Study of Technology and Engineering (CETE) in São Paulo.

1979 Creation of Industrial Technology Company (CTI) by the INT to transfer the technologies produced by the institute to the productive system.

—— III PND issued. Because of uncertainty created by the economic crisis, no quantitative targets are included in the plan, which emphasizes the explicit goal of reducing science and technology dependence.

—— Reorganization of SEPLAN. BNDE is placed under MIC jurisdiction.

1980   III PBDCT issued, stressing the same goals as III PND and
        likewise sector-oriented, emphasizing in particular energy
        and capital goods.
———   Creation of Superintendency of Institutional Programs (SPI)
        under CNPq to promote scientific development and the
        training of researchers.
———   Creation of Science and Technology Policy Advisory Group
        (GAPCT), a forum of mostly pragmatic antidependency in-
        tellectuals who serve as policy consultants, under CNPq.

Policy Making: Structure and Process (to 1979)

The top Brazilian policymaker is the president. It is likely that the institutions closest to him, the National Security Council, the National
Intelligence Service, and the Armed Forces General Staff, have played
an informal part in the policy-making process. The heads of the president's civilian and military cabinets have also been key policymakers.

The presidential agency for formulation of policy on economic development and science and technology at the highest level was the
planning secretariat, SEPLAN. SEPLAN's chief was the president's assistant and adviser on economic policy. He also functioned as secretary general of the Economic Development Council (CDE), the top
forum for economic development decisions, comprising the secretaries of finance, agriculture, interior, industry and commerce, works,
and planning and presided over by the president. The council set *overall* economic policies and as such strongly affected the actions of institutions such as the National Economic Development Bank (BNDE).

At the top of the industrial policy hierarchy was the ministry of industry and commerce (MIC) and its Industrial Development Council
(CDI). CDI was responsible for defining the industrial development
policy, selecting the priority sectors in accordance with the general
economic policies and plans, and setting fiscal incentives.[14] Formulation and implementation of industrial technology and technology
transfer policy was centered in MIC's industrial technology secretariat
(STI), which acted in accordance with directives of SEPLAN, the
CDE, and the presidency. Directly dependent on the secretariat were
the National Institute of Technology (INT), which performed R & D
activities and has been increasingly integrated into the productive

system; the Fund for the Support of Technology (FUNAT); and INMETRO, the industrial quality and standards organ.

Formulation of guidelines and creation and implementation of specific instruments of technology transfer were in the hands of the National Institute of Industrial Property (INPI), which was administratively divided into four bureaus: patents, trademarks, technology transfer, and technological information. In practice, INPI was responsible for renegotiating technology contracts, which were reviewed by advisory engineering enterprises and by INPI's council, with occasional consultations with the National Council of Scientific and Technological Development (CNPq) and such institutes as INT and the Technological Research Institute (IPT) to acquire technical information.

Implementation of technology transfer policy has not been solely in the hands of the MIC/STI/INPI system. For example, CACEX, the Department of Foreign Commerce of the Bank of Brazil, granted import licenses following the guidelines of the ministry of finance. With key decisions on technology transfer made by INPI, once it approved a contract, registration with the bank was practically automatic.[15] (Before INPI was created, the central bank was the ultimate authority over technology transfer contracts.)

SEPLAN presided over most of the institutions that formulated and implemented scientific and technological development policy (SNDCT), the key ones being CNPq, the Studies and Projects Financing Agency (FINEP), and BNDE. Since 1974 CNPq has had the role of formulating science and technology policy and executing it through long- and short-range planning; coordinating science and technology matters with the ministries and other government organizations; promoting research and training; and coordinating and centralizing scientific and technological information systems. Since 1980 it has articulated science and technology activities through the Programmed Actions, which mobilize resources, people, and institutions in pursuit of a specific science and technology goal.

CNPq's Scientific and Technological Council (CCT) comprised 30 members from all branches of the science and technology system and included 15 scientists and invited members from the national enterprises; CNPq's president presided. CNPq's vice president served as president of FINEP as well and coordinated the commission that prepared the science and technology plan. The commission, with basically the same representation as the council, was subdivided into subcommissions for industry, infrastructure, agriculture, social development, and scientific development.

FINEP administered the National Science and Technology Development Fund (FNDCT) as well as several programs for scientific and technological development of national enterprise, promotion of national consulting, and support to consumers of consulting services programs. It also functioned as the executive secretariat of the National Program of Executives Training and administered the training program of the Administration of Scientific and Technological Research and the Coordinating Commission of Articulation with Industry (CCNAI), which in turn oversaw the Nuclei of Articulation with Industry (NAI) in promoting the purchase and development of national capital goods by public enterprises. FINEP contained a Center of Studies and Research, participated in commissions of various ministries, noted technological needs, searched the national infrastructure for suitable national suppliers, and then decided what is worth financing. It received funds from SEPLAN, and additionally from business establishments for services rendered, allowing it to maintain a certain independence from federal budget fluctuations caused by changing economic conditions.

BNDE, also institutionally linked to SEPLAN up to 1979 through its Scientific and Technical Development Fund (FUNTEC), financed the promotion of national R & D and of national demand for domestically developed technology, as well as national production of capital goods.

SNDCT also included the Economic and Social Planning Institute (IPEA) and the Brazilian Institute of Geography and Statistics (IBGE). The main policy instrument of SNDCT, the science and technology plan (PBDCT), adhered to the guidelines set by its "senior counterpart," the economic development plan (PND). Through the planning process, SEPLAN connected the loose parts of the science and technology policy system, including those under MIC hierarchy such as STI/INPI. Until the restructuring of the science and technology policy-making system in 1974, the planning ministry was directly responsible for formulating PBDCT, with data and information from IPEA. Subsequently, CNPq formulated the plan.

The planning process consisted mainly of demarcating the areas to be considered for institution of the plan. This was based on the assessment of proposals from the ministries for their respective sectors, consisting of political directives, goals, priority programs and projects, activities, and budgets.[16]

Several ministries have had relevant roles regarding science and technology policy formulation in their own sectors, such as agriculture, mining and energy, communications, transportation, and the military ministries. In 1975 technology secretariats were placed in every ministry, all integrated to SNDCT.[17] Figure 6 summarizes the policy-making system.

Goals

The pragmatic antidependency approach implemented by Brazil lent itself to various interpretations and to a broad spectrum of values and beliefs, rejecting the free market on one hand and full autarky on the other. For the majority of the policymakers concerned, the explicit goal was self-determination or autonomy in science and technology. The so-called autonomy policy enjoyed a bandwagon effect as actors joined to fulfill some of their own political and bureaucratic needs. The military, for example, emphasized the nation value (even the "superpower-nation" value). For SEPLAN the central element was to produce Brazilian technology relevant to the Brazilian image and model of development. MIC/INPI stressed the "reality" of increased interdependence and the need to catch up through the absorption of foreign knowledge, or "real" technology transfer, while at the same time improving the balance of payments. And even the finance ministry, which has traditionally supported efficiency and growth over more nationalistic goals, acknowledged the importance of the strategic, economic, and social priorities that a national indicative plan would establish.

For many of the policymakers, autonomy meant the recognition, or as I PBDCT stated, the awareness, that economic and social transformation in modern society has come to depend on science and technology to such an extent that technological mastery is nowadays a basic factor in establishing the relative competitiveness of the various countries; or as II PND pointed out: "S & T [science and technology] in the present stage of Brazilian society represents a driving force, the channel par excellence of the idea of progress and modernization."[18]

Figure 6. Brazil. The Science and Technology
Policy-Making System (to 1979)

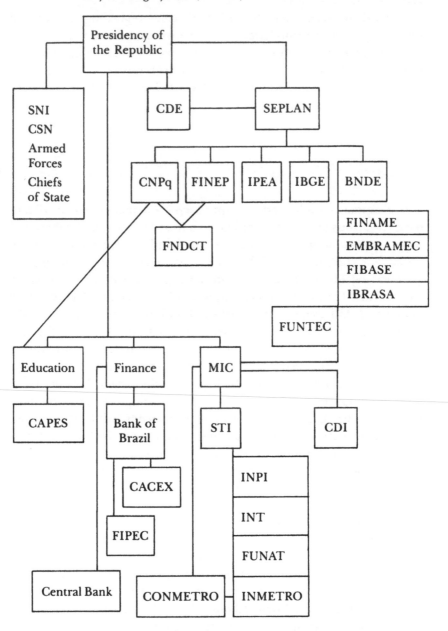

Note: Abbreviations include CDE for Economic Development Council; CSN for National Security Council; SNI for National Intelligence Service; and FIPEC for Scientific and Technological Research Incentives Fund.

Some technocrats and influential intellectuals, members of the pragmatic antidependency guerrilla group, were driven to the self-determination goal as well by the desire to create a national and cultural capacity to innovate. But this desire was, in most cases, not expressed in such a straightforward manner, and because the policymakers with power to effect change were mostly economists and engineers, objectives were presented in economic and technical rather than in cultural terms.

Behind the objective of self-determination was the fundamental drive to know more for the sake of production. As the distinctive Brazilian ideology of development evolved in the late 1960s and 1970s, policymakers such as Velloso, Pelúcio, and others came to regard development as a higher capacity of knowing, perceiving, and understanding. The science and technology policy, by aiming at increasing "national knowledge" and "national understanding," was expected to bring about a new stage of development, defined mainly in cognitive terms. Thus while many "felt" it but only a few explicitly recognized it, development became directly and intrinsically linked to consciousness, because development required that attention be placed on certain physical processes, on understanding and knowing and the choices these processes revealed.

The Brazilian national paper presented at the United Nations Conference on Science and Technology Development in 1979 reflected these feelings:

[Brazil] rejects the definition of development which points only to the developed countries as the target to reach, the state to achieve; the idea that developed countries exemplify the goals toward which to strive. The document proposes a definition which states that development implies a high capacity of knowing, perceiving, and understanding, in order to interpret the realities of the country, in order to maximize the achievement of the nation's objectives through the human and physical resources available. . . . This cognitive capability is the common element, the means of identification of advanced and developed countries. This is precisely what the nondeveloped countries lack. The cognitive capability will be equivalent to a concept of science and technology for development. The document underscores this cognitive capability, the idea of a critical mass, the idea of a platform of knowledge as a conditional element to the possibilities of development.[19]

Means

According to most Brazilian science and technology policymakers, more than the invisible hand was needed to guide the country to the goals outlined above. That is why control over the means of technology transfer has been steadily and incrementally increasing.

Knowledge-Generating Capital Equipment

Over the years the Brazilian industrialization process has generated large amounts of technology transfer through the purchase of capital equipment, such as machinery. Furthermore, the large presence of Brazilian engineering firms in the technology market and increasing government incentives for locals to handle the maintenance and operation of equipment have allowed for considerable effectiveness in technology transfer. At the same time, as the domestic capital goods industry developed, a desire to reduce embodied technology transfer and to acquire know-how became evident.

The policy with regard to imports in general and capital goods in particular has been directly linked to the balance of payments question. Thus, for example, in the high-growth period the complex and confusing imports controls were simplified, but with the balance of payments worsening again by 1974, they were progressively reinstituted. "The measures included steeply increased duties on a wide array of products, the reintroduction . . . of prior deposits . . . , and the extension of the requirement for a *guia* [in effect, a license to import] for virtually all imports." Among the new measures, a previous maximum permissible financing period of eight years for imports of capital goods was abolished.[20]

Also, tariffs that had decreased between 1966 and 1973—for example, machinery, from 48 to 38 percent, and electrical equipment, from 114 to 56 percent—began an upward trend, notably on steel, paper, and certain petrochemicals.[21] In March 1980 Brazil increased duties on some two thousand items, mostly chemicals and capital equipment. At the same time other nontariff measures were instituted, such as a requirement that governmental and quasi-governmental agencies purchase as much of their needs from local services as possible.[22] At the end of 1979 the Law of Similars suffered a setback when it began to be applied only to companies seeking any of the still-available concession incentives such as those offered under the Export Benefits Program (BEFIEX). But one year later the law was reinstated in its original form. And a central bank resolution (#638) in September 1980 making virtually all imports of capital equipment, including components and spare parts, contingent on obtaining foreign financing created one more roadblock for capital goods imports.[23]

The share of imported capital goods in the projects approved by the Industrial Development Council (CDI) for the industrial sector in general and the capital goods sector in particular went down drastically. For example, the 1971 share of 68.1 percent for the industrial sector and 76.0 percent for the capital goods projects decreased in 1979 to 28.9 percent and 38.7 percent, respectively. In 1973, under the participation agreements approved by the Department of Foreign Commerce of the Bank of Brazil (CACEX), 52.7 percent of the capital goods were bought in the domestic market, in contrast to 82.4 percent in 1979.[24]

Brazil also instituted stiff local-content rules. Thus, the automobile has achieved a national content close to 100 percent by weight, and local-content requirements for such capital goods sectors as steel making, railroads, and electrical energy have been increased to as much as 80–90 percent. Table 8 shows the nationalization index of

Table 8.   Brazil. Nationalization Index of Equipment in Projects Financed by BNDE (in %)

Sector	1974–1975	1979–1980
Steel	22%	70–80%
Hydroelectric	50	85–90
Cement	50	95
Petrochemical	60	70–80
Paper and cellulose	50–60	80–85
Nitrogen fertilizers	60	70
Phosphate fertilizers	70	90
Mining	50–60	80–90
Railroad equipment	60	90–95

Source: José Serra, Working document presented to ECLA (São Paulo, 1981), table 53.

equipment as reflected in payments financed by the National Economic Development Bank (BNDE). We see that the nationalization index went up radically between 1974 and 1980.

Technical Assistance, Industrial Property, and Know-How

In the realm of technical assistance, patents, licenses, and trademarks, technology transfer policy since the beginning of the 1960s has aimed at the "real" transfer of technology. The steps taken have been deliberate and consistent, beginning as early as 1958 with Law 3470, which limited to 5 percent the amount of royalties for trademarks, patents, and technical assistance that could be deducted for tax purposes.

But it was Law 4131 of 1962 that produced the first set of strict technology transfer controls. Its basic concern was to restrict the remittance of earnings abroad, but at the same time it attempted to foster the absorption of technology, creating a structure of incentives differentiated according to industrial sectors.[25] As a result of this law, those who wished to make remittances abroad for any technological activity had to submit the contracts to the Superintendency of Money and Credit (SUMOC), which would verify, if necessary, whether technical assistance really took place. Those who made the transfer had to furnish some proof that their privileges abroad were still standing. Article 14 of the law prohibited remittances between subsidiaries and their home companies or between an enterprise whose capital was in the hands of the royalties recipients abroad. Any transfer remittances above 5 percent in a five-year term were considered taxable earnings. Article 14 was to remain unalterable across all legislations and political changes, as were disincentives for the registration of patents without property and for the use of a foreign trademark.

The antitrust law (4137), also of 1962, considered abusive all practices that tended to take over national markets or tried, totally or partially, to eliminate competition. This law has been interpreted as the normative mechanism for the prevention of abuses in technology transfer contracts.

Decree 53451, which regulated Law 4131, limited to five years the period in which contracts with foreign patents could be allowed to generate remittances and restricted the total remittance to 2 percent of the manufactured product cost or of the product's gross earnings. Law 4390 of August 1964 (regulated by Decree 55762/1965) removed these restrictions but retained the other provisions of Law 4131 and also added to the 1962 provisions a supplementary income tax on remittances.

After two short-lived Industrial Property Codes, the one instituted in 1971 introduced modifications to the rules regarding technology transfer registration: technology agreements (technical assistance, license for patents and trademarks) had to be approved and registered with the National Institute of Industrial Property (INPI), and documentation had to be provided justifying the acquisition of technology from outside the country, stating the advantages for the contractee and the national economy, estimating time of absorption, and describing the technology.[26] INPI wanted such information as the value of the technology to be transferred, method and duration of transfer, whether the licensor would train employees of the licensed firm at its own plants, whether the licensee planned to set up its own R & D facilities or use existing ones in Brazil, and whether the licensee would be entitled to the technology and techniques it developed from the technology being acquired. INPI would not accept licensing agreements restricting or controlling the licensee's exports or stipulating the sources of materials or components to be imported for the manufacture of the licensed product.[27] According to the code, no minimum royalties were to be accepted, but it was possible to include minimum production and sales figures upon which the royalties would be based. Fixed payments were accepted for technical, engineering, and consulting services, but the maximum payment for technical personnel entering Brazil to render technical services was set at $6,000 per month.[28]

The Industrial Property Code set a patent duration limit of fifteen years from the date of filing the application. Except for priority rights under international conventions, companies would have to file a patent in Brazil within one year of obtaining rights abroad. Pharmaceutical, medical, and food products and processes; chemical products; metallic mixtures and alloys; and atomic materials and the alloys used in producing them were declared unpatentable. If a patent was not worked in Brazil within three years of issuance or work was interrupted for more than a year or was not considered sufficient to supply demand, INPI might release it from compulsory licensing to interested parties, who had to begin to work it within twelve months. Patent rights were forfeited automatically if not worked in the first four years by the patent holder or the first five years if licensed.

Remittability of royalties and fees to unrelated foreign licensors was limited to between 1 and 5 percent of gross sales, depending on the essentiality of the product. Payments for technical assistance were not deductible expenses for multinational corporation subsidiaries, and the remittance of patent and trademark royalties by the subsidiaries to their parent companies was prohibited.[29] "The 1971 Brazilian Patent Law," one source concluded, was considered "the toughest and most sophisticated piece of legislation for the control of technology in existence in the hemisphere."[30]

In 1973 a regulation was instituted to the effect that foreign technical service companies in the fields of engineering, architecture, and agriculture could work in Brazil only if they formed consortiums with Brazilian firms. This regulation came on top of a 1969 decree restricting governmental technical service contracts to companies with at least 51 percent Brazilian ownership and technical staff.[31]

Normative Act 15, enacted in 1975, classified (with the purpose of registration) technology transfer agreements into five categories according to their purpose: (1) license contracts for patent rights; (2) license contracts for the use of a trademark; (3) contracts for the supply of industrial technology; (4) contracts for technical-industrial cooperation; and (5) contracts for specialized technical services. It stipulated that "the provisions that apply to each purpose are quite distinct and must correspond specifically to one single contract in the respective category."[32] For the first time a government document identified know-how as being separate from other technology transfer means, implying the opening of the "technological package."[33] Act 15 also differentiated among three types of dealings between capital-linked parties—total control, majority participation, and minority participation—which opened a path for the distinction between technology transfer to fully foreign-owned enterprises and to joint ventures.

In 1978 INPI issued Normative Act 30, which barred payments by automotive subsidiaries to foreign parent companies for R & D on new models and for administrative or marketing assistance. Normative Act 32, enacted the same year, stipulated that a draft of the technology transfer agreement (in Portuguese and in a second language chosen by the foreign party) must be submitted to INPI for study. This act was aimed directly at increasing the bargaining power of the Brazilian parties. Three years later, other important directives were approved: Normative Act 53 established procedures for review and control of contracts in the area of data processing, oriented to defend domestic production; Act 55 made consultation on matters concerning technical assistance and contracts for specialized technical services compulsory; and Act 60 regulated the acquisition of specialized technical services abroad.

INPI officials reviewed contracts according to the following principles: veto purchases of technology that the country already possesses; know the technical needs of national industry; and link national enterprises (demand) with national research institutes (supply). The policy criteria used varied according to the nature of the enterprise involved. Officials have been very "liberal" with national enterprise in approving technology contracts, but they have been careful to see that the technology was indeed transferred. The policy has been more exacting with regard to joint ventures, even those with national capital majority, because if the foreign company in the venture (in spite of its minority capital status) provides the technology without the transfer, the "control" is assumed to be foreign. Thus INPI compelled renegotiation when there was no transfer. With regard to foreign enterprises, INPI flatly prohibited remittances between subsidiaries and their home companies.

I will now review some data on technology transfer since the Brazilian science and technology policy began to take shape. However, because data collection mechanisms are deficient, the following figures must be regarded as merely suggestive.

The number of authorized technological agreements did increase in the 1970s as compared to the second half of the 1960s. Between 1963 and 1970, 2,429 contracts were registered, compared to 9,600 between 1972 and 1979. The last few years of the 1970s, however, saw a decrease in the number of contracts registered annually (see Figure 7).

Technology payments of $221 million were authorized in 1972, $282 million in 1973, $375 million in 1974, and $225 million in 1975. In 1978 and 1979 the figures increased to $402 million and $519 million, respectively. But the next two years saw a sharp decrease (to $294 million and $282 million), probably reflecting an improvement in the control machinery and the worsening of the balance of payments crisis.

The average growth of technology payments was 20.9 percent in the second half of the 1960s but declined to an average 16.7 percent

Figure 7. Brazil. Authorized Technology Contracts, 1972–1980

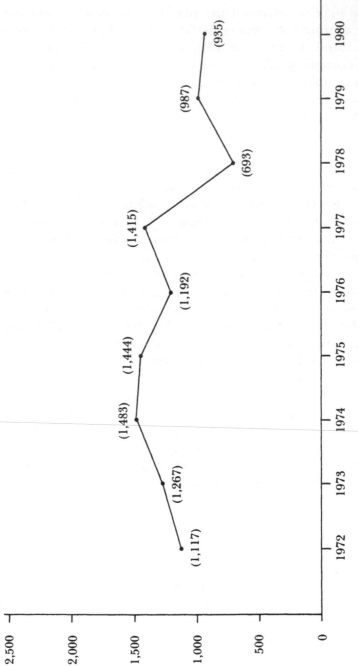

Sources: Francisco Almeida Biato et al., *A transferência de tecnologia no Brasil* (Brasília: IPEA, 1973), 27–29; UNCTAD, *Legislation and Regulations on Technology Transfer: Empirical Analysis of Their Effects in Selected Countries*, TD/B/C.6/55 (Geneva: United Nations, Aug. 1980), 9; *Planejamento e desenvolvimento* 5, no. 59 (Apr. 1978): 61; and MIC, STI/INPI, *Relatório de atividades 1980* (Brasília, 1980), 10.

between 1970 and 1976.[34] Their share of the GNP was only 0.06 percent between 1962 and 1964, 0.16 percent in 1965 and 1966, 0.31 percent in 1971, and during the 1970s it was maintained at approximately 0.2 percent,[35] indicating that the GNP and technology payments were growing at approximately the same pace. According to Fabio Erber, the ratio of technology imports to the Brazilian GNP is not substantially higher than in the developed countries; the main difference lies in the ratio between technology imports and domestic R & D expenditures. Whereas countries like France, for example, spent more than ten times the amount for technology imports on local R & D, the Brazilian factor was less than three.[36]

Remittances abroad for technology transfer increased markedly: 8.5 times between 1962 and 1976 and 3.75 times between 1968 and 1976. Furthermore, the industrial property share of total remittances fell from almost 40 percent in 1962 to 7.8 percent in 1973, while the share for know-how and services increased from 61.3 percent to 92.2 percent (see Table 9). The changes in legislation (Law 4131) with regard to remittances for patents and trademarks explain why the companies preferred to remit earnings as know-how and services.

In 1980, 58.4 percent of contracts brought for consideration were for specialized technical services, 17.6 percent for industrial technology, and 24 percent for patents and trademarks. Of the 90 percent approved, 12 percent belonged to foreign enterprises, 24.5 percent to joint ventures, and 63.5 percent to wholly owned Brazilian companies (of these, almost 50 percent were public enterprises).[37]

Between 1965 and 1973 there was a small increase in registration of process patents and a slight decrease in production patents. Between 1950 and 1973 only 18 percent of a sample of 48,212 patents (94 percent of all patents awarded during this period) went to Brazilians; the rest went to developed countries, mainly the United States and West Germany. In 1980 only 14.7 percent of the 4,204 patents awarded went to domestic enterprises.[38]

Direct Foreign Investment

Since 1964 Brazil has had a favorable attitude toward foreign investment. Laws 4131 and 4930 have regulated not so much the conditions of investment as the conditions under which capital and profits can be remitted abroad. The 1962 law specified that once registered in

Table 9.   Brazil. Technology Remittances, 1962–1976

	Industrial Property		Know-How and Services		Total	Index
	US$ mil.	% of total	US$ mil.	% of total	US$ mil.	(100 = 1956)
1962	12	38.7%	19	61.3%	31	91
1963	1	14.3	6	85.7	7	21
1964	—	—	8	100.0	8	24
1965	1	2.3	42	97.7	43	126
1966	3	6.5	43	93.5	46	135
1967	8	12.7	55	87.3	63	185
1968	7	10.0	63	90.0	70	206
1969	7	7.7	84	92.3	91	268
1970	8	7.7	96	92.3	104	306
1971	10	7.6	122	92.4	132	388
1972	6	3.9	148	96.1	154	452
1973	13	7.8	153	92.2	166	488
1974	—	—	—	—	194	571
1975	—	—	—	—	223	656
1976	—	—	—	—	263	774

Source: A. L. Figueira Barbosa, *Propriedade e quase-propriedade no comércio de tecnologia* (Brasília: CNPq, 1978), 141.

Brazil, foreign capital was to be accorded the same legal treatment as domestic capital. Foreign investment was prohibited in the areas of oil exploration, extraction, and refining; domestic airlines; communications; publishing; and coastal shipping; and it was permitted on a partial-participation basis in mining, fishing, hydroelectric power, banking, and insurance.

Foreign investment had to be registered in the central bank, and profit remittances abroad were limited to 12 percent (16 percent when not counting 25 percent withholding tax on the remittance) of registered investment. Beyond that limit, a supplementary income tax up to 60 percent was to be levied on the remittance. At the same time, "a foreign subsidiary cannot obtain long-term local financing from government sources and government guarantees against international loans, unless the investment is considered by the CDI . . . as high priority for the national economy."[39]

The Economic Development Council (CDE) passed guidelines for priority areas, namely capital goods, raw materials, and mining industries, in which preference was to go to local enterprises. If the locals could not do the entire job, then foreign investment could be called in to associate as a minority partner.[40] According to the Business International Corporation, multinational corporations were "finding it all but impossible to obtain [CDI] fiscal incentives for projects in areas that local companies have staked out as their own preserve, especially in capital goods."[41] The government has likewise been pressuring foreign firms to take in local partners when contemplating new ventures or new product lines. Indeed, foreign-owned engineering firms have not been able to get government business unless they merged on a minority basis with local companies. After 1976 the Special Agency for Industrial Investment (FINAME) no longer financed the sale of capital goods produced by foreign-owned firms. The CDI was given veto power over all proposed foreign investment, whether incentives were being sought or not, and the access of foreign firms to local credit was limited.[42]

Foreign investment increased quite dramatically. As of December 31, 1979, the central bank tallied $15.9 billion in direct foreign investments, with new investments reaching $1.42 billion, up from $1.36 billion in 1978.[43] In fact, foreign investment grew faster than the GNP. "The average annual growth rates between 1970 and 1973 [were] 99.7 percent for foreign investment, 16.3 percent for gross total capital formation and 12.5 percent for GNP respectively in real terms."[44]

The lion's share (about 75 percent) of foreign investment has been in manufacturing, with the heaviest concentrations in chemicals, transportation, metallurgy, and mechanical, electrical, and communications equipment. Of the top two hundred companies in Brazil in terms of sales, 39 percent were majority foreign-controlled, accounting for 32 percent of total sales, although foreign sources accounted for only 8.3 percent of total investments.[45]

A study undertaken on the financial impact of multinationals in Brazil between 1965 and 1975, using eleven of the top two hundred companies as examples, showed that their $298.8 million capital investment was turned into a $1.467 billion surplus generated in Brazil. Of this, $774.5 million was remitted abroad, $272.1 million, an amount roughly similar to the original investment, in payment for technology. One multinational, Volkswagen, accounted for 76.7 percent of the technology remittances; its own technology remittances in turn were 75 percent of its total. The eleven companies studied remitted an average of 35.1 percent for technology matters.[46]

Joint ventures.   Associate investment of foreign enterprises with national—state or private—enterprises is not new, as there were similar cases before World War II. But only after the 1960s did the Brazilian government establish it as policy for achieving a real transfer of technology: "The government encourages joint ventures between foreign and local investors as a means of transferring technology into Brazilian hands. It therefore wants the local partners to be capable of absorbing the technology." Foreign companies are thus under pressure to establish joint ventures, especially in the manufacturing sector such as in communications equipment. Standard Electric, ITT, Ericsson, NEC do Brasil, Siemens, and Olivetti have done so in communications, and General Electric in heavy electrical equipment.[47] Government enforcement of joint venture investment has been particularly strict in the petrochemical sector. In 1980, for example, the government blocked Dow Chemical from going ahead with a large project of its own, reminding Dow and other enterprises in the sector that the joint venture policy was still in effect. A study found that among 315 sampled foreign enterprises, 61 percent were wholly owned (95 percent or more), while the rest were joint ventures, of which 27 percent were foreign majority owned (50—94 percent) and 12 percent were foreign minority owned (6—49 percent).[48]

Management of Knowledge and Information

The management of knowledge and information became the cornerstone of the Brazilian science and technology policy as planned innovation and adaptation superseded the idea of a free market of ideas. No longer was random contact between foreign and Brazilian scientists and engineers to be counted on to trigger innovation and set trends. Scientists and engineers remained free to pursue their scientific instincts, but the financial rewards and other types of incentives were such that many scientists and engineers chose to serve the ideas of economic development and industrialization. Thus training and education became a means of cultivating what was considered one of the most valuable components of Brazilian progress: human resources—bearers of the scientific, technological, and managerial know-how.

Training and Education

In 1967 the government undertook what was called Operation Return to bring back to Brazil scientists working abroad and to integrate them into the process of building an autonomous science and technology capability. Duties for returning scientists were reduced in 1969 and wages were increased significantly.

One of the key measures taken to improve the level of education and to train experts was the 1968 university system reform.

> According to the new legislation, central institutes and departments were to be established in all universities, abrogating their traditional structure as federations of independent professional schools, and providing a place for research. A common basic course was created for broad areas of knowledge, thus postponing the students' decisions about their professional careers; entrance examinations for the universities ceased to be administered by each school and became unified for each geographical region and field of knowledge, in an approximation of the French *baccalauréat;* the courses ceased to be organized around a rigid succession of "years" and were replaced by a system of "course credits." At the same time, the universities lost their freedom to refuse unqualified students if they had places for them, and the number of

university enrollments increased pronouncedly. . . . The decision
to open the gates of higher education to a much larger proportion of the applicants, without waiting for the effects of university reform, was carried on mostly through the new, separate and
privately owned professional schools which were allowed to be
created. From 1968 to 1969, first year enrollment in these schools
more than doubled, rising from 39,000 to 85,000. In consequence, in 1971 more than half of all university students were
enrolled in private institutions, compared to a third in 1964.[49]

In 1964 a total of 142,386 students were enrolled in undergraduate
studies, of which roughly 40 percent were first-year students; in 1977
these figures were an estimated 1,117,000 and 41.5 percent.[50]

Significant measures were also taken by the government in coordination with faculty and scientists to build a national system of graduate studies. "The new graduate programmes, established very often in
isolation from the undergraduate schools, appeared to be alternative
places where some serious educational and research activities could be
carried on: they started to command a growing amount of funds and
to attract the best of the available talent in Brazilian higher education."[51] From 1969 to 1973, 18,777 students participated in the graduate programs (of which only 17 percent graduated).[52] Between 1975
and 1981 the number of graduate students increased from 22,245 to
40,209, and graduate courses increased 60 percent.[53] By 1974, 2,451
master's degrees and 1,910 doctorates had been awarded, and 335 individuals held graduate degrees in the biosciences. More than 40 percent of all master's degrees were in engineering,[54] but at the doctorate
level engineering represented only 11.5 percent. The largest percentages of doctorates were in physics, the social sciences, and the humanities. In 1975 a national Graduate Studies Plan was approved with the
goal of awarding 16,800 new master's degrees and 1,400 doctorates
between 1975 and 1979.

Worthy of note among the graduate studies programs is COPPE
(Coordination of Graduate Programs in Engineering), created in
1965 on a BNDE/FUNTEC initiative and by 1978 the largest engineers' training program in Latin America.[55] COPPE is a fine example
of the connection between economic development and the idea of
building human resources for the sake of technological autonomy:
note that a national development bank promoted and supported
what became the largest graduate engineering school in Latin America. COPPE's programs cover eleven fields of specialization, including

nuclear, systems, biochemical, and chemical engineering and computers, with approximately 100 faculty members and 1,500 graduate students.[56]

One of the most important tasks of the National Research Council (CNPq) in the area of human resources has been the awarding of fellowships, a traditional activity since its beginning in the early 1950s; 41,117 were awarded between 1972 and 1978. The value of fellowships and grants for the training of human resources increased 50 percent between 1974 and 1975. An even larger increase—127 percent—occurred between 1975 and 1976; and there was a 66 percent increment per year from 1976 to 1978.[57]

The organization of knowledge for training experts was given special emphasis by the first and second science and technology plans (I and II PBDCT). Plan I allocated approximately 22.3 percent of the 1973 budget and 21 percent of the 1974 budget for the development of scientific personnel, graduate training, and scientific development—or approximately $72 million and $80 million, respectively. Plan II allocated 27.3 percent of the budget for scientific and technological training in 1975, 26.2 percent in 1976, and 25.6 percent in 1977, totaling about $185, $201, and $211 million, respectively.[58] II PBDCT was very explicit in this area: among the principal concrete objectives were contracting of foreign professors and creation of fellowship programs; participation of Brazilian professors and researchers in scientific meetings in Brazil and abroad; supplying of laboratory material; and financing of fundamental research and graduate studies.[59]

Between 1970 and 1978, 41 percent of the funds of the National Science and Technology Development Fund (FNDCT) were released for scientific development and training. The funds were processed through CNPq, BNDE/FUNTEC, MEC/Department of Education, and the Brazilian Academy of Sciences, all for graduate courses at home and abroad. Physics received almost 40 percent of the financial resources, which the Studies and Projects Financing Agency (FINEP) attributed to the heavy emphasis on nuclear research.

Research and Development

Brazilians are constantly expressing concern and frustration about the transmission of the fruits of the R & D process to the productive system. But R & D has to be seen in the perspective of long-range, incremental progress. In fact, notwithstanding the complaints, a relationship between R & D and industrial development *has* evolved, which cannot be said about Argentina or the other Latin American countries.

An idea of the tremendous task that had to be performed can be gained by looking at the supply-and-demand situation of domestic research at the beginning stages of the explicit science and technology policy. An Economic and Social Planning Institute (IPEA) study carried out between 1967 and 1969 concluded that of the 46 research institutions investigated, only those with the largest number of technological activities were involved in real innovation and that the bulk of work was in technology adaptation.[60] Only a quarter of these institutions had an active relationship with the industrial sector, and only one-third of their activities were initiated by it.[61]

A Studies and Projects Financing Agency (FINEP) study, which looked at enterprises with some need for the work of research institutes, found that over 55 percent of the research jobs performed in these institutes were of the applied and of the development and adaptation kinds, and not merely routine—an indication of some, but still precarious, technological progressive endeavor.[62] Since the late 1960s the advance toward science and technology policy objectives has been steady, if slow, although it is on the whole still too early to assess the activities and the results of the programs being implemented.

One means of linking R & D and the productive sector is through the creation of foundations and companies that sell research services. For example, COPPE's faculty members created the Foundation for the Coordination of Graduate Programs in Engineering (COPPETEC) with the aim of providing services to private and public enterprises. "COPPETEC was officially described as an entity destined to intensify the linkage between the program and the enterprises, allowing the participation of COPPE's professors and students in the development of Brazilian technology and thus attracting to the university projects of value for the progress of the nation. COPPETEC is COPPE's scientific unit oriented toward the [solution of] technology problems."[63]

Other examples of such research institutes include the Technological Development Company (CODETEC; University of Campinas), created to transfer the results of R & D (up to the level of prototype construction) to the market. The University of São Paulo possesses several foundations, such as the Carlos Alberto Vanzolini Foundation, part of the Production Engineering Department of the Polytechnical School, and the Foundation for Engineering Technological Development (FDTE), an evolution of the Library of Digital Systems (ISD). The University of Minas Gerais created a Development Research Foundation (FUNDEP) to work on contracted research and technical assistance.[64]

Two research institutes of major importance for the development and adaptation of technology in Brazil are the Technological Research Institute (IPT) in São Paulo and the National Institute of Technology (INT) in Rio de Janeiro. IPT is one of the largest Brazilian research institutes: it has one laboratory devoted solely to the production of capital goods and another for metallurgical physics; it conducts research in such areas as solar energy, coal, hydrogasification, and alcohol; provides technical assistance to infrastructure works and to industry; participates in engineering; holds more than one hundred contracts with state and private industries; consults with and trains experts for research; performs tests and chemical analyses; and is involved in industrial metrology and pilot plan experimentation.

INT, with a foundation that has set up a company to sell services, also commercializes the products of its laboratories. Although the INT central laboratory is in Rio de Janeiro, it works in several regions and states as well. Its research concentrations are paper and cellulose, chemistry (technology), food, construction technology, pharmaceuticals, and industrial developments for small and middle-sized industries. The INT also undertook the task of developing alcohol out of biomass as part of the Pro-Álcool Program. In addition, it holds agreements with many national and international institutions such as UNIDO, OAS, the Canadian Research Council, and the Denver Research Institute of the United States.[65]

Other important institutions are the Minas Gerais Technological Center, which emphasizes metallurgy, minerals, and transportation; the Bahia Research and Development Center (CEPED), working on the control of metals, oil derivatives, pollution, and minerals; the Rio Grande Foundation, working in the areas of minerals and coal; and the Aerospace Technical Center in São José dos Campos, which develops technologies in the conversion of engines to alcohol, aeronautical engines and equipment, and solar and wind energy.[66]

FINEP has been stimulating R & D since the beginning of the 1970s, which has resulted in the intensification and diversification of applied research. It has acted according to the sectoral priorities of the economic and science and technology plans, financing R & D for basic research and up to product development by the national enterprise, in such sectors as energy, chemistry and pharmaceuticals, fertilizers, oceanography, material science, computers, and biophysics and health sciences. It supported the development of new technologies (including nuclear and space technologies, ocean resources, and alternative energy); industrial technology; agricultural technology; environment, health, and education; human resources training for R & D; and support activities for scientific and technological development. FINEP is thus a major instrument in Brazil's quest for technological self-determination, linking technology and the industrial sector (the latter accounted for 56.5 percent of the value of the operations contracted by FINEP in 1980). FINEP's Industrial Technology Program covers applied research aimed at generating technologies in the direct interest of the productive sector: "FINEP's activity is directed toward the expansion and strengthening of a network of technology production centers, in its aspects of physical infrastructure and of human resources."[67] As a corollary of these measures, several regional industrial technology centers were set up between 1970 and 1973 in Rio Grande do Sul, São Paulo, and Bahia. FINEP has also cooperated with INT and the Aerospace Technical Center.

CNPq has supported research activities through fellowships and grants, offering sectoral, regional, and institutional programs considered important for science and technology and for economic and social development. Beginning in 1970, the Integrated Programs—multi-institutional scientific endeavors coordinated by CNPq—have dealt with endemic diseases, genetics, oceanography, the continental shelf, computation, semi-arid and humid tropics, flora, urban and housing development, and nitrogen fixation in the tropics.

Later, CNPq created the so-called Technological Innovation Nuclei (NITs), units implanted in research institutions and universities to link research with industry. The NITs offer help to interested parties in obtaining R & D project financing, technology transfer or commercialization, or intercourse with other state science and technology institutions. Between 1972 and 1978 CNPq awarded 8,848 fellowships for research, and it has several dependent institutions performing R & D, including the National Institute of Amazonian Research (INPA), the Institute of Space Research (INPE), and the Brazilian Center for Physics Research (CBPF).[68]

Data comparing R & D performance show a significant change

during the last decade. The number of scientists and engineers engaged in R & D almost tripled between 1974 and 1978, from 0.8 to 2.1 per 10,000 population, as did R & D expenditures as a percentage of GNP between 1971 and 1979 (from 0.24 to 0.65).[69] In 1979 R & D expenditures on basic research amounted to 16 percent of the total, on applied research 31.9 percent, and on experimental development 52.2 percent. This last figure is very impressive, given comparable figures for France and Japan of 44.5 and 58.4 percent, respectively. Also particularly significant is the fact that 52.2 percent of the science and technology expenditures is accounted for by the productive sector (public and private).[70]

In 1977, 18.8 percent of Brazil's scientists and engineers were engaged in engineering and technology, which, although not remarkable, is higher than Argentina's 14.3 percent in 1976. Brazil spent 14.4 percent of its total R & D expenditures for 1977 on industrial development and 26.9 percent on the advancement of knowledge. Corresponding figures for Argentina in 1976 were 11.5 and 10.0 percent, respectively. In 1978, 35.4 percent of Brazil's 24,015 scientists and engineers engaged in R & D were involved in integrated and nonintegrated R & D and 64.6 percent were in higher education. In the same year, 31.0 percent of the expenditures went for integrated and nonintegrated research, 26.7 percent for higher education, and 42.3 percent for general services. The differences between these data and those for Argentina in 1976 are striking: 7.7 percent for integrated and nonintegrated R & D, 31.5 percent for higher education, and 60.8 percent for general services.[71]

Retrieval Systems and Data Banks

Since the beginning of its explicit science and technology policy, Brazil has placed heavy emphasis on the development of information retrieval systems and data banks. Some early steps were taken by Petrobrás, which had a Documentation and Patent Division. In 1970 the Targets and Bases Economic Plan suggested a national technological information network, and in 1973 that idea was incorporated into the new National Science and Technology Information System (SNICT), under CNPq's jurisdiction.[72] The initial components of the SNICT as specified by I PBDCT were the subsystems of Scientific Information, Free Industrial Information, Patented Technical Information, Patent

Bank, Information of the Infrastructure and Services, Agricultural Information, Social and Economic Information, and Collection and Distribution of Information from Abroad. II PBDCT appropriated approximately 1.5 percent of the 1975–1977 science and technology budget for the information system.[73] III PBDCT did not include quantitative targets, but SNICT's share of the science and technology budget in 1981 was 1.43 percent.[74]

A strong belief in the importance of information systems and data banks led in 1976 to the creation of the Brazilian Institute for Science and Technology Information (IBICT), evolved from the Brazilian Institute of Bibliography and Documentation (IBBD) attached to CNPq. IBICT collects data about information transfers between Brazilian and foreign institutions; archives scientific and technological information in Brazil, allowing access to the documents; records the Brazilian scientific and technological literature; trains researchers in information science; provides individual researchers direct access to the international data bases; and provides information for science and technology decision making. It also supports the information and documentation centers, providing complete sets of published material and facilitating regional distribution. IBICT is thus truly an information retrieval and dissemination institution.

A Document Center of Technological Information was created in 1978 in INPI, the first such center in Latin America; it includes a patent bank open to the public. By 1980 the patent bank comprised more than 14 million items, derived mostly from patent documentation—an increase of 22 percent since 1978.[75] A network of computer terminals with access to a central source of information was being considered. IPT and INT also have their own information retrieval systems.

Role of the State

Brazil's economy has been a mixture of market mechanisms, state intervention, and planning; the productive sector has combined private (domestic and foreign) and state enterprises. The state has played an important economic role through guidelines and planning, incentives and controls—establishing the objectives and the means to achieve progress. It has increased public spending; its development banks have financed education and production; and some of its enterprises are among the largest in Brazil. It follows, then, that decision making, both public and private, is affected directly and indirectly by a government that allows independent action, but only within the limits of objectives aimed at maximizing the nation's worth: this is the Brazilian mechanism for the allocation of values.[76]

In 1979 federal and state governments accounted for 48.6 percent of expenditures for science and technology, and 42.3 percent came from universities and financing agencies, many of which are also state-owned. The productive sector spent another 6.1 percent, and foreign sources 3 percent. From a sample of 46 research institutions that in the late 1960s worked on industrial technology, 82.6 percent were public (federal and state government); by 1977, 97.5 percent of R & D activities were performed by the state.[77]

The main role of the Brazilian state in science and technology has been one of guidance, in which planning mechanisms such as the economic development and science and technology plans play a part. This role also encompasses others, however, such as educating and facilitating, and promoting.

Educational and Facilitative Role

As a corollary of its explicit policy, the state performs many educational and facilitative functions. It has aided both private and state enterprises in acquiring foreign licenses and patents when the required knowledge was not available domestically. It has supported private enterprises in bargaining situations to assure that "real" technology transfer took place. It has worked to strengthen human resources and the scientific and technological infrastructure simultaneously in order to facilitate selection and use of foreign innovations, which policymakers view as a prerequisite for the achievement of self-determination. FINEP's efforts to improve the domestic technological capacity, together with the work of INT and IPT, for example, were consistent with this educational and facilitative role.

The state has not been entirely autonomous in these activities, however. Although the bulk of funds for science and technology has come from domestic sources, some are provided by international organizations such as the United Nations, OAS, the Interamerican Development Bank (IDB), and the World Bank. Relations with these entities and with other countries are managed through CNPq's Superintendency for International Cooperation (SCI), SEPLAN's International Economic and Technical Cooperation Secretariat, and a science and technology information agency of the foreign ministry. I PBDCT stated, for example, that the science and technology "contribution made under official international and foreign programs should attain annual values on the order of $25 million. Of outstanding importance within this program, on account of the volume involved, are the United States, the cooperation provided by the USAID, and the contribution of the United Nations Development Program (UNDP), which totaled $14 million in 1972."[78]

One way to see how international assistance and the domestic educational and facilitative effort cooperate is to focus on a program such as the Training Program for Scientific and Technological Research Administration, created by II PBDCT to promote the creation of technical groups to oversee or coordinate R & D activities, both public and private. Between 1974 and 1978, with the participation of 90 technicians from 15 Latin American countries, about 720 technicians were trained and 20 fellowships for study abroad were awarded. UNESCO supported the program with $330,000 in technical assistance budgeted through the UNDP, and the OAS also contributed. Related to the program were various seminars on science and technology questions—for example, one on technology transfer in which Argentina, France, Japan, the United Kingdom, and the United States participated.[79]

CNPq has also had a close cooperative relationship with UNESCO, the Organization of Economic Cooperation and Development (OECD), the International Council of Scientific Unions, the International Center for Theoretical Physics, and the Latin American Association of Physiological Sciences.[80] The World Bank has aided scientific and technological development, assisting the National Institute of Amazonian Research (INPA) in initiating research in the fields of ecology, medicine, and climatology and at the same time sponsoring several priority projects in the Amazon.[81]

IDB has also had extensive relations with Brazilian state institutions, such as BNDE and FINEP. For example, a program of industrial technology known as PROGRAM MINIPLAN/CNPq/BID, set up in 1971, involved an estimated first-stage expenditure of $58 million, of which IDB agreed to finance 55.2 percent ($32 million). The contracts between FNDCT and IDB were signed in 1973. IDB also participated in FINEP's beginnings in 1965 with a $5 million loan, representing 45.1 percent of FINEP's initial resources.[82]

The State as Promoter

The Brazilian government has fulfilled its role as promoter in several ways. Having set broad priorities for industrialization, it first identified the sectors that would be closed to foreign investment (see p. 177) and invited foreign companies to contribute to the development of selected industrial sectors, especially manufacturing. At the same time, the state managed its own enterprises in development projects and devised fiscal incentives to encourage investment in this development. All this was done with the greatest discrimination, according to the needs and capabilities of individual and of national industry as a whole.

Many of these incentives were managed by CDI, who determined which sectors were eligible for exemptions. In 1975 there were eight sectors, covering hundreds of products; the list was constantly revised as priorities and economic conditions changed. When a sector approached full development, new projects were no longer covered by incentives.[83] Decree 1137, which created the CDI, also stipulated that the tax exemption would apply only to imported capital goods without national counterparts. Decree 77065/1976 introduced new fiscal incentives, such as a 50-percent tax reduction on industrial products necessary for the manufacture of machinery and instruments not domestically available.[84] The incentive scheme was extended in 1977 to include components for capital goods made by foreign companies that join a CDI program for increasing the "nationalization index."[85] CDE also passed a resolution in 1977 establishing priority investment areas, namely the capital goods, raw materials, and mining industries. Local capital would be favored, with foreign investment invited only if no domestic company could do the job.[86] In 1979 the investment incentive program was severely curtailed, however, mainly to increase tax revenue.

With regard to R & D incentives, Decree 76063/1975 exempted capital and consumer goods purchased abroad for research from import tariffs; Law 4506 of 1964 and Decree 76185/1975 allowed expenditures for R & D to be deducted as operational; and the finance ministry further allowed R & D expenditures to be deducted from taxable income on a case-by-case basis. The CDI policy was to provide incentives when technological processes adequate for regional and sectoral development were used or when foreign capital was willing to contribute to local R & D. Dow Chemical, for example, built a $5.2 million R & D system as a supplement to its $500 million investment in Brazil.[87] Incentives did not always work, however. When a 1974 CDI resolution singled out the pharmaceutical sector to encourage technological research, giving preference to studies using available domestic resources,[88] Johnson and Johnson invested $2 million in its center for the study of endemic diseases, and American Cyanide and the Beecham group contributed some to Brazilian R & D; but all in all the response was very poor.[89]

Technology exports have also benefited from tax incentives. The Bank of Brazil has issued bonds to cover the loss of contract bids and failure to fulfill operation specifications for exported projects. The bank has also occasionally granted pre-export financing. Available data on technological exports estimate that between 1966 and 1980 Brazil built close to 60 infrastructure projects abroad, 36 of which totaled $3.049 billion, and 88 industrial projects, 12 of which amounted to $110.9 million; and that 10 out of 29 consulting services have rendered $8.4 million. Brazil's direct foreign investment between 1971 and 1978 amounted to approximately $400 million. Among the infrastructure projects have been oil pipelines, ports, airport expansions, and hydroelectric plants, and industrial projects have included plants and processes for the production of steel, food products, machinery and equipment, chemical products, paper, consumer durables, and alcohol (the latter almost 30 percent of the total). Sometimes the sale abroad of turnkey plants was accompanied by direct foreign investment.[90]

The state has also played a significant role in the technology market, mainly through its enterprises, which between 1972 and 1975 accounted for 33.9 percent of all agreements and 75.9 percent of all approved payments. Petrobrás alone had all the agreements in the oil sector, with approved payments of $290.7 million, about 26.3 percent of total approved payments. The Brazilian government held 94.2 percent of the extractive sector payments, 46.3 percent of manufacturing, and 89.3 percent of service. (The majority of the payments by the government were for technical assistance and engineering services.) The government share of agreements, as opposed to payments, was less during this period, but still impressive at 33.9 percent.[91] In 1977, 67 percent of the value of technology contracts was in nuclear energy.[92] However, although state enterprises have played a relevant role in the technology market, they have been exercising more independence with regard to technological purchases, payments, and services than some policymakers deem appropriate.

The State as Guide

In its role as guide, the state assumed a position at the center of a host of complex relationships and interrelationships (see Figure 8). To guide science and technology policy into the capture of new foreign technology and the creation of a national capacity to innovate, the state (1) attempted to control multinational corporations through guidelines for technology transfer and payments; (2) reinforced the science and technology infrastructure, focusing on the education system and technological adaptation and innovation; (3) created the mechanisms to promote national industry, both private and state-owned, identifying priority sectors directly connected with science

Figure 8.   Brazil. The Science and Technology Triangle: The State as Guide

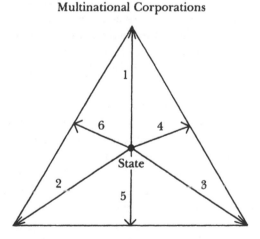

Multinational Corporations

Science and Technology Infrastructure

National Industry (public and private)

and technology autonomy, such as capital goods, and providing financial assistance to firms for R & D and technical, consulting and engineering services; (4) guided the relationship between the multinational corporations and national industry by encouraging joint ventures; (5) guided the relationship between the science and technology infrastructure and national industry by creating mechanisms, mainly through FINEP, STI, and BNDE, to link domestic supply and demand for science and technology and by centralizing science and technology information services; and (6) promoted multinational investment in local R & D operations (with little success).

It is interesting to compare this set of relationships with the Sabato triangle. Whereas the latter prescribed a dependency-reducing regime that deliberately ignored the multinational corporations (see pp. 58–59), in the Brazilian case a science and technology policy was designed to reduce dependency given the very market presence of the multinationals with their capital and technology. The state therefore placed itself in the triangle's center, guiding not only each of the elements involved but their relationships as well. Indicative plans established public fund allocations for science and technology within well-defined industrial and social priorities. Information retrieval systems were set up to allow informed choices on local adaptations to be made and to increase the bargaining power of the state and of Brazilian private enterprises—this was largely the domain of INPI from the mid-1970s. In addition, the state set priorities regarding the training of human resources and the creation of an indigenous R & D capacity and indicated methods by which to achieve them.

I have already described state actions in the areas of technology transfer and the national capacity to innovate. Thus I will conclude with a description of the financial part of the state's guidance and indicative planning role.

Science and technology as a share of the total national budget (created as an explicit budget item only in 1976) has been steadily increasing, at well over the rate of inflation: from 0.84 percent in 1970, to 2.20 percent in 1979, 2.11 percent in 1980, 2.31 percent in 1981, and 3.64 percent in 1982.[93] Between 1973 and 1977 the total programmed expenditures for science and technology (I and II PBDCT) according to large sectors approached $3 billion. Moreover, it is worth noting that the roughly $700 million appropriated by I PBDCT was more than twice the financial aid granted by the United Nations to all developing countries for science and technology development ($320 million).[94] II PBDCT alone allocated $2.27 billion for 1975–1977, an unprecedented expenditure on science and technology in a developing country. III PBDCT did not have financial commitments, but the 1981 budget was approximately $500 million. Furthermore, the fact that 33.6 percent of this amount was for technological research, about 20 percent for graduate studies, 8.5 percent for nuclear power, and only 5.1 percent for scientific research demonstrates Brazil's desire for quick results in the technology area.[95]

FNDCT has financed all the stages of scientific and technological knowledge creation and assimilation. Between 1970 and 1980 it channeled Cr 87.6 billion (1980 prices), or approximately $1.52 billion, into science and technology development (see Table 10). Although only 8 operations were approved by FNDCT for financing in 1970, 53 were approved in 1974, 72 in 1976, and 141 in 1978. Table 10 shows the value of FNDCT operations in three basic items: higher education, research institutions, and national enterprises. Although the 1979 value of operations was 60 times larger than that of 1970, the 1980 figure was almost 40 percent less than the 1979 value.

In real terms, the value of operations, increasing up to the mid-1970s, began a downturn in 1976. Thus FNDCT's share of the GDP quadrupled between 1971 and 1975, from 0.032 percent to 0.128 percent, but in 1976 it went down to 0.074 percent and continued to decrease to 0.06 percent in 1979 and 0.04 percent in 1980. FNDCT's share of the resources allocated to science and technology in the national budget declined as well, from 35 percent in 1978 to 32.2 percent in 1979 and 30 percent in 1980.[96] Table 10 shows also that the share of contracts with national enterprises grew markedly up to 1978, when it represented almost 50 percent of the value of operations; then a downturn began.

FINEP, the industrial technology secretariat (STI), and BNDE/FUNTEC (Scientific and Technical Development Fund) have financed programs oriented toward linking the supply of domestic technology with the demand. "The financial assistance provided by FINEP has been directed largely toward priority areas within the capital goods industries. Programs funded have involved activities such as new product development, the testing of prototypes, and the construction of pilot installations. FINEP has traditionally funded preinvestment studies which have allowed industry groups to study potential projects

Table 10. Brazil. FNDCT Operations, 1970–1980 (1980 values, in Cr 1,000)

	1970	1971	1972	1973	1974	1975	1976	1977	1978	1979	1980
Higher education	137,580 78%	598,780 55%	965,676 24%	1,013,065 17%	946,269 17%	4,423,069 41%	2,861,153 25%	3,961,953 33%	4,200,681 23%	4,248,030 39%	3,002,173 44%
Research institutes	28,221 16%	391,929 36%	1,891,116 47%	3,337,155 56%	2,996,517 51%	4,114,483 38%	3,237,620 28%	2,749,110 23%	5,640,914 30%	3,186,023 30%	2,119,181 31%
National enterprises	10,583 6%	97,983 9%	1,166,859 29%	1,621,873 27%	1,895,189 32%	2,314,980 21%	5,380,239 47%	5,212,509 44%	8,812,115 47%	3,360,230 31%	1,656,898 25%
Total	176,384	1,088,692	4,023,651	5,972,093	5,837,975	10,852,532	11,479,012	11,923,572	18,653,710	10,794,283	6,778,252

Source: FINEP/CEP, Internal Memo CEPO24/81 (May 19, 1981).

before taking final action."[97] FINEP has also been providing substantial support for national consulting, especially in the technology area. For example,

> FINEP and FINAME signed an Agreement of Financial Collaboration to support those enterprises which develop their own product engineering, thereby benefiting the capital goods industry. FINAME will give financial aid to enterprises on special conditions, and establish the index of nationalization to be attained. The funds allocated to the Agreement will come from the Program of Support for Technological Development of National Enterprises generated by FINEP (interest 2 percent to 4 percent annually, inflation adjustment of 10 percent annually, under 3 years, amortization over 9 years) and the Special Program of FINAME.[98]

FINEP's expenditures for technological development of national enterprises and for domestic consulting can be seen in Table 11. For example, it financed 83 percent of the total support cost for technological development of national enterprises in 1973 and 86 percent in 1978. Its share of total capital support for 244 operations in 1978 was over 76 percent. Industry in general received 54 percent of FINEP's total contribution.[99]

Following its creation in 1964, BNDE/FUNTEC funded education and applied and technological research (of which almost 50 percent was for engineering). Since 1974 it has concentrated on promoting and guiding autonomous industrial technological development at the company level.

> FUNTEC operates in this manner from two subprograms: . . . Demand and Utilization of Technology, "which seeks to establish attractive conditions for Brazilian industry to make good use of technological innovations as a factor of modernization, progress, productivity, and competition," the object being to substitute the importation of technology; [and] . . . Generation and Supply of Technology, which "has as its objectives to enlist research institutions and universities and to participate directly in the technological efforts of enterprises to search for innovations appropriate to the realities of the national economy."[100]

Also—parallel to FUNTEC—the Special Agency for Industrial Investment (FINAME), Brazilian Mechanics, Inc. (EMBRAMEC), Basic Goods, Inc. (FIBASE), and Brazilian Investments (IBRASA),

Table 11. Brazil. FINEP, Contracted Operations: Summary According to Type of Activities, 1967–1978 (1978 values, in Cr 1,000)

	Support for the National Enterprise Technological Development[a]			Support to the Consulting Services Customers[a]			Support to National Consultants			Total		
		Value			Value			Value			Value	
	Operations	FINEP	Total Cost	Operations	FINEP	Total Cost	Operations	FINEP	Total Cost	Operations	FINEP	Total Cost
1967	—	—	—	03	83,118	83,760	—	—	—	03	83,118	83,760
1968	—	—	—	18	234,454	271,731	—	—	—	18	234,454	271,731
1969	—	—	—	15	71,871	135,744	—	—	—	15	71,871	135,744
1970	—	—	—	41	97,435	116,929	—	—	—	41	97,435	116,929
1971	—	—	—	25	156,730	280,842	—	—	—	25	156,730	280,842
1972	—	—	—	76	247,043	295,071	—	—	—	76	247,043	295,071
1973	02	3,051	3,676	120	1,025,452	1,918,964	08	21,050	21,904	130	1,049,553	1,944,544
1974	14	137,544	190,434	110	779,506	1,020,578	30	117,504	236,215	154	1,034,554	1,447,227
1975	20	134,253	159,942	93	3,123,792	4,432,345	26	230,829	467,706	139	3,488,874	5,059,993
1976	40	935,994	4,387,797	59	531,648	911,672	15	54,492	462,947	114	1,522,134	5,762,416
1977	87	909,334	1,410,227	76	1,339,158	1,729,901	19	110,856	110,856	182	2,359,348	3,250,984
1978	125	1,609,616	1,870,764	111	1,293,603	1,935,271	08	37,487	39,603	244	2,940,706	3,845,638
Total	288	3,729,792	8,022,840	747	8,983,810	13,132,808	106	572,218	1,339,231	1,141	13,285,820	22,494,879

Source: FINEP.
[a]Includes direct operations and those performed through financial agents.

have been channeling resources for domestic capital goods and related high-technology programs. BNDE appropriated almost Cr 75 billion through these four institutions between 1974 and 1977; the FINAME share, which supports domestic manufacture of machinery, was 88 percent of that amount and represented 37.6 percent of the total BNDE operations.[101]

Indicative planning. The main priorities for I PBDCT—as part of the guidance role of the state—were new technologies, such as nuclear energy, space activities, and oceanography; infrastructure technologies such as electricity, oil, transportation, and communication; and industrial technology. The latter specified 27 priorities, including nonferrous metals, electronics, capital goods, petrochemicals, and food industries, and the minicomputer industry was singled out for special development.

II PBDCT emphasized energy, especially alternative sources, agricultural technology, and regional and social development. In the industrial sector it followed its II PND counterpart's focus on metallurgy, chemicals, electronics, mechanical agriculture, and food. III PBDCT's priorities were energy, agriculture, education, and welfare, and since energy and agriculture require a large input of capital goods, this determined the special emphasis in the industrial sector.

The State of São Paulo

To conclude this description and analysis of the Brazilian science and technology policy I should add that São Paulo, Brazil's most industrialized state, possesses something close to a science and technology policy of its own. The Science and Technology Program developed by the State Council of Technology (later replaced by the secretariat of culture, science, and technology), involves various projects in the areas of science and technology policy, technology marketing, research administration, technological information, metallurgy, quality certification, and food technology. The São Paulo State Development Bank, in charge of the program's execution, finances the operations with resources from the state and from FINEP and FUNTEC, and the State Fund for Science and Technology Development coordinates the financing enterprise with the State Council of Technology.

The Science and Technology Program includes a system designed to orient and develop the capacity of technological research institutions, create extension services and financial incentives to encourage the use of existing Brazilian technology, establish new technological services, and develop the technology that is supported by public funds. Thus the São Paulo science and technology system has become, in a way, a microcosm of the federal system.[102]

An Image of
the Future Takes Hold 8

The key to Brazil's success in conceiving and implementing a science and technology policy and in building the bureaucratic machinery to serve it lies in part in the country's post-1964 political stability and the economic miracle that produced the resources to be invested in development. But it also, and in interaction with that, lies in the existence of a general ideological consensus about industrialization as the way to progress, in the proliferation of antidependency ideas during the 1950s and 1960s in places of intellectual and political influence, in the basic nationalist view of scientists, and in the ad hoc alliance between the military and technocrats in pursuit of a common nationalist goal.

Actions might not have happened at all or might have taken a very different course had it not been for some of the actors and processes I will review in this chapter. The success of those who made the policy possible, who conceived and then built the science and technology "system," is more significant than meets the eye. Their main accomplishment has been to infuse a political system characterized by different and sometimes conflicting political and economic objectives with the awareness that science and technology are important for progress (as defined by most of the Brazilian actors) and that technological innovation and the control of technology transfer are linked to the development of industry, employment, welfare, and security. The major achievement of the guerrillas has been in turning their image of the future into something taken almost for granted.

Ideological Background: Ideas, Sources, and Carriers

Continuity of the Ideology of Development

Any outside observer of postwar Brazilian economic policy-making cannot fail to be struck by the special way in which many issues were approached. Important assumptions, attitudes, and even "data" were taken so much for granted that there was little discussion of such "self-evident" truths. This is perhaps the best indication for the prevalence of a widely held ideology.[1]

Among the most important elements of the ideology of development and the consensus built on it was the perception that Brazil had a bright future but that it would only come about through industrialization and modernization, to be achieved by developing basic industries to provide a domestic supply of steel, oil, and chemicals[2] and a capital goods industry to provide domestic machinery and equipment. In this sense national achievement and economic modernization were more important than simply raising per capita income.

Another element was the belief that the Brazilian state and capitalist economic planning had to play a strong role.[3] It can be argued that from the cognitive point of view industrialization and modernization undertaken in a changing international environment would only create more complexity and uncertainty. But it was understood that the state would deal with these problems by formulating plans and directing market forces toward national objectives. Still another element of this ideology was pragmatism. As Leff (quoting Bonilla) pointed out: "'Brazil is a nation of pragmatists. Unembarrassed by rigid commitment to ideas or principles.' This pragmatic approach, however, operated only *within* the framework of the ideology. There it was possible only because of the consensus and the consequent delimitation of issues and solutions."[4] The decline of traditional export producers in national politics permitted the growth of such an ideological consensus on economic development. Its statist nature was aided by the failure of national private entrepreneurs to develop an independent ideology.[5]

After 1964, however, the consensus suffered a major setback. Whereas some advocated the achievement of Brazilian objectives through opening the economy to the international economic system, dependentistas advocated just the opposite: that the ISI process toward self-sufficiency be continued. The rift has been large, but both groups still shared the broad objectives of national development with heavy reliance on state intervention, and both have been very nationalistic.

Even Maria da Conceição Tavares, well known for her outbursts against Brazil's "selling out" to the multinationals, believes that "Brazil has been Latin America's most nationalist country, even during a 'sell-out' period such as that of the previous [Médici] government." She argues that it is unquestionably

one of those countries which, in spite of its liberal rhetoric and its rhetoric in favor of foreign capital, has systematically used its bargaining power, i.e., the bargaining power of its dominant classes, of its government technocracy, and of its national entrepreneurs, to resist. I have not been informed that in this country any large national capital, big national bank, large national landowner, or big national technocrat has been liquidated. Even if the national technocrats consider themselves transatlantic, consider themselves liberals, in practice they have increased State intervention, have increased the strength of State enterprises augmenting Brazil's political control.[6]

The Intellectual Carrier and the Brazilian Subversive Elite

Leff has argued that intellectuals became prominent in the political system because of the political leaders' perceived need for specialized technical knowledge following the collapse of the coffee economy in the Great Depression and the desire to achieve the high levels of administrative practice (development, rationalization, and practice) being developed in the advanced countries.[7] It can also be argued that Brazil has been strongly influenced by positivistic, mainly Comtian, philosophical thinking, which created a fertile ground for intellectual scientific elites. "Positivism was taken more seriously in Brazil than practically anywhere else. Even at the end of the nineteenth century, most of the teachers of the Escola Politécnica in Rio were positivists, as were those of the Escola de Medicina. Comtian positivism was pervasive in the Brazilian army and was the outlook of a substantial part of the political leadership in the states of Rio de Janeiro and Rio Grande do Sul."[8] This positivistic mentality was probably linked with the desire for higher levels of administrative management practice, modernization, and, later on, technological development.

The egalitarian nationalist intellectuals, through their studies and their work as technocrats, became carriers of the development ideology. From this point of view, the 1964 "bureaucratic-authoritarian" coup[9] had a reinforcing function, because it brought intellectuals, technocrats, and planners to the service of the state, which in turn bred groups of young subversive-elite intellectuals. At the same time, the expansion of the public sector provided the opportunity for patronage appointments to the policy-making circle—but through knowledge, not politics or class. After 1964 reliance on the intellectual rather than on the "político" was the consensus among the ruling elites. Key policymakers such as Roberto Campos, Mário Henrique Simonsen, and João dos Reis Velloso publicly expressed their conviction that technocrats rather than politicians were best suited to make the Brazilian productive machinery work.[10]

Postwar institution building created the physical environment for these carriers, and for their ideas to develop and "grow" to political relevance. The Getúlio Vargas Foundation became not only a center for economic problem solving but also an ideological factory. Later the Economic and Social Planning Institute (IPEA) would fulfill a similar, and crucial, role, and the Superintendency of Money and Credit (SUMOC) and the National Economic Development Bank (BNDE) would also become "strategic places within the government where these individuals could develop their careers and their influence on policy."[11]

Many of the Brazilian technocrats working in the science and technology area seem to have shared the basic belief that coercive dependence should be reduced or eliminated and that more autonomous policies should be implemented. They have differed significantly about how to do this, usually along the lines of the pragmatic versus structural antidependency debate. However, few structural dependentistas have worked for government institutions, most being employed at the universities. The majority of technocrats, then, have been pragmatic dependentistas, favoring a capitalist solution to Brazil's economic development and equality problems. They have strongly advocated strengthening national entrepreneurship, with active state involvement. Although they favored redistribution, they opted to join forces with a government that was overwhelmingly concerned with stabilization. For even though the pragmatic dependentistas opposed the strong military government, what it stood for, its repressions and its Orwellian connotations, they knew that its services were necessary in the interests of national economic development and equality, when the objective was to reduce dependency and increase their nation's self-reliant potential.

The Scientific Community and the "Other" Community

One of the distinctive ideological origins of the movement for a science and technology policy came from among the scientists themselves, mainly the physicists. As one prominent scientist put it, the

> development of physics in Brazil in the last period was a cultural event of major transcendence. . . . The cultural importance of the physicists' work—especially that of César Lattes, which had a large international repercussion—is something that needs to be analyzed in more depth. Even in areas that were not directly linked to physics, it represented a certain sense of self-confidence about Brazilian capabilities. . . . Thus even if this cultural movement did not accomplish much, it ended up exerting a strong ideological influence. In general, the Brazilian scientists of that period, not only because of the relevance of research, acquired a high national prestige and were always linked to nationalist positions.
>
> I believe also that Oswaldo Cruz [a famous biologist] and the scientists of the past were nationalists, yet without a conscious nationalist ideology. Let's say they were patriots. But contemporary scientists have a nationalist ideology. They were linked, for example, to the oil campaign and other campaigns that defended the national wealth. . . . We cannot adopt an excessively technologist point of view: we have to take into account the large interaction between the technologist and the ideologist. From this point of view, the most eminent Brazilian scientists of the period always provided strong ideological contributions.[12]

One major difference between the nationalism of the scientists and that of the economists and engineers is that the scientists have usually opposed short-range solutions. They view Brazilian science from a long-term, global perspective of Brazilian development in all areas of human endeavor. Said a biologist, "The conception behind financing research in Brazil cannot be subordinated to a pragmatic ideology of the immediate. . . . It is essential to include the perspectives of immediate and long-range problems within a certain ideological conception. This applies to any area, such as energy, transportation, housing, health, and so forth." [13] At least some scientists did not share the pragmatism of other intellectuals; we will encounter the sequel to this disagreement in the making of science and technology policy after 1968.

The engineers have been ideologically closer to the economists than to the scientists. One reason for this is their academic background: "Concerning the prominent role of engineers in economic policy formation, we should add that the Brazilian engineering schools always taught some economics. The professional self-definition of Brazilian engineers also tends to be much broader than, for example, is the case with engineers in the United States. In the discussion forums of their professional clubs and in their journals, general issues of national economic policy are often raised." [14] Moreover, engineers, like economists, believe in the state. [15] Whereas scientists tend to be cognitively detached from production, from policy making, and from the multinationals and their activities in Brazil, engineers are in direct contact with the multinationals, learning and copying. And working for the state, it is the engineers who, together with the economists, think about how to control them. Their point of view was to be crucial for the future development of science and technology policy, as Schwartzman elaborated:

Alongside the scientific community in Brazil there exists another community, historically and currently important, which we can call the "technical-scientific and planning community." Historically, this community can be linked to the tradition of engineers graduated from the Polytechnic School of Rio de Janeiro, the School of Minas de Ouro Prêto, and some other institutions, as well as to the Armed Forces. It has to do with an elite group of men of action, concerned with practical results, linked to public administration. . . . Since the 1950s, the banks and the development and economic planning agencies have been organized and stimulated by this community. It is in this environment that ideas have been generated for economic and social planning, for the implementation of comprehensive regulations, and for the administration of the public sector. . . . For this community, scientific activity has meaning only when it is integrated with a planning project of broader scope. This type of scientific policy is upheld principally by big government corporations, starting with the Armed Forces. . . . It is superimposed then, in part, on the technical-scientific and planning community, but deals with a sector of this community endowed with strong organizational structure and sizable administrative economic resources. The object is the acquisition of operational autonomy and national and institutional control of technical strategic resources.[16]

ISEB: "Fábrica de Ideologias"

Among the social science institutions created in the 1950s we will focus on the Higher Institute of Brazilian Studies (ISEB), whose strong ideological legacy was described in Caio Navarro de Toledo's *ISEB: Fábrica de ideologias* (Factory of Ideologies).[17] ISEB's impact lay not only in the ideas it produced before the military government shut it down in 1964, but also in the influence of those ideas on subsequent political action, including the development of a nationalist science and technology policy.

ISEB, created in 1955 by President João Café Filho, became internationally known through the work of such intellectuals as Hélio Jaguaribe and Cândido Mendes. It was characterized by two clearly defined nationalist currents: a moderately egalitarian one represented by Mendes and Jaguaribe and a more radical one represented, for example, by Álvaro Vieira Pinto and Wanderley Guilherme.[18]

The common cognitive ideological element in all of ISEB's "strands" of thought was nationalism. The differences that arose had to do with the intensity of the egalitarian dimension, with opinions about dependency generally divided along the structural and pragmatic lines.

The radical intellectuals could not conceive that foreign investment would promote development in Brazil. So structural dependentistas saw no alternative but emancipation from foreign capital. As Vieira Pinto put it:

It is no longer a question of promoting development per se, but of bringing about *national emancipation* based on development. . . . *The old nationalism* could be identified with simple development needs; the new one . . . demands more, demands that development should be understood not only as quantitative progress but above all as a transformation of the essence of the objective process, which would have to be free from any foreign influence.[19]

For the more moderate Cândido Mendes, foreign capital had to be disciplined and controlled but not repressed or dispensed with if the national development objectives were to be achieved. Mendes (and later Cardoso) believed that dependency and development could occur simultaneously—the key objective was autonomous national capitalism. He and other pragmatic ISEBians were firmly convinced that capitalist development could be achieved on a strictly autonomous basis. Thus,

> it was believed that significant sectors of the industrial bourgeoisie were motivated by national or "native" interests. An ideology, upon being formulated, would respond to the function of informing and organizing these sectors . . . and of convincing other segments of the dominant class of their objective interests in promoting economic and political policies that would be more rational and clear.[20]

These ideas of the pragmatists with regard to foreign investment, autonomy, and capitalism versus socialism had a direct impact on state officials as they began to formalize a conception of autonomous technological development.

Escola Superior de Guerra (ESG)

The development and functions of the Higher War College (ESG) have been treated elsewhere.[21] My aim is to explore its effect—in terms of ideas and ideology—on the military officers who took power in 1964, who had a significant role in supporting a national science and technology policy. Stepan credits the ESG, created in 1949 by President Dutra, with the development of a "new professionalism" within the armed forces, involving increased military intervention in the polity. In 1963 the ESG announced that its mission was "preparing civilians and the military to perform executive and advisory functions especially in those organs responsible for the formulation, development, planning, and execution of the policies of national security."[22] Furthermore, the ESG was developing a world view, known as *segurança e desenvolvimento,* that included both the role of the military and ideas about development and security.

It was no coincidence that the ideas of the military government were also those of one of the main ESG ideologues—Col. Golbery de Couto e Silva, who until 1981 stood behind five presidents as the strongman of Brazil. Other top policymakers who actively participated in the ESG were former presidents Humberto Castello Branco and Ernesto Geisel and his brother Orlando Geisel. In the early days of the ESG the critiques of Brazilian society by military intellectuals seemed academic. But by the 1960s these critiques had evolved into calls for action and change, made imperious by the perceived threat of communist aggression.[23] Economic development was thus conceived as a means to enhance national security, and later, when the threat did not materialize and following the economic miracle, it became the means, through industrialization and scientific and technological development, to achieving an "ultimate" goal: the status of superpower—security and development, *segurança e desenvolvimento*, had become inseparable.

This brings us to the science and technology issue. The military have generally regarded both development and technological independence as necessary to Brazil's becoming a major military power— thus the drive for a policy of autonomy in science and technology. What ESG publications euphemistically called "national objectives" could be achieved only by such a policy. As a high-ranking military officer, trained as an engineer and serving also as a high official in a technological institution, explained:

> It is today an uncontested fact that Brazil exhibits a sum of highly favorable factors that permit the country to aspire to the status of world power. It is also recognized that one of the main obstacles to this achievement is its dependence on external sources for technological material. Were the flow of imported technology to cease, the Brazilian industrial park would rapidly become incapable of competing in the outside market, for lack of the internal capacity to prevent technological obsolescence. . . . Consequently, it would be desirable to have a national consciousness of these circumstances, which civilian and military leaders should take into careful consideration. With respect to military power specifically, it can be affirmed that to this day history has not recorded a single nation of global import whose military industrial park was supported by foreign enterprise and by the importation of technology. On the contrary, history shows that just the reverse has happened: the great powers rarely admit external partners in their military industry.[24]

The Evolution of an Idea

The idea of technological autonomy evolved as part of the more general concept of dependency, which became a stage in political consciousness—a perception and a cognition of social causes and effects in domestic and international relations. In other words, science and technology and dependency became linked issues in Brazil once politically relevant individuals were alerted to underdevelopment and inequality at the international level and to the ascendancy of certain production models and ways of life.

It is impossible to trace all the ideological connections between the individuals and institutions that over the years contributed to the idea's evolution. Ideas are shared, copied, and passed along from individual to individual and from institution to institution; their origins and paths of development become lost. However, it is possible to point to certain key individuals and organizations that have had a noticeable and strong impact on the way the idea of technological autonomy evolved. These actors fought for their ideas, created institutions, and influenced other actors, sometimes even inducing them to change their ideological stance. As Dionísio Dias Carneiro, a former high official in the science and technology system, said, "What they did is invent a system." Also, as we will see, because of them the evolution of the idea proceeded in spite of the fact that economic development policy took an apparently different ideological direction. Two of these institutional actors were the National Economic Development Bank (BNDE) and the Economic and Social Planning Institute (IPEA); two of the individual actors were José Pelúcio and João dos Reis Velloso.

BNDE, Pelúcio, Velloso, and IPEA

From the science and technology point of view BNDE is the institutional embryo, the place where many of the ideas regarding development were conceived and found a reinforcing environment. It was established in June 1952 as a federal *autarquia,* "a type of independent agency with the characteristics of a government corporation," [25] with the objective of financing the development of the Brazilian infrastructure, and later of financing industrial development, by giving help to private investors and thus guiding and controlling the industrial development of Brazil. "BNDE performed a useful function in reconciling decisions and establishing orders of priority, as well as in connection with the execution of the projects financed."[26] BNDE also became a tacit center for the analysis of governmental programs.

In the 1950s BNDE devoted most of its financial resources to the development of the infrastructure and heavy basic industry. As a corollary of the Target Plan (covering the years 1956–1960; see below), BNDE became very much involved in financing the development of the capital goods industry, granting loans, and guaranteeing loans from abroad. Its share in the gross formation of capital grew from an initial 1.78 percent to 7.29 percent in 1967. During the first year of its existence 100 percent of its funds went to utilities services in the public sector, but the distribution gradually shifted; in 1965 productive industry (both public and private) received 95.8 percent, and by 1972 78.2 percent was going to the private sector.[27]

It is relevant from our point of view that besides playing a financial role, BNDE became a mechanism for setting development priorities and for judging and controlling industrial development. Even more relevant is the fact that it also became an institutional source of nationalist and egalitarian-nationalist economic ideas and plans. Its Department of Economics was in charge of research, undertaking studies of monetary policy, the balance of payments, and national income. "That this planning role of the Bank is a conscious one is indicated in its official reports: 'The relationships of the Bank with the government require, obviously, a direct participation in planning and in the execution of the governmental policy in the area of economic development.' It is apparent, then, that the Bank has developed not only as a continuing mechanism for the execution of plans but also as a planning agency proper." It is significant that BNDE signed an agreement with the Economic Commission for Latin America (ECLA) in 1953 "which provided that a joint team of economists from the Bank and from ECLA would make a complete study of the Brazilian economy"— from which the Target Plan was drawn.[28]

José Pelúcio's going to work for BNDE in 1963 was a fundamental event in the evolution of the idea of scientific and technological autonomy. Pelúcio acted as a "carrier," giving the idea access to political power and changing the minds of policymakers; ultimately in 1968 a policy with the goal of technological self-determination and autonomy was formulated—the idea's day had come.

In 1964 a Brazilian physicist, J. Leite Lopes, published *Ciência e des-*

envolvimento (Science and Development),[29] which presented the thesis of scientific dependency and the "myth" of technology transfer as a disguise for new forms of dependency. In the preface of a second book, *Ciência e libertação* (Science and Liberation), he wrote:

> Many of us had the privilege of getting specialized scientific training abroad. Upon returning to our country we were impelled to participate in the so-called economic development efforts. Little by little, we were confronted with obstacles that prevented us from doing our work: insufficient financial support for scientific inquiry, the absence of a national policy to stimulate science and technology, universities lacking adequate structures. We thus discovered the absence of a national technology that could be the basis for our industries, the absence of projects for the development of diverse branches of applied science, necessary for technological research, and most of all, the lack of access, or impeded access, to health and education for the majority of our population. We discovered then that these problems are part of what is called underdevelopment and result from historical and internal factors, among them the primary factor that the economy and culture of Latin American countries were and continue to be dominated, dependent.[30]

Pelúcio admitted that he was greatly influenced by Leite Lopes's first book, which was published at about the time Pelúcio began his association with BNDE. "The physicists," he said, "were the first to become conscious of dependency; later came the economists."[31] He said that although the physicists and economists developed their ideas separately, by the middle of the 1960s the ideas had converged to create an awareness of science and technology dependency. The economists' emphasis on the linkages between technology and economic development was particularly important. Pelúcio also acknowledged that both *ideas cepalinas* and ISEB had considerable effect on his subsequent work in the science and technology field.

When Pelúcio started with BNDE, his ideas were pragmatic; he looked toward self-determination but remained aware of the need for real technology transfer. Together with his friend Alberto L. Galvão Coimbra he devised the idea of graduate studies in engineering in Brazil, which BNDE adopted as its "pet program"; in 1965 COPPE was created. Coimbra became its director and until his resignation in 1973 was its guiding force. Also as a result of Pelúcio's efforts, BNDE set up FUNTEC to finance science and technology development.[32]

Pelúcio's and BNDE's ideas were incorporated into the science and technology section of the Strategic Development Plan of 1968. "From that date," said Pelúcio, "the question of science and technology development became routine." Pelúcio remained active in the policy-making process until 1979, as president of the Studies and Projects Financing Agency (FINEP) and vice-president of the National Council of Scientific and Technological Development (CNPq). (He returned to activity in science and technology, after a six-year hiatus, in 1985.) He was also credited with conceiving of the economic and political "trick" of keeping science and technology funds independent from the national budget, which provided the scientific and technological community with a partial, but nonetheless important, shield from economic and political pressures. The following exchange between São Paulo scientists Z. Vaz and Rogério C. Leite shows how scientists felt about Pelúcio:

Vaz: When Rogério and others went to Campinas [a university near São Paulo] they knew they had nothing, no buildings, no equipment, not even tables to sit around.

Leite: We had Pelúcio, at FINEP.[33]

Pelúcio exerted his influence not only on the scientific community but also on one man who in the 1970s would become a strong figure in the Brazilian political system—João dos Reis Velloso. Velloso founded IPEA and was its president; in 1968 he became general secretary of the ministry of planning, and in 1969 minister of planning; and he served as minister-in-chief of SEPLAN from its creation in 1974 until 1979. It was mainly through Velloso that Pelúcio diffused his ideas into the policy-making and implementation processes. Fábio Stefano Erber, a well-known economist who specializes in science and technology, put it this way: "Velloso was in the business of buying ideas; Pelúcio was in the business of selling them. . . . Behind Velloso and his actions, there were many ideas."[34]

When Velloso became minister of planning, he invited Pelúcio to create a group to work on science and technology policy for a government development plan. Pelúcio believes that he and his coworkers were invited because of BNDE's influence in creating favorable conditions for industrialization and economic development, and because the military, propelled to power in 1964, understood the importance of science and technology. According to Carlos Lessa, another economist and a colleague of Pelúcio and Erber, the military, too, were in the business of buying ideas. Many of Pelúcio's and Velloso's ideas were especially attractive because at the time they fit the military strategy of development.[35] In addition, the military were influenced by certain publications that reinforced their nationalistic and proindustrialization views. Lessa cited two books in particular: Jean-Jacques Servan-Schreiber's *The American Challenge*, which stressed the relevance of mastering production techniques; and the controversial *Toward the Year 2000* by Herman Kahn and A. J. Wiener,[36] which to the consternation of the military and other political leaders predicted that in the year 2000 Brazil would have only a $500 per capita income (this elicited a direct response from Velloso and a challenge in the form of the book *Brasil 2002*, by Mário Henrique Simonsen).

Velloso was fascinated by the role of knowledge in the development process, envisioning intensive technological industries aided by universities and research centers—which only the state could promote.[37] But Velloso was critical of dependentistas and had strong ideas about Brazil strengthening its relations with the rest of the world. "It is natural," he said, "that a market economy would actively maintain its relationship with the external world. . . . A certain degree of economic interdependence is also natural, mainly among the countries of the West."[38]

Therefore, Velloso and Pelúcio, for their own reasons, favored giving a push to Brazilian industrial strength and domestic science and technology capabilities. Velloso had the political power needed to make decisions and turn them into action; Pelúcio was his right hand on science and technology matters, the organizer and executor. In addition to giving explicit credit to Pelúcio and other economists for having influenced his thinking about science and technology, Velloso cited a series of Organization of Economic Cooperation and Development (OECD) studies published in the late 1960s dealing with the technological gap and various Organization of American States (OAS) studies on development in Latin America, most of which were presented from the pragmatic antidependency perspective.[39] He also mentioned ECLA's studies and books on the brain drain, but said, "It was only when the economists started becoming interested in the area of science and technology that I realized the importance of having a clearly defined policy to accelerate scientific and technological development in certain areas." As an example he referred to Pelúcio's work in BNDE: "BNDE was a pioneer in this area. I learned a great deal via Pelúcio. The economists started thinking about science and technology—that was the main point."

Other Actors

The development of science and technology policy after 1968 was the work of hundreds of individuals employed in public institutions. For example, between 1970 and 1972 Francisco Biato and other egalitarian-nationalist IPEA technocrats carried out a series of studies about science and technology, including one on the problems of technology transfer and another on Brazil's technical research potential.[40] The studies, "squeezing" an ideology into their arguments, illuminated the deficiencies and problems arising from not having an explicit science and technology policy.

Another guerrilla stronghold has been FINEP's Center of Study and Research. Under its first director, Erber, the center published many policy-oriented studies on science and technology. Erber himself is credited with two of the most influential, one on the demand for national scientific and technological services, which complemented the two IPEA studies, the other on the absorption of technology into the capital goods sector.[41] Before these studies were published there was no clear idea about technological supply and demand or about concrete problems of technology transfer.

The pragmatic antidependency guerrillas have certainly penetrated CNPq, with Pelúcio as its vice-president until 1979 and several scientists in the Scientific Consulting Group expressing pragmatic antidependency views. In 1980 a new organ, the Science and Technology Policy Advisory Group, was created to advise CNPq on the elaboration of policy and to strengthen its programs of education and research.[42] The group's first participants were Amílcar Herrera, Eugenio Lerner, Fábio Erber, Fernando Garcia, Francisco Biato, Federico Gomes, Gabriel Cohn, José Tavares, and Sergio Baptista Zacarelli, many of whom have been the "elite" of the science and technology subversive elite in Brazil.

Policy Continuity: Seizing Opportunities and Overcoming Obstacles

Implementation Hurdles

The pragmatic antidependency guerrillas had to fight to change the minds of politicians, industrialists, and scientists in favor of a policy of autonomy—in short, they had to create new awareness. They also had to bargain, compromise, and coalesce with other agencies and contend with groups firmly opposed to the policy. Thus many bureaucratic and political obstacles hindered the policy-making process, its coherence, and sometimes its efficiency, but despite setbacks, the main policy thrust was maintained.

Science and technology actors, and particularly the pragmatic anti-dependency guerrillas, had to overcome strong opposition from various political sectors. Some saw the science and technology policy as one more attempt by the state to flex its muscles and dominate socio-cultural and economic affairs; others, because of either liberal ideological viewpoints or political and business ties to multinational corporations, saw it as threatening Brazil's economic progress. Resistance could be found all over the bureaucracy, but it was especially evident within the ministries of finance and of industry and commerce (MIC)—mainly within the Industrial Development Council (CDI)—and even within the National Institute of Industrial Property (INPI) and CNPq.

When Antônio Delfim Netto became head of the planning secretariat (SEPLAN) in 1979, bringing his own people with him, the science and technology system had to contend with a new planning philosophy, one very different from Velloso's. But although Netto de-emphasized the importance of planning and eliminated quantitative targets, the commitment to science and technology planning was too deep for even Netto to extinguish completely. Neither did he reverse the main thrust of the policy, autonomy: by 1979 technological autonomy was not just an idea in the minds of some individuals as in 1964; it was a full-blown ideology, espoused by powerful governmental agencies, institutes, research centers, the armed forces, and influential individuals—forces to be reckoned with.

Part of the policy's difficulties have had to do with the less-than-total success of science and technology institutions in centralizing the policy-making process. For example, a narrow sectorial preference of foreign goods over national on the part of a few ministers could be enough to block specific efforts toward technological autonomy. This bottlenecking was aided by the fact that part of the science and technology budget went directly to the ministries, which were in full control of the funds.

An additional hindrance to implementation has been the opposition of some national enterprises to INPI regulations involving the renegotiation of contracts to effect actual technology transfer. Because small and middle-sized enterprises were unable to equip themselves to absorb and adapt imported technology, they saw science and technology policy as a threat to their existence. To overcome the problem, INPI created working groups to coordinate private enterprise and government institutions. For example, during 1981 a capital goods working group was formed with the participation of the Brazilian Association for the Development of Basic Industry, the Coordinating Commission of Articulation with Industry (CCNAI), the Department of Foreign Commerce of the Bank of Brazil (CACEX), BNDE, and FINEP. Moreover, consultation between INPI and research institutions and engineering associations has increased in recent years, also easing some of the implementation hurdles.

State enterprises have hindered implementation as well. Being relatively independent from other government structures, and consequently from the policy formulation and implementation process, and in spite of having great recognized potential for autonomous technological development, these enterprises (particularly the larger ones such as Petrobrás and Eletrobrás) have not always followed the SEPLAN and CNPq directives. For example, state enterprises have gone against policy by contracting with foreign instead of national consulting and engineering firms: a 1975 FINEP study found inconsistency in contracting in the electrical sector and a predominance of foreign consulting firms in the metallurgy sector.[43] Petrobrás set regulations to encourage the use of Brazilian engineering firms, but usually the national firms were employed as minor parties to the foreign ones. Inasmuch as state enterprises have been generating 70 percent of the national demand for capital goods, their behavior could create a large implementation obstacle.

Another political obstacle has been the basic disagreement of scientists, mainly pure scientists, regarding the increasing preponderance of the technologists and planners in the policy-making process and in policy outcomes.[44] This disagreement concerns not the main thrust of the autonomy policy but rather the definition of its objectives and means. Scientists have been troubled by what they considered the policy's large bias toward economic development and planning, technology and production, ignoring the questions of basic science. They have also argued that they are being alienated more and more from CNPq as the economists and engineers become more prominent. Both groups, the scientists on the one hand and the technologists and economists on the other, have been engaged in a continuous process of bargaining and copromise, but none has been totally satisfied with the end result. This was particularly true after SEPLAN proposed a new reorganization of the science and technology system in 1981, which most of the scientists deplored. Schwartzman has proposed that "the most important government action to support scientific activity in Brazil would be to give science the conditions for its development in all its plenitude according to its own logic of growth, and not to pay direct attention to the demands by the educational, technological, economic, or military sectors."[45]

The growth of the policy-making system to unmanageable, even ridiculous, proportions has further aggravated the policy implementation problem since 1968, leading to deficient coordination between the apparatus that regulates technology transfer and the one in charge of technological innovation and adaptation and to increased separation and redundancy of the BNDE/FUNTEC and the SEPLAN hierarchies. The size of the bureaucracy has likewise negatively affected the chances for ideological cohesion. But differences have been linked more to *institutional* ideological or personal considerations than to basic disagreements about progress, economics, and/or scientific and technological development per se. The science and technology plans (PBDCT) attempted to overcome some of the bureaucratic obstructions by granting all sectors representation and accommodating various interests in advisory groups. But the plans did not overcome political conflicts over who should get what and how much. For example, I PBDCT, originally conceived as a means to aggregate and match forces in diverse sectorial levels of government, wound up because of conflicts as a large indicative program with a global budget.[46]

In sum, the making of a science and technology policy of autonomy and its battle for survival have not been easy . . . which makes it all the more remarkable that the policy has come as far as it has.

The Political Environment

President João Goulart's regime, hounded by economic crisis, domestic opposition, and outside forces, collapsed in March 1964. The military regime instituted to replace it remained in power for the entire period covered by this study, during which time Brazil had five presidents. The regimes of the first three—Humberto Castello

Branco (1964–1967), Artur Costa e Silva (1967–1969), and Emílio Garrastazú Médici (1969–1974)—were authoritarian and repressive. However, it is significant that Médici used the "ablest manpower from the earlier military administrations, and thus provided for maximum continuity in the work of the economic planning technocrats. In addition, new talent uninvolved in the rivalry between followers of Castello Branco and Costa e Silva was brought into the government and its communication with the public greatly improved."[47] Thus, there was considerable stability and continuity in the key ministries in the following years, which contributed to the "stable" evolution of science and technology policy.

A new government under Ernesto Geisel was installed in 1974. Velloso carried over from the Médici administration, and Mário Henrique Simonsen was brought to the finance ministry to replace Netto. President Geisel started to gradually open up the political system, allowing it to organize itself. The process came to be known as *abertura* (opening) and was taken up even more strongly by the fifth president, João Baptista Figueiredo, who took office in March 1979.

Abertura has had an effect on science and technology policy. As the political system was freed, as dissident voices began to be heard and opponents of the military government returned to Brazil, the nationalist rhetoric and the egalitarian-nationalist and Marxist attacks against foreign investment increased, pressuring elites and government agencies to take a more nationalistic stand.

The Economic Environment

Policy for science and technology was made explicit at the end of the 1960s. By then the economy was experiencing its "economic miracle," with growth at a very high rate and resources available for investment in the future. The GDP, the industrial growth rate, exports, capital goods production, growth of manufacturing industry, all reached impressive levels. For example, between 1968 and 1973 GDP grew at an average yearly rate of 10.1 percent. Industrial production growth rates were even higher, with consumer durables and the construction industry at the top.[48] By 1975 the Brazilian manufactured value-added represented almost 20 percent of that of all the developing countries combined, and it was about 25 percent of Brazil's GDP.[49] But even more remarkable was the real growth of capital goods output, at a 1968–1973 average of 20.8 percent.[50] Not only did the value of total exports rise dramatically, more than quadrupling between 1968 and 1974, but the share of manufactured exports rose from 10.5 percent to 28.5 percent, with the share of all industrial exports climbing from 20.2 to 40.7 percent during the same years.[51]

This period of relative financial bonanza generated confidence among domestic and foreign investors and international financial institutions, reinforcing the image of a bright future. The economic environment thus aided and promoted the ideas of the pragmatic antidependency guerrillas. Growth was good for science and technology policy, and the policy was perceived to be good for growth.

In 1974 the oil crisis hit oil-dependent Brazil. It was aggravated by the fact that thanks to the "miracle," Brazil was living beyond its means, spending on consumption and investment some 7.8 percent more than its domestic production at current prices. By 1975 the foreign debt had exceeded $21 billion, and inflation was approximately 30 percent. Oil as a share of total imports rose dramatically, and the value of total imports went up as well, owing not only to price increases, which accounted for 51 percent of the rise, but also to a 34 percent increase in the quantity of goods imported. Indeed, the FOB (free on board) value of capital goods imports almost doubled between 1972 and 1974.[52]

Exports, which increased from an FOB value of $6,199 billion in 1973 to $8,670 billion in 1975, were not enough to level either the balance of trade or the overall balance of payments, which was $936 million in the red in 1974, $950 million in 1975. Industry continued to grow at an average rate of 9 percent between 1974 and 1976, manufacturing industry at 7.8 percent. Capital goods production grew at lower levels than in previous years, and manufactured exports as a share of total exports, after rising in 1974 and 1975, came down slightly in 1976.[53]

As a response to these economic developments, the policymakers decided to put the blame on external factors,[54] in other words, to continue the process of development, industrialization, and diversification; this of course reinforced pragmatic antidependency in science and technology in the mid-1970s. Emphasis was placed on the energy sector, primarily oil, with the objective of becoming energy independent. An import substitution policy for a number of basic and capital goods products was established to decrease dependence on imports such as, for example, machinery and equipment. Whereas in 1973 national industry supplied 36 percent of the machinery and equipment for projects approved by CDI, in 1977 it supplied 67 percent.[55] According to Velloso, the stimulation of national industry in these sectors was directly linked to the capacity to assimilate and adapt foreign technology and to the strengthening of national technology. At the same time, an outward-oriented policy featured a new line of manufactured exports, among them computers and microprocessors, farm machinery, vehicles and parts, roadbuilding equipment, steel products, and power tools.[56]

Since 1977 an economic crisis caused by domestic and foreign events has made long-range investment more difficult. Between 1976 and 1979 the GDP went up at an average rate of approximately 6.4 percent, although 1977 was a particularly bad year for both the GDP and industrial growth. Imports and exports continued to rise, being almost equal in 1977 but with imports outnumbering exports in the next two years. However, exports increased more than 30 percent in 1980. The balance of payments—positive in 1976 and 1977—was well in the red in 1979 and 1980, and inflation rose to 55.6 percent in 1979 and to 100.3 percent in 1980. Manufacturing industry grew only 2.3 percent in 1977, but manufactured exports as a share of total exports rose to 40.4 percent in 1979. In 1980 the GDP grew at a "healthy" 8.0 percent, but it dropped dramatically in 1981 because of recession, as did industrial growth.[57]

Notwithstanding this uneven economic growth, the capital goods sector accounted for 25 percent of the industrial output in 1981 and 8 percent of the GDP. Its total production was $3.6 billion in 1979, as compared to $350 million in 1970. Also while in 1970 national and foreign capital goods production (in dollar value) had been about the same, in 1980 national was double foreign.[58] The balance of payments deficit persisted, the energy independence goal was upheld, and government pursuance of a policy aimed at strengthening both private and state entrepreneurship and promoting the establishment of a self-sufficient basic goods industry also strengthened the science and technology policy. Thus, in spite of the internationalization of the economy since 1964, Brazilian economic developments have promoted the survival of the pragmatic antidependency policy.

Many Brazilian economists have complained that there is no conceptual coherence between Brazilian economic policy, which has increasingly been outward-oriented and pro—foreign investment, and the promotion of technological autonomy. They have argued that

without an economic policy explicitly attuned to technological autonomy (that is, inward-oriented and anti–foreign investment) the attempt to achieve such autonomy is futile. They therefore want not
only a self-reliant science and technology policy but also a more comprehensive self-reliant economic development policy.[59] The argument
has its merits, for multinational corporations have been dominant in
Brazil's industrialization drive, but it is important that the "internationalization" of the Brazilian economy be kept in the right perspective. Despite the expansion of exports, Brazil, relatively speaking, is
still to a great extent producing for a domestic market. Whereas in
1973 Korea's ratio of manufactured exports to manufactured value-
added was 82.46 (a country that makes its living from exports) and
India and Mexico's ratios were 18.5 and 14.84, respectively, Brazil's
ratio that year was 7.64; it rose to 13.4 in 1978.[60] True, this situation is
more than some nationalists and egalitarian nationalists would want,
but it is a far cry from "hard-core" interdependence.

Also, from the point of view of the ruling elites, the mixture of exports and foreign investment incentives on the one hand and the support for national entrepreneurs, national basic and capital goods, and
national technology on the other, made sense. Both maximized their
desire that Brazil become a developed and industrial nation and a
world power. Because the elites understood that they could not achieve
their goals without the multinationals, at least for the time being, they
increased Brazil's interdependence; at the same time, however, they
invested in a future based on an efficient domestic capitalism and on
scientific and technological knowledge applied to production—in
short, in an independent future.

III

Argentina's Aborted
Venture into Computers
in the Mid-1970s 9

The following two chapters are ambitious: they attempt to clarify further the cognitive and ideological elements involved in the development of policy and to shed more light on the groups that through their influence and access to the power structure succeeded in effecting political and economic changes. Involved is a single choice—whether to develop a national computer industry or continue to rely on foreign suppliers of data-processing equipment—that is intrinsically related to the linkages between scientific and technological development and economic development and to the need to manage interdependence and reduce technological dependency.

Given that the developed countries such as the United States and Japan indisputably dominate the computer field, the choice for Brazil and Argentina in the 1970s was not between dependency and absolute self-reliance. Rather, it was between dependency and starting the *national* manufacture of micro- and minicomputers, technologically and economically much easier to substitute than the medium-sized and large computers, thus paving the way to increased self-reliance and managed interdependent relations. Those who favored the latter course knew that it was the only realistic alternative and that the technological and economic, managerial, and political learning to be ex-

tracted from the experience could positively influence the development of larger and more sophisticated computers in the future.

The decision to set up a national computer industry grew from
technological developments in the international computer industry [1]
and the trend away from large, expensive systems toward smaller and
cheaper ones. The availability of relatively inexpensive integrated circuits, along with the possibility of obtaining technology under license,
helped Brazil shift its technological dependence from the older computer hardware market dominated by market giants [2] to the dynamic
semiconductor market, and the foreign components and software
know-how available from small new companies. Joseph M. Grieco, in a
rich study of the Indian computer industry, found such developments
to be crucial to the Indian decision to set up a domestic industry. [3] He
also identified key state institutions in assertive countries, such as
India, that promoted the computer projects against both opposing
domestic interests and the multinationals. As he pointed out, the establishment of such an institution in Brazil, matching the experience
of India, was critical to the relative success of subsequent choices.

But as the cases of Argentina and Brazil will demonstrate, much
more than technological change and the creation of state institutions
is involved in the establishment of a new industry. To find out how
these and other factors affected the Latin American experience, we
must disaggregate the notion of state, even institutions, and explore
the development and evolution of ideologies and their impact on the
actors involved in the political and economic processes. We must study
the actions of ideological groups such as the pragmatic antidependency guerrillas and their influence, or lack of it, over political power.
And we must look at the relationships between science and technology
in general on the one hand and the development of the national computers on the other.

The cases I will present have a clear comparative appeal, especially
because Brazil and Argentina met, at approximately the same time,
such different fates in their attempts to establish a domestic computer
industry. Argentina, after a prototype had already been built, decided
to halt the experiment and rely on the market's "efficiency"—which
meant dependence on foreign computers. Brazil, after a process of
consciousness-raising and political give-and-take, proceeded with the
state-promoted and state-supported computer industry, reckoning on

some inefficiency in its beginning stages, because they saw that they could not afford to remain dependent on multinational corporations for computers.

What has happened in the 1970s and early 1980s in Argentina and Brazil is not the end of the story; it is just part of these countries' "journeys toward progress," one sequence of choices and changes, one frame of a motion picture. From a long-range perspective, an Argentine computer industry can still be developed and flourish, and that of Brazil can fail or succeed. The two countries can also choose to cooperate and share their expertise in order to develop and strengthen their potentialities in the computer field. But the future, whatever happens, will be affected by the choices made in the 1970s. These choices and their timing have not only irreversibly altered the direction these countries will take toward progress, but they have also altered the nations' relative capabilities and attitudes.

These choices have also affected the multinationals, their markets, and their strategies. Brazil learned that the multinationals were adaptable, and the multinationals learned that those who do not adapt pay a price. And other countries and actors may have learned from these interactions and experiences, broadening the effect of the choices even more. By incrementally changing the images and perceptions national actors have about multinationals and, more important, about themselves and their ability to develop in the context of international dependency, bold national choices may have the propensity to transform the relative capabilities of national and multinational actors and their relative bargaining power vis-à-vis each other, thus leading to change at the level of international interactions.

The story of the Argentine computer concerns an idea that did not become reality—not much can be said about something that did not actually occur. The data are sketchy. They are based on studies of the electronics industry;[4] on an article written by Eugenio Lahera Parada in 1976 on FATE (Argentine Tire Mfg. Co.), the enterprise that was to produce the computers and finally decided not to;[5] substantially on interviews with past and present members of the firm and with persons from the science and technology establishment, government officials, and business figures close to FATE; and on events surrounding the decision to commercially develop a national computer.

Electronics and the Computer Market

Electronics in Argentina

To illustrate the background of the decision we must first look at Argentina's electronics industry, focusing on the first half of the 1970s, when the major developments took place.[6] In the 1950s and 1960s the industry had grown at high levels in terms of the physical value of production, higher than the growth rates for the manufacturing sector as a whole.[7] Production concentrated at first in consumer goods for the domestic market, but started to change slowly at the end of the 1960s, when the field of electronic instruments and components began to acquire some dynamism. But the importation of active and passive components remained dominant. Many factories just assembled products for the domestic market, relying on a large amount of imported materials.

In 1973 and 1974, surveys of the majority of electronics industry enterprises (the remainder were too small and insignificant to affect statistics) found 285 enterprises in 1973, with a total of 21,392 employees, of whom 1,869 were technical personnel and 683 professionals. Labor in this industry was highly qualified compared to other developing countries such as Korea.

Total production value in 1973 was approximately $285 million. One year later it was higher; the United Nations Conference for Trade and Development (UNCTAD) figure is $593 million. During 1974 the electronics industry's per capita gross product was 41 percent higher than that of the manufacturing sector as a whole. Significant drops in production, though, occurred in 1975 and 1976, with a drop of 18 percent in 1975 alone.[8] One study concluded that although Argentina's electronics production lagged behind Brazil's, it was far more advanced in technologically sophisticated equipment.[9]

Among the components manufactured in Argentina were electronic tubes, integrated circuits, and semiconductor manufacturing equipment. The foreign firms involved in this subsector were IBM, Texas Instruments, and Olivetti. In 1974 multinationals controlled about 30 percent of the electronics sector production and accounted for 90 percent of the exports,[10] most of it in data-processing equipment and parts.

In 1973 the electronics industry exported less than it imported;

among the exports were ceramic and plastic capacitors, transistors, cathode ray tubes, and magnetic ceramics.[11] Electronic products exports enjoyed some benefits, such as drawbacks, reimbursements, fiscal benefits, and special financing systems, but in 1974 these were reduced to only reimbursements of up to 15 percent according to the type of product, with 5 percent additional reimbursement in the case of opening of a new export market. The orientation was clearly to supply the internal market.

Because production was geared largely toward domestic consumption and was mostly in the consumer electronics sector, it required a relatively low technical level. Most of the capital goods and technologies for the professional sector were imported. But technological backwardness was not the main factor hampering dynamism of the national enterprises. Indeed, one study has shown that in 1978 R & D expenditures in the electronics industry amounted to 9.3 percent of total production value, with data-processing and office equipment accounting for one-third of those expenditures.[12]

Between 1971 and 1974 the industrial electronic instruments subsector (worth $10 million), with only 10 percent of the firms producing through foreign license, achieved product quality and a price structure competitive with imports from the United States. Something similar happened with medical electronic instruments, where production was based on local design and engineering without the use of licenses.[13] Furthermore, several laboratories and research centers were working on silicon crystals and the implantation of ions in semiconductors,[14] and it was believed that by 1976 prototypes could be built, with manufacture soon to follow. At the universities teams were working on computer hardware and software. Petrocolla et al. have shown the constant effort taken to adapt designs to both local conditions and the new components entering the market. They concluded that this type of technological modification and adaptation effectively broadened the market, either because of lower prices or because new products increased demand.[15] Thus, while in general the industry could by no means match IBM and Olivetti technologically, there was some local production and adaptation, mainly in the industrial and medical electronic instruments subsectors, with computer technology only a few steps ahead.

But there were other problems, and they were not technological. Argentina was in a state of economic and political crisis, and the electronics sector did not receive the government's attention and support as a future leading sector. According to the Argentine Association of Electronic Industries (CADIE), the difficulties the industry encountered in the first half of the 1970s were due primarily to the following factors: price control instituted between 1973 and 1976; a delay in infrastructure work that led to divestment and the flight of human resources; the use of personnel in nonproductive jobs; a strong contraction in demand since the end of 1975; and an increase in the real cost of domestic and foreign components. After the Videla government came to power, the main problem was the lack of protection for the industry.[16]

The Computer Market

The office machine and data-processing equipment subsector showed a consistent rate of growth compared to the electronics sector as a whole. Production increased steadily between 1970 and 1976 (the period when the domestic computer idea was born and partially implemented), from $14.1 million to $117.9 million (1978 dollars).[17] Also, this subsector accounted for a majority of the electronics industry exports (over 60 percent in 1973) and, in contrast to the overall electronics sector, enjoyed a positive trade balance between 1970 and 1974.[18]

There were an estimated 356 computers in Argentina at the end of 1969; 20 percent were used by the government, 27 percent by heavy industry and manufacturing, 17 percent by banking and finance, 10 percent by data service business, 7 percent by education and research, with the rest scattered among other users. By 1973 there were about 500 computers, worth $130 million.[19] Some 70 percent of the computers in use were installed in the second half of the 1960s; virtually all of this equipment had been imported and most of the hardware supplied by the multinational corporations. Imports of computer hardware in 1969 reached $5.3 million, and that of peripherals $2.9 million. By the end of 1969 IBM held 65.9 percent of the market, while NCR had 15.1 percent, Bull/GE 10.1 percent, Burroughs 6.7 percent, Univac 2.1 percent, and others 0.1 percent. Seventy-five percent of the computer installations were leases, preferred over purchase because of expected support from the leasing firms and a belief that the equipment was subject to rapid obsolescence.[20] Prior to 1974 three companies were producing data-processing and office machine equipment: IBM, Olivetti, and FATE. They employed 1,340 individuals, including 145 technical personnel and 142 professionals.

IBM Argentina, which started its activities in 1923, manufactured computer hardware. In 1973 it was the thirty-second-largest Argentine firm working to full capacity, with a total of 610 employees, of whom 140 were technicians and professionals. All but 5 percent of its production was for export, half of which went to the United States, Canada, the United Kingdom, Japan, and Sweden and the rest all over the world.[21] IBM exports increased from an FOB value of approximately $22 million in 1974 to $29 million in 1975, but dropped back to $22 million in 1976. In 1980 and 1981 exports were approximately $57 million and $97 million, respectively.[22]

Olivetti Argentina was ranked thirty-fifth among the largest Argentine companies in 1973, when it employed 230 people and worked to only 30 percent of capacity. It operated under a foreign license, assembling electric accounting machines since 1962, electronic calculators since 1969. In 1973 it produced 8,000 "Logos" calculators.

Texas Instruments produced transistors. In 1974 it began to market pocket calculators (imported from Brazil), selling 70,000 in six months, a 60 percent market share for that year. But then import restrictions lowered that share to 20 percent during the first semester of 1975, and Texas Instruments decided to strike a bargain with the Argentine government: in exchange for lifting the restrictions, the American company offered domestic production of calculators. Argentina accepted, and Texas Instruments began to produce 5,000 calculators a month.[23]

FATE, a private, 100-percent-Argentine-capital company, was set up as a tire manufacturing company in the 1940s, but by the end of the 1960s it began to diversify into electronics, more precisely electronic calculators and printed and integrated circuits, and into an aluminum venture with a company named Aluar. FATE soon rose to the "big league" in Argentina (ranked forty-second), owing at least partially to its technology policy. After some successful technical assistance contracts with General Tire in the United States, FATE became more interested in technological assimilation, training technicians and engineers, providing space for university researchers, and getting involved in the production of R & D–intensive products. The creation of an electronics division and the manufacture of desk calculators beginning in the early 1970s fit this pattern.[24]

FATE Electronics and
the National Computer That Never Was

The history of FATE's computer starts with the production between 1956 and 1959 of Argentina's first experimental computer, known as CEFIBA,[25] and with the work of scientists in the early 1960s at the University of Buenos Aires, which had a Mercury-Ferranti computer that was used for research in the components, digital techniques, automation, and industrial electronics areas. These activities produced a cadre of qualified scientists in the computer field, who later were recruited by FATE.

Three groups of scientists in the engineering department of the University of Buenos Aires, specializing in computation, semiconductors, and process control, were the seed for what was about to come. They were under the supervision, and intellectual guidance, of Humberto Giancaglini and Alberto Bilotti, the latter an expert in microelectronics and solid states. Roberto Zubieta was an engineer who worked at the department's semiconductors laboratory and who became the leading intellectual and active force in the development of an Argentine computer. Other scientists involved in electronics R & D during the "golden age" of Argentine science, and who would become part of the FATE team, were Hector Abrales, Carlos Duro, Horacio Serebrinsky, and Pedro Joselevich, director of the department's electronics application laboratory. But all these efforts at the university stopped in 1966 with the "night of the long clubs." Many scientists left the universities, some leaving the country and others going to work for the multinationals; Zubieta went to Texas Instruments.

Oscar Varsavsky, a physicist, became the other key actor, with Zubieta, in FATE's development of electronics and the computer. Varsavsky strongly favored technological autonomy in Third World countries in general, and Argentina in particular, and wrote extensively on the subject.[26] One of the most renowned Argentine scientists of his time, he was listened to with respect. It was Varsavsky who, at the end of the 1960s, convinced Manuel Madanes, FATE's owner and a strong nationalist, of the benefits of developing electronics within FATE. At that time FATE was in excellent financial condition owing to its production of tires and its diversification into aluminum with Aluar, which had taken the company into more capital-intensive and technologically sophisticated investment. In the venture with Aluar, Madanes had explicitly set out to acquire some measure of technological independence by creating an excellently trained scientific group within the company. Now, Varsavsky suggested, with money available, the time had come for intelligence-intensive enterprises.

We should remember that this was the time when Onganía was on the way out, nationalism was growing, and the mild antidependency science and technology policy was already under way. It was also quite relevant for FATE that Gelbard held a significant amount of Aluar's shares, so that the General Economic Confederation (CGE), and after 1973 the ministry of economy, were very supportive of FATE and its enterprise. This "link" would later be one of the causes of FATE Electronics' downfall.

Madanes accepted the idea of vertical rather than horizontal diversification, getting involved in digital electronics first with the production of calculators, then "going up" toward computers. Varsavsky approached Zubieta, and both started to "look for people with technological autonomy consciousness." Zubieta brought in the core of the university groups: Bilotti, Joselevich, Abrales, Duro, Serebrinsky, and many more; Zubieta and Bilotti were placed at the head of the technological operation. According to several sources, Varsavsky brought to FATE the best Argentine minds in the electronics of his time.

In July 1969 Zubieta and Bilotti began to organize the technological structure, the heart of FATE Electronics. Zubieta was given almost total autonomy to make allocation decisions, and he became the operation's spearhead, selecting goals and using the means at his disposal to achieve them. His colleagues admitted that he was an excellent leader, with a good deal of political sensibility and a spirit of opportunism. His core ideas were basically three: technological self-reliance is possible, it should be the goal, and the state should help in bringing it about.

Now Zubieta was in command of a group of antidependency-minded scientists within a company headed by someone close to the Peronists, to Gelbard, and to the CGE, when all were on the rise. He envisioned FATE as an "island" and example of technological self-reliance. According to one of the highest-ranking scientists of the FATE group, "there was a challenge kind of attitude." As long as FATE was financially solvent, and its political contacts among the very best, the picture looked bright.

The engineering department at FATE Electronics became the enterprise's most dynamic element, the engineers by far outnumbering the employees in commercialization. Fifteen percent of the personnel were involved in R & D, all in engineering and development, as there was no basic or applied research.

FATE's strategy was to develop, copy, and adapt technology. Although some components, such as chips, had to be imported, FATE Electronics used no foreign licenses or any other trademark. It searched aggressively for nonproprietary technological information, such as that available in journals (it subscribed to 80), and visited foreign plants and international fairs; it sent technicians to study at MIT; and sometimes it consulted foreign experts and bought foreign machinery, equipment, and components to train domestic personnel. FATE's technological style clearly matched a pragmatic antidependency ideology. This is also why it spent 7 percent of its gross sales on R & D (fixed cost), a good share for such an endeavor in Argentina. FATE also received direct help in R & D mainly from the National Atomic Energy Commission (CNEA), the National Institute of Industrial Technology (INTI), and La Plata University.[27]

FATE Electronics received operational financial aid from the World Bank, and their position in the marketplace was boosted when President Lanusse fixed a high tariff for the importation of electronic calculators and a somewhat lower (but still high) tariff for mechanical-electronic calculators, of the kind Olivetti was producing in Argentina. FATE also received import exemptions and, as did the multinationals, draw-backs and subsidies to encourage exports.

FATE developed a new line of products almost every year. It produced four calculator models in direct competition with Olivetti, with the brand name CIFRA: #311 (1150 integrated circuits [IC]); #211 (7 IC); #121 (3 IC); and 100/13 (1 IC). The calculators were of FATE's own design; circuits were first purchased in foreign markets, then built abroad according to FATE's specifications. Finally in 1974 FATE itself began to produce about 15–20 percent of its demand for integrated circuits. The first calculator took fifteen months to build. Production grew considerably between 1971 and 1975, from 500 to 134,000,[28] and FATE's share of the market rose from 1 percent to 50–55 percent (helped by protection). Importation of components decreased from 70 percent in 1971 to 40 percent in 1975.[29] FATE's success led to the building of a semiconductor manufacturing plant with capacity for 1,400 workers, and some in the company began to consider establishing plants in Mexico and Brazil. By 1972 FATE Electronics was making money.

These developments had a clear positive-feedback impact. Zubieta was confident that technological change was not a problem: FATE Electronics was meeting all challenges, and it had practically put Olivetti out of business. As one ex-member of the Zubieta group put it, this was "a period of delirium."

The next stage, the computer, came right along. Madanes had been approached by the military, who told him how important computers were for the armed forces. The scientists at FATE told him the enterprise was viable. Madanes then advised the military that he needed government support, which the military promised to provide. The computer idea branched into two: a medium computer (which some said would be similar to the IBM 360; others, to the IBM 370[30]—in any case, it would be simpler and faster than either of these), to be called Serie 1000; and a microcomputer, Sistema 75.

Sistema 75, the second project, was built under the supervision of Bilotti and successfully commercialized. With a memory of up to 16K, it was designed for cards and later for disks. Manufacture began in 1974, and a considerable number were reportedly sold.

Development of the Serie 1000 started in 1972, and by December 1973 its specifications were complete. According to one report, Serie 1000 was to have a microprogrammable processor, with up to 2 megabytes memory capacity. Memory and microprocessors were to be MOS/LS, designed with the help of an American company, Macro-Systems. The peripherals included disk units, magnetic tapes, a linear printer, a CRT terminal, and a data-entry terminal. Serie 1000 was to work with an advanced operating system, multiprogrammable and with realtime extension, and be compatible with COBOL, FORTRAN, BCPL, and other languages. The system was designed to diminish costs, using a "spooling system" (*impresión diferida*) for printing and a low-cost terminal with direct-to-disk data entry. It was also designed to interface with larger systems.

Building of Serie 1000 began in 1974. The hardware was assembled with mostly foreign parts, with the intention of adding Argentine components as soon as they were developed. By the end of 1975 a prototype, basically different from the IBM 360, was complete and working. Software development proceeded at a much slower pace; it turned out to be one of the computer's major obstacles.

FATE's technicians believed that Serie 1000 would be on the market by 1977, competitive with foreign firms in terms of price; others doubted the capacity to produce it commercially. Although one engineer involved in marketing argued that the problems ahead had to do mainly with production, the consensus was that marketing was being disregarded and would become another major bottleneck.

A CNEA scientist with close links to the project argued that it was workable. The problem, he felt, was that "there was no awareness within the government about computers. The effort being made was not institutional or governmental, but personal." Some of the military, especially air force personnel, expressed interest in the development of domestic computers, but they were not then in power, and the military who came to power in 1976 had a completely different, pro-liberal ideology. FATE did receive political support from the Armed Forces Center for Scientific and Technical Research, the CGE, and Gelbard, but the latter two lost political power at the end of 1974 (when Isabel Perón and López Rega took control of the Peronist movement), and soon thereafter the import exemptions and export incentives were ended.

FATE's electronics venture prompted contrasting responses from IBM and Olivetti. IBM did not pressure the government against FATE Electronics computers because it did not feel threatened. Most of its Argentine market was in large machines for the public sector. Furthermore, its disbelief in the Argentine capacity to produce domestic computers led it to take a wait-and-see attitude during the Peronist period, preparing itself not to sell its products if things should turn sour.

Olivetti, however, had lost much of its calculator market to FATE. Skeptical in the beginning about FATE's chances, Olivetti decided to put up a fight when FATE Electronics began to grow. At the head of this effort was Edgar C. Bustos, an engineer who became the main lobbyist against FATE's venture. He said that since FATE was copying technology and using imported elements, the "self-reliance" policy was in fact a fake—a very inefficient fake—and that without high protection the project was doomed. He applied the most pressure on the technical staffs of the ministries, but other companies were also putting pressure on the government not to continue to protect FATE. As the head of the Chamber of Office Machines Manufacturers, Bustos also helped push entrepreneurs against FATE's venture. The argument was efficiency. As Bustos saw it, FATE, by failing to influence the armed forces and the bureaucracy sufficiently, finally lost to the lobby campaign waged against it.

During 1975 FATE's financial situation was bad. Zubieta's critics ar-

gue that although the goals were viable and the engineering department a success, he failed in management, specifically in the allocation
of funds and in marketing. In the same year FATE and its owners became involved in an economic scandal concerning Aluar, which attracted public attention because of Gelbard's association with Aluar.

The process of deciding to terminate calculator and computer production took place between November 1975 and August 1976, at the
peak of the political and economic chaos of the Peronist interlude. The
timing could hardly have been worse for FATE Electronics. Madanes,
by then burdened by the national turmoil, the financial difficulties of
the company, the Aluar scandal, and the lack of government support,
decided to bring a new general manager to FATE, R. Bargagna, whose
business background and liberal outlook were a signal that the end
was near.

As Bargagna said, "Madanes under these conditions would not
think of investing even more on the computer." The end came late in
1975, and Bargagna was the "key" decisionmaker with power dele-
gated by Madanes. According to Bargagna, the nationalists among the
military and the government were consulted and told, "'If you want a
national computer show me the purchase orders.' Up to that moment
we had spent $2.5 million, and we needed another $2.5 million to continue with the project. They did not respond positively and the decision was simple."[31] Bargagna felt that the ISI process had already
been exhausted and that a new approach was needed. It was time, he
said, that industrialists understand that they have to modernize and
compete rather than rely on overprotection. Bargagna's concerns
were efficiency and market considerations; he deplored inefficient
and overprotected ISI projects and saw technological change in the
international computer field as being so rapid that the Argentine
computer would be obsolete by the time it was marketed. In Brazil,
however, this same change was a decisive motive for embarking on the
manufacture of national minicomputers, with the ultimate goal of
autonomous technological development. In pure market terms, of
course, Bargagna was right.

The timing of the decision—*before* the Peronists were ousted by the
military—was not directly connected with a political regime change,
but the decision was anticipatory. "Everybody knew there was going to
be a coup and that the Peronists would be out; the question was when."

Bargaining continued in the first months of 1976, but Zubieta was
in no position to make deals. He and his collaborators were identified

as militant Peronists and classified by military intelligence as a focus of subversion. When he and his group left FATE Electronics, the Serie 1000, the microcomputer, the international venture, the semiconductor plant, the calculators, were all terminated. FATE Electronics became an assembler of foreign products and the Argentine representative of Nippon Electronics Company. And as Lahera wrote in a postscript to his article (and a postmortem to the computer idea), "the enterprise decided to concentrate in the sector it knows best—tires . . . which presented the best options." [32]

Some time later, Zubieta fled to Brazil, where he became a manager of one of the most important Brazilian microelectronic enterprises, Elebra, a subsidiary of Doças de Santos. The insight gained from time and experience brought him to the conclusion that a country can reduce dependency, as Brazil had in computers and Argentina in nuclear technology, only with almost full government support and freedom to develop a self-sufficiency strategy—the lack of which had doomed the Argentine computer industry to failure.

FATE started the electronics project on the premise that self-reliant development was possible and that the company could benefit from it, but Argentina lacked a systematic science and technology policy, institutions to support the development of an electronics technology, and government awareness of the industrial relevance of computers; and so it failed. FATE also suffered from the political and economic turmoil of the time, which personally involved its shareholders and led Madanes to reject any further investments. Madanes then named a general manager whose ideology was totally different from that of the group that had developed the computer. Bargagna knew that the military was returning to power; he believed that efficiency and the market, not protection, were the answer for Argentine industry; and he was convinced that technological change in semiconductors would make FATE's project inviable—so he decided to kill it.

In the last analysis, the computer project succumbed to bad timing; the disintegrating Peronist government lacked the means to successfully pursue it, and the new military government set goals of efficiency and modernization that led to the dismantling of much of Argentina's industrial capabilities. As Marcelo Diamand commented, FATE and the computer became the first victims of the change of government. They also fell prey to a lack of awareness of the dynamics of development, of the fact that a successful technological project might, despite short-term inefficiencies, have more important national payoffs.

In 1980 the Argentine government started to show some interest in microelectronics, which was intensified after the Falklands (Malvinas) war. An informatics subsecretariat within the planning secretariat was set up and an Informatics National Commission planned. But political repression had led to mistrust and a considerable brain drain. Meanwhile, Argentina has asked Brazil, which has acquired significant expertise in the last decade, for assistance in developing its computer industry.

Whether the Argentine computer experience was "good" or "bad" for Argentina's development in general is an ideological question. So is the subject of the "right" strategy for scientific, technological, and economic development. Nationalists view the loss of capacity to innovate and to adapt to what is becoming a crucial technology as a failure; from a market perspective there was no failure—efficiency decided what had to be done.

Interviewees were asked what happened to the knowledge and experience generated within FATE: *where have all the scientists gone?* "Gone to IBM and other multinationals, every one," was the response, and implicit in this answer was the question, when will they ever learn, when will they ever learn?

Brazil's Domestic
Computer Industry   10

The Data-Processing
Market, 1970–1982

Data-processing systems, ever more complex and comprehensive, are turning into the nervous system of modern societies. Mastering their technology will be increasingly essential for a nation willing to know itself and consequently to maintain or to develop its own decision power in order to exercise its political independence, setting by itself its degree of economic interdependence with other nations. For Brazil, abstaining from this effort of national technology creation means at a minimum to give up having in our territory—more precisely in the minds of our technical teams—the key element for setting up a data-processing industry based on national interests, and not just on the interests of companies linked to foreign powers, which rarely fit Brazil's interests in the international political-economic game.[1]

Many Brazilians involved in the development of the computer industry saw control by a few international giants, such as IBM, as a threat to Brazilian independence. This perception grew from an awareness—as Simon Nora and Alain Minc put it so graphically in their report to the French president—of "the computerization of society," an awareness that not only is the computer a technological innovation of recent years, but it also constitutes "the common factor that speeds the development of all the others [and that] will alter the entire nervous system of social organization."[2] This chapter describes how the cognitions and ideas of a small group of scientists and technocrats, while at work for assertive political and professional institutions, ended up generating broad political consciousness and irreversibly affecting industrial policy and industry.

In the early 1970s the Brazilian computer market was already the twelfth largest in the world and, behind Japan, the second fastest growing with annual growth rates of 30–40 percent (compared with a world rate of about 20 percent). Growth rates were still high, at 20–30 percent, in the mid-1970s. By 1975 Brazil's data-processing equipment market had climbed to tenth place, and one year later it was worth about $1.4 billion, or approximately 1 percent of the GDP.[3]

The value of installed computers in Brazil was $2.8 billion by 1982.[4] In dollar terms the computer industry grew 64 percent between 1979 and 1980, 26 percent between 1980 and 1981, and 51 percent between 1981 and 1982 (the latter after adjusting for 100 percent inflation). While the 1979–1980 growth reflected the entrance of new domestic enterprises to the market, the 1981–1982 figure reflected a real growth in sales. It was expected that the market would reach $5 billion by 1985.[5]

Growth in quantity of installed computers between 1970 and 1982 is shown in Table 12, broken down into the six categories used by the Brazilian Special Secretariat of Informatics (SEI), which has been in charge of computer policy since 1979.[6]

Between 1970 and 1978, the number of computers in the country grew almost fourteenfold. Even discounting micros, they increased by 270 percent between 1973 and 1978 and by 673 percent between 1973 and 1982—71 percent in the 1981–1982 period alone.

The market changed dramatically between 1970, when small and medium-sized computers accounted for 99 percent of the total, and 1978, when micros and minis made up 71 percent. By 1982 this latter figure had jumped to 87 percent. Because micros and minis were now doing what small and medium-sized computers had done in the past, and because large and very large computers were still unmatched, the medium-sized computer market was compressed whereas the

Table 12. Brazil. Installed Computers, by Size, 1970–1982

	1970	1971	1972	1973	1974	1975	1976	1977	1978	1979	1980	1981	1982
Micro	a	a	a	586	1,514	2,143	3,131	3,846	4,290	4,791	4,722	8,756	17,702
% of yearly total				38	54	56	60	64	62	60	53	61	73
Mini	a	a	a	19	81	173	265	356	656	1,015	1,675	2,719	3,571
% of yearly total				1	3	4	5	6	10	13	19	19	14
Small	378	403	454	639	775	1,057	1,309	1,296	1,378	1,494	1,688	1,858	1,950
% of yearly total	75	70	68	40	27	27	25	21	20	18	19	13	8
Medium	122	163	184	250	288	327	338	353	370	377	388	408	400
% of yearly total	24	28	28	16	11	9	7	6	5	5	5	3	2
Large	2	2	10	45	72	82	99	122	166	226	248	374	544
% of yearly total	0	0	1	3	3	2	2	2	2	3	3	3	2
Very large	4	10	19	33	42	61	72	87	93	97	123	134	172
% of yearly total	1	2	3	2	2	2	1	1	1	1	1	1	1
Total excluding micros	506	578	667	986	1,258	1,700	2,083	2,214	2,663	3,209	4,122	5,493	6,637
Overall total	506	578	667	1,572	2,772	3,843	5,214	6,060	6,953	8,000	8,844	14,249	24,339

Sources: SEI, Boletim informativo, no. 5 (Aug./Sept./Oct. 1981): 9; no. 8 (July 1982): 4; no. 11 (June/Sept. 1983): 6.

ᵃAvailable information unreliable.

extremes grew significantly: minis at 903 percent from 1977 to 1982 (combined rates) and micros at 128 percent, and large computers at 346 percent (51 percent in 1980–1981 alone). The "slow growers" were small, medium, and very large computers, which increased at combined rates of 51, 14, and 98 percent, respectively, between 1977 and 1982.[7]

Before the Brazilian computer policy was formulated and began to be implemented during 1975, Brazil's computer requirements were met by multinational corporations such as IBM, Burroughs, Hewlett-Packard (H-P), Honeywell Bull, Data General, Digital, and Olivetti. Imports increased from $13.3 million in 1969 to $99.8 million in 1974, and to $111.9 million in 1975.[8] IBM, Burroughs, and H-P also manufactured in Brazil, to meet domestic needs as well as their global production requirements. By 1980, IBM do Brasil, the largest computer company in the country, held 53.8 percent of installed computers value and was IBM's fastest-growing subsidiary, generating about 50 percent of the company's business in Latin America.[9] It produced medium and large computers, tape drives, CRT displays, printers, and data-entry equipment in its Sumaré plant. Burroughs, the second-largest company, with approximately 15 percent of installed computers value in 1980, manufactured medium, large, and very large computers in Brazil.[10]

Once Brazil decided to enter the domestic computer market, the industry developed rapidly. Within only two years domestic companies were producing hardware and software systems, peripherals, terminals, modems, and special terminals (banking and lottery). The dollar value of installed domestic computers grew from 2 percent of the total in 1978 to 19 percent in 1982, by which time 67 percent of installed computers had been produced by domestic companies (see Figure 9).

By 1983 about one hundred domestic computer companies, with 18,000 employees and sales of $687 million, existed, most of them founded after 1976 under the guidance of the computer policy.[11] In 1982 they accounted for 67, 91, 13, and 1 percent, respectively, of installed micro, mini, small, and medium computer value.[12] Cobra SA, the largest, and the only state company, ranked third in sales by June 1982, with 36.2 percent of installed minicomputers. The other important large national minicomputer manufacturers by that time were Labo, with 18.4 percent of the market; SID, with 7.6 percent; Edisa, 23.3 percent; and Sisco, 5.0 percent. Cobra, Dismac, Edisa, and Prológica held approximately 72 percent of installed microcomputers.[13]

Brazil's main goal in developing its own computer industry was to achieve a measure of self-sufficiency in this high-technology sector. The contribution of domestic R & D toward this end was considerable.

Domestic computer companies invest a relatively high share of their sales in R & D. In 1980 those using indigenous technology spent an average of 14.4 percent of their sales on R & D, and those working under foreign licenses spent 7.9 percent. The total industry's R & D average was 8.7 percent—2.6 percent more than the American computer industry spent during the same year.[14]

The pragmatic antidependency policy, with its emphasis on R & D, has been especially successful in increasing the use of domestic technology and reducing imports in the data-processing equipment industry. Between 1979 and 1981, the industry's share in sales of prod-

Figure 9.   Brazil. Domestic Computers, as Percentage
of All Installed Computers, 1980–1982

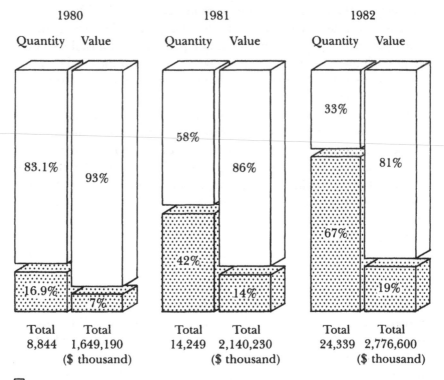

1980		1981		1982	
Quantity	Value	Quantity	Value	Quantity	Value
83.1%	93%	58% 42%	86% 14%	33% 67%	81% 19%
16.9%	7%				
Total	Total	Total	Total	Total	Total
8,844	1,649,190	14,249	2,140,230	24,339	2,776,600
	($ thousand)		($ thousand)		($ thousand)

▣ Domestic Computers

Source: SEI, *Boletim Informativo,* no. 11 (June/Sept. 1983): 10.

Table 13.   Brazil. Technological Dependence Reduction by
National Data-Processing Corporations, 1979–1981

Equipment	1979		1980		1981	
	Local Tech-nology[a]	Imports[b]	Local Tech-nology[a]	Imports[b]	Local Tech-nology[a]	Imports[b]
Systems (hardware and software)	28%	29%	41%	18%	60%	7%
Peripherals	—	111	4	48	6	36
Terminals	100	8	100	8	100	3
Modems	10	21	37	22	50	13
Special terminals	100	6	100	22	100	14
Total Equipment	31	29	39	20	53	8

Source: United Nations Center on Transnational Corporations, *Transborder Data Flows and Brazil* (New York: United Nations, 1983), 223–25.

Note: Total sales figures include exports.

[a]Equipment produced with local technology as percentage of total dollar sales for a given year.

[b]Imports as percentage of total dollar sales for a given year. Because corporations may import to increase inventories, percentages may be higher than 100.

ucts based on domestic technology (technology not obtained under licensing agreements or only slightly adapted from recently expired licenses) rose from 31 percent to 53 percent, whereas that based on imports fell by a factor of almost four (see Table 13). The shift toward computer products based on local technology is evident in the case of four of the industry's products: terminals, special terminals, systems, and modems. Terminals and special terminals are now entirely domestic, and between 1979 and 1981 the share in sales of systems based on local technology rose from 28 percent to 60 percent, and that of modems rose from 10 percent to 50 percent. During the same period, the share in sales of imports in these last two areas fell from 29 to 7 percent and from 21 to 13 percent, respectively. However, in 1981 peripherals were still being produced almost entirely with foreign technology. Imports as percentage of sales accounted for by foreign corporations rose from 28 percent to 40 percent between 1979 and 1981.[15]

Some domestic computer companies have now reached a level of technological sophistication and economic efficiency that allows them to produce for export. Cobra, Microdigital, Prológica, and Elebra have been the domestic export leaders, with the most popular items being central processing units (CPUs), printers, magnetic disks, and video terminals. Cobra has exported micros and minis to Argentina, and Elebra components to the United States. Logus Computadores Ltda. has sold microcomputers to the United States that can run up to fifteen separate programs at once, something that previously only large computers were able to do. At least one domestic software firm has sold its merchandise to West Germany, and a sale of two thousand microcomputers to China was under negotiation throughout 1984.

Development of the Brazilian Computer Industry

Cobra: The Early Days

The first attempt to build a Brazilian computer was made in 1961 when a group of four engineers developed a prototype, known affectionately as "Zezinho." Mário Ripper, one of the engineers and later an important actor in the development of the Brazilian computer industry, said, "While we were in Paris in 1961 visiting Machines Bull, we were struck by their effort toward developing their own technology. . . . The fact that today we know that Bull failed is irrelevant for the history of Zezinho. What is important is that the idea of developing our own technology impressed us. If France could do it, why not Brazil?"[16]

With help from the National Research Council (CNPq) and from several Brazilian companies, the computer was built and tested for 60 days. Ripper indicated that it was a success, but "we failed, apparently, in our naive objective of impacting the electronics industry in Brazil. A decade had to go by before the idea of a national computer could be taken up again, and then this effort was not a consequence of Zezinho. Why? Maybe, really, the idea was premature because in 1961 there were no technological conditions, nor was there a market for a computer industry. . . . Was the effort worthless? Certainly not."[17]

Only at the end of the 1960s did National Economic Development Bank (BNDE) planners identify the minicomputer field as a promising one for developing a Brazilian industry based on domestic technological capabilities. The technology was more accessible, investments were not as large as for the big computers, and there were no minicomputer manufacturers in the Brazilian domestic market. At the same time, the Brazilian navy was modernizing its ships and buying new ones, which had to be equipped with electronic gear, including computers. The navy's communication and electronics directorate concluded that the navy could not depend on foreign sources for a security-sensitive technology such as that required by computers. Therefore, early in 1971, when the navy decided to equip its vessels with English Ferranti FM 1600-B computers, it also initiated a project for the planning, development, and manufacture of a computer prototype suitable for naval operations, preferably in association with Ferranti.[18]

Naval officer José Luis Guaranys became more involved in this project than anyone else. First he contacted Pelúcio at BNDE, and in February 1971 what came to be called the Guaranys Project was institutionalized by a special commission, Special Working Group (GTE)/FUNTEC 111. The Scientific and Technical Development Fund (FUNTEC) represented BNDE and financed the project, which was to start with a 60 percent contribution from BNDE and 40 percent from the navy.

The GTE decided to follow two courses of action: to promote and finance the development of a domestic minicomputer prototype; and to set up a company with both state and private participation and the participation of a foreign partner that would not impose technological restrictions and would be willing to transfer its technology to the company—in other words, *tripé*, a three-member partnership of the state, private industry, and multinationals. In 1972 E. E. Eletrônica was selected as the private domestic firm, which then with GTE looked for a foreign partner to complete the tripé. The most serious candidates were Ferranti, preferred by the navy, and Fujitsu, preferred by BNDE.

In April 1973, with capital from BNDE, Petrobrás, Telebrás, and the finance ministry, a holding company, Eletrônica Digital Brasileira, was created, which was then to create two computer companies, one in association with Ferranti for the military market and the other with

Fujitsu for the civilian market. The former was called Digibrás (Brazilian Digital Enterprise); the latter did not materialize. In July 1974 Digibrás became Cobra SA, to cover both the civilian and the military markets, and Eletrônica Digital Brazileira soon changed its name to Digibrás.

When the GTE was eliminated in 1975, Digibrás assumed full responsibility for the national computer project, thereby becoming an industrial promotion agency to approve projects and set up research centers and companies to develop the sector. Digibrás concentrated on three areas: (a) *the market*—studying the market and assessing its change as computer technology changed; (b) *incentives and state involvement*—developing production lines and incentive means to help the participating national enterprises and providing consulting services and support for the incipient industries; and (c) *technology*—identifying R & D priority sectors, evaluating Brazilian potential, and promoting R & D.

Cobra became one of Brazil's main instruments for developing the minicomputer industry. Investment in the company was a tripé venture involving the Brazilian state, E. E. Eletrônica, and Ferranti, although the latter was allowed very limited participation, being relied on mainly to provide only the technology for Cobra's first computer. This arrangement reflected a pragmatic position. Those involved thought that domestic technology would progress most rapidly if Cobra used foreign technology to develop the national computer, but only if the foreign firms committed to full technology transfer, with technology then to be absorbed by Cobra.

> Ferranti was chosen because of its positive response to Cobra's demand for technology transfer, rather than because of the performance of its products. Cobra's experience in shopping for computer technology showed that international market leaders are seldom willing to sell expertise without linking it to investment or to other mechanisms of control. Digital, the world's largest minicomputer manufacturer . . . was involved in licensing negotiations with Cobra, but no agreement was reached because the U.S. firm insisted on taking a majority share in the Brazilian computer company. Negotiations with Data General, the world's second largest independent minicomputer manufacturer, broke down because Data General would not transfer its ownership rights to the technology to Cobra.[19]

Ferranti was also chosen, of course, because the Brazilian navy was using Ferranti computers for its ships' control system. From this venture resulted the first Brazilian-assembled minicomputer, the 700 series.

Digibrás had initially planned to produce and commercialize a business computer for commerce and industry that was domestically manufactured with minority foreign investment. Data General seemed the best candidate to transfer minicomputer technology to Cobra until it proved unwilling to accept Brazil's condition that patents, blueprints, and other know-how be transferred to Cobra at the end of the license period. Attempts to bring in Fujitsu and Nixdorf (West Germany) also failed. Instead, Digibrás studied the technology market and finally, in 1976, chose a relatively small American company, Sycor, Inc., that agreed to technology transfer. This effort produced the smaller 400 series minicomputer for business and accounting, with a 64K main memory and a 20 megabyte disk memory. The use of foreign technology for the two series was relatively successful because it "substantially reduced the time required to begin local production of minicomputers and helped to avoid mistakes both in product and process designs that would probably have occurred had Cobra relied initially on local technological sources only."[20]

Meanwhile, the hardware for Brazil's first domestic computer, the G-10, affectionately called "*patinho feo*" (ugly duckling), was being developed at the Digital Systems Laboratory (LSD) of the Polytechnic School of São Paulo University, and the software at the Pontifical Catholic University of Rio de Janeiro (PUC/RJ). First planned as a scientific computer, when patinho feo was transferred to Cobra it became more of a general-applications computer.[21] The work itself at LSD and PUC/RJ had a payoff beyond the G-10: the project employed two hundred technicians and professionals and provided an important learning experience for all of them.[22]

The G-10 got help too from Serpro (Federal Service for Dataprocessing), created in 1970 as a public enterprise under the supervision of the finance ministry. Its manufacturing unit developed the STV-1600, a video terminal known as the "*telinha*" and used in the model that Cobra later commercialized, and a terminal remote control (intelligent terminal) called the "*telao*." The telao and a computer terminal developed at the Federal University of Rio de Janeiro (UFRJ) were transferred to Cobra. The result was the G-11 minicomputer, which had greater speed and capacity and accepted more peripherals than the Cobra 400. L. Pereira, Cobra's hardware manager, said the many improvements made on the original model resulted in "a product at the level of the other computers in the market, the difference being that this one has been totally conceived and designed in Brazil, by Brazilian technicians."[23]

Cobra products included TR-100 and TR-200 terminals, TD-100 and TD-200 data-entry terminals, and the SC-300 microcomputer. It supplied computer equipment and peripherals to the army and navy, and a small group worked on military computer applications in its so-called military systems division, consisting of three technical groups: software, products, and analysis. Although it was not a sound investment, this project was politically necessary.[24]

Cobra's financial situation contrasted sharply with its relative success in R & D and technology transfer. Despite the fact that its capital increased more than tenfold in three years (from Cr 30 million to Cr 350 million), Cobra was usually in financial distress. By 1976 things were not going well for the company, which, since there were no purchase requests from the private market, was selling only to government institutions such as Digibrás and Embratel, and of course to the armed forces. It was discovering the difficulty of competing with multinationals, with their tremendous advantage in technology, access to supply, economies of scale, and overall finances. Pressure mounted when it became clear that IBM was planning to introduce its minicomputer, System 32. Planners began to fear that the Brazilian minicomputer industry would disappear even before it was born.

But in mid-1977 a consortium of eleven banks, including some of the largest, such as Bradesco and Itaú, realized a need for electronic automation and decided to invest in the national venture, acquiring 39 percent of Cobra stock.[25] BNDE invested approximately $3.3 million and gave Cobra guarantees for foreign loans valued at approximately $1.3 million,[26] and Digibrás and Ferranti each invested $1.5 million.

With the help of the banks, Cobra was able by 1980 to issue its newest line of computers, the Cobra 500 series, replacing the 400 series. These descendants of the G-10 and G-11 were the first computers to be designed totally in Brazil, using 92 percent locally developed components. The Cobra 530, for example, had a 500K memory, about the same as the lowest-powered IBM 4331. The two other models were the Cobra 520 (128K) and Cobra 540 (up to 1M). Cobra phased out its

TD-100 and TD-200 terminals, replacing them with the Cobra 305 microcomputer, which is faster than the terminals for about the same price.

Computer Policy and CAPRE—
The Institution Behind the Change

The Commission for the Coordination of Electronic Data-Processing Activities (CAPRE) was created by presidential decree (Médici) on April 5, 1972, in response to an exposition by Velloso concerning heavy imports of computers, the increasing national scientific and technological capabilities in the sector, and the need to create government incentives for the computer industry. "CAPRE was given the task of managing data processing within the federal government, maintaining statistics on the national market, and developing a strategy to encourage a local industry. However, at its inception CAPRE did not have the power to regulate computer imports or to control the Brazilian manufacturing activities of foreign computer firms."[27] Its main policy-making body, the council, had representatives from the armed forces, the ministry of finance, BNDE, the Brazilian Institute of Geography and Statistics (IBGE), the Modernization and Administrative Reform Secretary, and CAPRE's executive secretary, although this composition was later changed.[28]

CAPRE's first decision, on January 12, 1973, was to create a Permanent Working Group, with BNDE, CNPq, the Studies and Projects Financing Agency (FINEP), and the ministry of education participating, that would be responsible for a National Program of Data-processing Centers. Headed by Ricardo A. C. Saur, CAPRE's executive secretary, its explicit objectives were to achieve economies of scale, reduce imports, promote the development of a national industry, prolong the life of equipment, and help in the process of technology transfer among research and education centers.[29] In 1973 CAPRE identified what it considered to be the sector's liabilities in terms of scientific and technological potential and as a remedy created a National Program for Computer Training with the help of the ministry of industry and commerce (MIC).[30]

As time went by, it became increasingly clear that the balance of payments deficit was growing and future prospects worsening. In 1974 the Geisel government moved to limit imports of consumer goods by federal agencies, and computers came to the attention of the high decision-making level during 1975. As a result CAPRE acquired additional financial support[31] and new power that allowed it to act on four fronts: to limit imports, foster domestic supply, attempt to reserve the minicomputer market for Brazilian firms, and continue to develop its activities for generating awareness within the government, the science and technology infrastructure, and domestic industry of the importance of developing a national computer industry. With the aim of reducing imports, CAPRE was "able to raise import duties on computers and to require that foreign computer firms deposit the value of imported systems with the Bank of Brazil for one year without interest; these two actions raised Brazilian prices for U.S. computers from about 140 percent of U.S. prices to about 190 percent."[32]

But CAPRE also achieved something more precious: it became the "guardian of the gate." Decree 104 of the National Foreign Trade Council (CONCEX) in December 1975 commissioned CAPRE to review all contracts for data-processing equipment that involved imports. CAPRE set the following import limits: $110 million in 1976, $100 million in 1977, $130 million in 1978, $150 million in 1979, and $180 million in 1980.[33] According to Saur, CAPRE examined two thousand requests in 1976 and, of the $250 million requested, granted $115 million. Decree 105 of March 1976 expanded CAPRE's import control power.

Another major development came with Decree 77.118 of February 9, 1976, which gave CAPRE the task of studying and proposing a national informatics policy. It also delineated CAPRE's formal structure: an executive secretariat in charge of day-to-day policy and procedure, and a council whose function was to propose the national data-processing policy and draw up a plan of informatics, examine the executive secretariat's major policy decisions, and rule on certain specific cases. The council's members were, serving as president, the secretary general of the planning secretariat (SEPLAN), at that time Élcio Costa Couto; CNPq's president, then José Dion De Melo Teles; and representatives of the armed forces and the ministries of communications, education and culture, finance, and industry and commerce. A consulting commission created to assist the executive secretariat consisted of scientists and technologists from private and public institutions.

By mid-1976, IBM sensed where things were heading and decided to announce the manufacture of its System 32 minicomputer at its Brazilian plant at Sumaré, to be assembled from parts introduced as part of IBM's import quota. A blitz advertising campaign attracted nearly four hundred potential buyers.[34]

CAPRE's council Decision 01 of July 15, 1976, created the basis for reserving the micro/mini computer market to Brazilian companies. It came not only as a continuation of the policy initiated in the early 1970s but also as a response to IBM's challenge. Deciding to divide the market and industry in two, the council recommended that "the national informatics policy for the medium and large machines computer market should be based on investment rationalization and optimization of installed resources [i.e., on the market or, in effect, on foreign industry]." But it recommended that

> the national informatics policy for the minicomputer and micro-
> computer market, their peripherals, modern transcription and
> transmission equipment, and terminals should be oriented in such
> a way as to allow the control of initiatives aiming at achieving con-
> ditions for the consolidation of an industrial park with total con-
> trol of technology and decision within the country, while trying to
> prevent investment overlapping, waste, and losses [i.e., state inter-
> vention and market reserve].[35]

Decision 02, also of July 15, 1976, gave CAPRE the power to control the purchase of software and data-processing services by government agencies and enterprises.

Another indication that Brazilian high-level government was more responsive to domestic than to international pressure was the Economic Development Council's (CDE) Decision 05/77 of January 12, 1977, which established five criteria for fiscal incentives in the data-processing industry: degree of nationalization, export potential, extent of technology transfer, viability of the enterprises already in the market, and domestic capital majority.

In June 1977 CAPRE's council Decision 01/77 invited domestic and foreign companies to submit bids for producing minicomputers in Brazil, with proposals to be presented within 90 days and selections to be made according to the above criteria.[36] Among the sixteen companies that entered the race were seven multinationals and only two joint ventures (see Table 14). This made the decision to favor the nationals easier.

Table 14.   Brazil. Minicomputer Projects
Under CAPRE's Examination

Bidding Company	Leading Enterprise	Nationality	Technology Source
Sharp/Inepar/ Dataserv	Sharp Equipamentos	Brazilian	Logabax
Edisa SA	Procergs	Brazilian	Fujitsu
Hidroservice/J. C. Mello	Hidroservice Ltda.	Brazilian	J. C. Mello
Elebra SA	Elebra SA	Brazilian	Honeywell
Ifema SA	Ifema SA	Brazilian	Ifema SA
Protondata/Isdra	Isdra SA	Brazilian	Philips
Doças de Santos	Doças de Santos	Brazilian	NEC
Labo Eletrônica Ltda.	Grupo Forsa	Brazilian	Nixdorf
Maico Ltda.	Grupo Lucas Nogueira Garcez	Brazilian	Basic Four
IBM Ltda.	IBM	American	Own
Burroughs Ltda.	Burroughs	American	Own
Hewlett-Packard	Hewlett-Packard	American	Own
NCR	NCR	American	Own
Olivetti	Olivetti	Italian	Own
Four Phase	Four Phase	American	Own
TRW	TRW	American	Own

Source: Silvia Helena, "A indústria de computadores: Evolução das decisões governa-
mentais," *Revista de administração pública* 14, no. 4 (Oct./Dec. 1980): 98.

CAPRE's blow to the multinationals came at the end of 1977 when
it chose four companies: the government Cobra and three domestic
consortiums.

1. *Sistemas de Informática Distribuída SA* (SID) was to be created in
   January 1978 with participation of Sharp (51 percent), Inepar/
   Dataserv (39 percent), and Digibrás (10 percent). With technology
   supplied by Logabax of France they planned to produce 1,700
   units for the domestic market within five years. Their system was
   to be called SID 5200, with a memory capacity of 64–96K, or 128K.

2. *Labo Eletrônica Ltda.* was founded in 1961 and composed of two
   consortiums, Grupo Forsa (75 percent) and Brasilinvest (25 per-

cent). It was set up to produce the Labo 8870/1 and Labo 8870/2 with 64K and 96K of memory, respectively, both with Nixdorf technology, and turn out 260 units in the first year. It also planned to export.

3. *Edisa Eletrônica Digital* (Edisa) was organized in November 1977 with private investments from Rio Grande do Sul's firms and banks. It decided to acquire Fujitsu's technology and produce the ED 301 with two memory capacities—8K to 48K and 16K to 64K. It planned to produce a minimum of 60 units per year up to the third year and then 100 units annually to the fifth year.[37]

Later CAPRE approved a fifth company to produce minicomputers: *Sistemas e Computadores Ltda.* (Sisco), a new firm created out of the Hidroservice/J. C. Mello enterprise.

Under the terms agreed to by the approved companies, technology transfer had to be complete by 1982, and payments could not exceed 3 percent of net sales. The policy then was to allow only a few companies per market segment; the local firm could buy the technology of the foreign company only once, and the next model had to be developed locally. The minicomputer firms were to concentrate on CPUs to allow other companies to manufacture peripherals, with standard interfaces.

The minicomputer battle was only the prelude for the next: filling the slot between the mini and larger systems like the IBM 370/148.[38] The question became in part one of definitions. Was the 370/148 a large or a medium computer? IBM called it a large system, but it was defined worldwide as medium. Nevertheless, it was beyond the reach of the newborn Brazilian computer industry. Therefore, in November 1978 CAPRE approved a new version of the 370/148, as well as Burroughs' large system, the B-6800.

CAPRE feared that if the multinationals were allowed to produce medium-sized computers, they might later be down-powered and used to compete with the Brazilian minicomputers, and that the domestic industry might not have a chance to compete in the medium market once it was ready to. Thus the strategy was to defend what had been achieved and to fight for the next slot. There were then reports that Brazil was aiming at the production of medium-sized computers in three years,[39] substantiated by the fact that the Physics Institute of São Paulo[40] and Computer Nuclei of UFRJ were already at work on one.

In December 1978 CAPRE issued new criteria for the manufacture of CPUs and peripherals beyond the minicomputer range, including the need for assurances that the projects would not interfere with the minis and micros, local decision making, possibility of technology transfer, growing nationalization index, and export potential.[41] CAPRE thereby prohibited IBM and Burroughs from manufacturing medium computers in Brazil. "Capre charged that IBM's proposal to manufacture its 4331 and System 38 midicomputer models and Burroughs' plan to produce its 2800 medium-sized computers did not provide for either sufficient national content or effective transfer of technology to Brazilian companies."[42]

Then on October 8, 1979, a major political development transformed data-processing policy in Brazil. By Decree 84067, CAPRE was abolished and replaced by the Special Secretariat of Informatics (SEI), attached to the National Security Council, with the aim of centralizing computer policy. This move followed the change of the presidential guard in 1979, but as we will see later, there were other and more profound reasons for the change.

SEI and the Strengthening of the Domestic Market Reserve

SEI became Brazil's new normative instrument of computer policy. Octávio Gennari Netto, a former CAPRE executive, was its general secretary, and Decree 84266 of December 5, 1979, regulated its structure and attributes. Its main task was to advise the National Security Council on the formulation and implementation of a national informatics policy and plan. It was also charged with stimulating and participating in the development and absorption of technology, components, equipment, programs, and services; with promoting and protecting the technical and commercial viability of domestic manufacturers of computer systems and components;[43] and with coordinating realtime control systems, microelectronics, and national software policies.

Among SEI's various bureaus, the informatics commission (CNI), which met every two months, corresponded most closely to CAPRE's council. It was composed of SEI's general secretary, representatives of the ministries and agencies involved in informatics, and the joint chiefs of staff. An informatics activities fund was created to finance

SEI activities, having a budget of Cr 250 million in 1980, about $4 million at that time.

SEI's new authority allowed it to coordinate all the activities in the field of informatics, which, according to Gennari, CAPRE failed to do. Gennari claimed that SEI not only would maintain the domestic minicomputer market reserve but in the mid-range would approve only projects that guaranteed technology transfer. He also assured that the import restrictions would continue, that computers with Brazilian counterparts would not be granted import permission, and that SEI would try to match national and multinational interests. Thus, he said, the multinationals would concentrate their investments on the large computers for which Brazil lacked the technology.[44]

SEI's first actions showed its determination to maintain the market reserve, control the data-processing sector, and deal with the multinationals in a tough yet pragmatic manner. Its first normative act, of March 1980, set criteria to guide the approval of data-processing imports, stipulating that the analysis of requests would take into account availability of domestically produced hardware and software. SEI would approve imports only for priority sectors as determined by the economic development plans; equipment destined for state agencies; equipment for private use (for planning and production); equipment for independent services; intelligent terminals that would avoid the use of large equipment; and inexpensive peripherals. The act stipulated that once a decision on imports was made there would be no second consideration.[45]

Additional normative acts during the same year ordered that all data-processing equipment be registered with SEI, required that any domestic and foreign equipment purchases by the federal government receive SEI's previous permission, and called on the federal government to prefer national alternatives when purchasing data-processing services, a condition of SEI's approval of imports and technology transfer contracts.[46]

SEI also stated that new projects aimed at manufacturing data-processing equipment and imports of parts and components, including process control equipment, would be authorized only if Brazilian nationals guaranteed effective control and direction of manufacturing ventures and if domestically developed technology was used. It further required all government agencies to submit to SEI annually the so-called informatics directive plans, which account for all the developments and transactions in the informatics sector. These plans had to be formulated in the spirit of SEI's policy of promoting domestic technology, strengthening the national enterprise, and reducing imports.[47]

The first major test for the new policy arose in August 1980, when SEI gave IBM permission to manufacture in Brazil its medium-sized 4331 MG2 in limited quantities. At that time the midicomputer market was growing by 10 percent a year and demand could be met only with imports, which SEI would not allow. To make sure the 4331 remained a medium-sized computer, SEI restricted the minimum memory capacity to 2 megabytes. It also decreed that the nationalization index would have to be 85 percent, the level required by the Industrial Development Council (CDI) for the issuance of nationalization certificates.[48] Furthermore, SEI determined that the manufactured computers should be mostly for export but that for every three units exported, two could be sold in the domestic market—with restrictions: a maximum limit of 242 units in the next four years was imposed, and this only for the substitution of old IBM models and by permit from SEI. This meant that IBM was allowed to manufacture a maximum of 605 units during four years, although SEI also allowed IBM to produce magnetic disks for export, which would provide estimated revenues of $60 million a year.[49]

The approval of the two operations at the same time was the outcome of active bargaining between SEI and IBM. SEI approved the 4331 on the condition that IBM would produce its disks in Brazil, generating crucial savings of $30 million per year. With that added to the 4331 exports and the expected $60 million in disk exports, the deal was very good from the balance of payments perspective.

Burroughs was also given permission to produce 50 units of their large B6900 computer annually. H-P, by what it considered a loophole in the Brazilian controls of the micro- and minicomputer markets, gained authorization to produce a desktop microcomputer specifically for scientific applications, with 1,950 units targeted for the domestic market up to 1984 provided 6,020 units would be exported.

During 1981, SEI tightened its control and strengthened the domestic market reserve. Normative Act 16 made it more difficult to get approval for ventures to manufacture the reserved products. It reiterated the domestic control and technology conditions and added that in considering approval of manufacturing ventures SEI would take into account the enterprise's technological experience and whether the company intended to use domestically developed software, adapt products to the domestic market, and—in an open waiver of the domestic technology rule—absorb the technology. Another decree developed criteria for products not included in Act 16, namely, those on which, in principle, foreign companies could make a bid. The many criteria included preferences for Brazilian control, management, and technology; Brazilian-developed software; a high nationalization index; use of local suppliers and components; export potential; and total openness of patented and unpatented technology.[50]

A more recent pronouncement instructed that all R & D performed in the informatics sector receive SEI approval, and another allowed the federal government (40 percent of the computer market) to contract services from foreign firms only when no national company was qualified to render the service. The market reserve was broadened to include digital machinery used for testing and measurement as well as digital biomedical machinery, and a Software Registry was established within SEI for the registration of all domestically and foreign-developed software programs available in the domestic market. Although registry is not obligatory, SEI would not approve equipment imports or manufacturing projects whose software had not been registered in the Software Registry. Furthermore, failure to register might lead the National Institute of Industrial Property (INPI) to consider software imports as technology transfer. No enterprise not registered with SEI would be able to remit payments accrued from software sales.[51] SEI also decided to strengthen the software policy by denying permits to microcomputer projects that did not use locally developed software.

When SEI was slightly restructured in 1981 and an advisory council created consisting of both private and public sector representatives, data processing controls were extended to process control, teleinformatics, data banks, transborder data flows, and microelectronics. With the goal of decreasing Brazilian dependency on imported microelectronics devices and/or on the ones manufactured in Brazil by foreign firms, SEI established a component import control policy and began to coordinate the R & D activities of various institutions. To this end it created the Microelectronics Consulting Group and called for the establishment of a specialized research center, the acquisition of semiconductor technology, and the creation of a microelectronics market reserve. The resultant Informatics Technological Center was inaugurated in May 1984 in Campinas, near São Paulo. SEI selected two domestic private firms, Itaú and Doças de Santos, to set up plants nearby for the manufacture of microelectronics products. The center became one of SEI's highest priorities, with full support from President Figueiredo and particularly from Gen. Danilo Venturini, who as secretary-general of the National Security Council was the key policymaker on these matters. He called the institute "one of our most vivid aspirations." [52]

During 1984 SEI decided to move the market reserve "up" into the 32-bit "supermini" computers—minicomputers with the power and speed of a mainframe. Eight domestic companies answered the call, three of which—Labo, SID, and Cobra—committed themselves to developing the supermini with local technology; the other five—Edisa, Elebra, Itautec, Sisco/Hidroservice, and ABC Sistemas—requested permission to manufacture superminis with imported technology. All eight companies pledged to effect technology transfer and a high nationalization index when their proposals were accepted. SEI decided first to accept the domestic technology projects, hoping some of the others would merge, but when they had not done so by June 1984, it approved all five supermini projects using foreign technology. Labo, SID, and Cobra, unable to compete, then had to halt their projects and buy foreign technology.

In October 1984 Brazil's computer policy was further protected from domestic and international pressure when the Brazilian congress, strengthened after the 1982 congressional elections and the imminent return to full democracy, enacted a National Informatics Law. This legislation gave the domestic computer industry eight more years' protection from foreign competition, provided fiscal incentives to stimulate local firms, and established a National Council on Informatics and Automation (CONIN), attached to the presidency. The 18-member council will be on an equal footing with the National Security Council and will take over from the security council control of SEI's policies. Multinational corporations already producing computers in Brazil will be allowed to continue their operations, but new foreign investment will be restricted to products for export. [53]

The Pragmatic Antidependency
Guerrillas at Work

Why and how has Brazil succeeded in implementing a computer policy with the explicit aim of reducing technological dependency on outside sources? And why and how did it establish a domestic computer industry and begin taking control of national needs by excluding international giants such as IBM from Brazil's lucrative micro- and minicomputer markets?

Many factors were involved, starting with the growing demand for computers in Brazil in the early 1970s following the "economic miracle," when much of the capital necessary for technological development became available. Also, technological changes in microelectronics and computer hardware gave developing countries the opportunity to develop and produce nearly state-of-the-art computer hardware. Thus, a domestic industry in conjunction with the ongoing industrialization and capital goods substitution processes would allow Brazil to reduce its dependence on international companies and thereby its imports, which were causing a serious balance of payments problem. The country's political stability and continuity since the 1964 military coup, the cooperation of the technocratic leadership with the military in setting economic and technological goals for Brazil's development, and heavy state intervention in the economy all contributed to Brazil's ultimate success in establishing its computer enterprise.

But these factors by themselves are not sufficient explanation. The growing Brazilian computer market, the large multinational involvement in that market, and the change toward smaller machines did not necessarily have to work in favor of indigenous computers. The growing demand could have been met with imported computers or computers manufactured in Brazil by multinationals. In fact, the technological changes favored the multinationals, with their advanced technology and larger capital assets. Furthermore, the balance of payments crisis was neither at the root of the development nor the main consideration at the ministry of planning, the National Economic Development Bank (BNDE), and the navy: when the computer initiative started, Brazil, in the midst of the economic miracle was experiencing a balance of payments surplus.

To gain a full understanding of the development of Brazil's computer industry we must look at processes, at choices that cannot be taken for granted; and we must look at those who made the choices, the organizations they represented, their ideologies, and the impact of ideology over choices. Although cognitive and institutional resources were not single or sufficient causal factors, they nevertheless were crucial in turning an economic, technological, and political propensity into a technological and industrial reality.

Development of Local Scientific and
Technological Data-Processing Potential

Mário Ripper, the engineer involved in the development of the first Brazilian computer (Zezinho, in 1961) and who later became executive director of the Federal Service for Data-Processing (Serpro), rightly suggested that the Zezinho effort was premature. Ten years later BNDE/FUNTEC (Scientific and Technical Development Fund) and FINEP (Studies and Projects Financing Agency) were established; not only was Brazil experiencing the economic miracle, but a carefully designed science and technology policy was already strengthening the Brazilian capability to adapt technology and innovate. As Velloso said, computers were not just industry, they were industry and technology policy both—and technology transfer was the goal.

Pelúcio (first at BNDE, then at FINEP and CNPq), BNDE, and the other autonomy-oriented institutions and planners not only developed (in conjunction with the navy) the idea and reality of the domestic computer industry, but they also established the scientific and technological system and its financial support, provided the means to educate and train cadres of professionals in computer science, and gave them the resources necessary to accomplish their work. The improvements in the computer industry's scientific and technological infrastructure produced a critical mass of experts adequate for "the government to adopt an aggressive policy of technological independence in the sector aimed at setting up a minicomputer industry, based on national firms with the commitment to absorb technology and create the bases for a national technology."[54]

FINEP supported hardware, software, and process control development; it also helped finance several university projects, including the development of the first computers at Brazilian universities. The National Council of Scientific and Technological Development (CNPq) organized a special commission of experts to produce computer science research guidelines. This resulted in the Integrated Program of Computation Technology, which provides fellowships and research support to institutions, assists a microelectronics project, and has organized a task force that coordinates the policy planning of future data-processing technology.

Graduate studies in computer science were instituted at universities in São Paulo, Minas Gerais, and Rio Grande do Sul. By the mid-1970s, when the computer policy began to take shape, students who had gone abroad to study were beginning to return, strengthening their institutions and universities. Although prior to 1972 professional training depended heavily on "free courses" offered by the multinationals by 1977, 40 undergraduate and graduate university courses were being offered.[55]

Brazil has 19 universities, 450 research scientists organized into 74 groups, and 12 government research centers working on computer technology. In 1981 the data-processing equipment industry workforce numbered 14,646, of whom 31.5 percent were university graduates. It has been estimated that in 1985 there will be 87,000 technical personnel: 30,000 operators, 32,000 programmers, and 25,000 systems analysts—certainly adequate to satisfy Brazil's data-processing needs.[56]

Ideology and the Policy-Making Process

While the ministry of planning and BNDE were pursuing plans for a domestic computer for reasons of economic development, the navy wanted it for strategic reasons; an ad hoc alliance for the development of a national computer resulted. According to a former high government official, Saur and BNDE used the navy and the navy used BNDE, each to achieve its own objectives. Also, the fact that BNDE was working together with the navy helped Saur convince the high political echelons to go along with the domestic computer project.

Organized pressure began with the creation of the Commission for the Coordination of Electronic Data-Processing Activities (CAPRE) in 1972. CAPRE became not only an organization for a particular sector's industrial development but also the home of an ideologically assertive group—a veritable guerrilla headquarters. Thus, in addition to discharging the duties entrusted to it by presidential decree, CAPRE (more explicitly, its executive secretariat) became involved in the selling of ideas and consciousness raising—that is, guerrilla work. According to one CAPRE official, the institution developed according to Saur's initiative. Saur himself argued that they sought consistently to raise consciousness and to find additional uses for computers in all sectors of the society and the economy.[57]

The pragmatic antidependency guerrilla members within and outside CAPRE, known in Brazil as "the group" (among them Saur, Ivan da Costa Marques, Mário Ripper, Arthur Pereira Nunes, and Claudio

Zamitti Mammana), formulated an implicit doctrine, known as "the model," which would later be adopted by SEI as well. "The model" had two key features: wholly domestically owned companies, and one-time purchase of foreign technology. "The group," working as teachers at the universities and as technocrats at the government agencies, also infused the science and technology community and the political system with optimism that "the thing can be done." According to some members, the positive results achieved in scientific and technological activities had strong positive reinforcement for their work. With the creation of the Seminars of Computation at the University (Secomu), which became another forum for airing guerrilla ideas, computers, industry, politics, and the university became interwoven.

For example, Secomu IV in 1974 explicitly called for the development of national computer technology and exhorted the government to actively protect the market. Secomu VII, held in Florianópolis in 1977, concluded that "real technological autonomy was dependent on the existence of 100-percent-national capital enterprises with the capacity to develop appropriate technology," and recommended that CAPRE not let foreign companies into the market and not approve projects that would involve foreign technology.[58] Secomu VII and its successors called on the government repeatedly to maintain and nourish the incipient computer industry and to continue with support for the development of domestic technology. Secomu thus became a realistic pressure group to which the government had to listen.

The Brazilian science and technology community found another valuable forum in the First Technology Transfer Seminar, held in 1975 in Rio de Janeiro, when particular critical attention was focused on the Digibrás proposal of a tripé venture with Fujitsu and/or Nixdorf. As the government began to consider giving Nixdorf permission to produce minicomputers, the seminar decided to call for improvement of local technology's capacity for development, especially in small computers, and it urged the government to actively participate in this development, to formulate an "informatics" policy, and to pass legislation that would protect local technology. Pressure also came in monthly doses through the pages of *Dados e Idéias*, a data-processing magazine issued by Serpro, which readily adopted "the group's" pragmatic antidependency position and became a valuable source of ideas, criticism, and consciousness raising.

The active search for a foreign partner undertaken by Special

Working Group (GTE)/FUNTEC 111 and Digibrás, and the minicomputer project description in the first Basic Plan for Scientific and Technological Development (I PBDCT) show that Velloso and other high policymakers were not at the time considering a totally national computer industry, but rather had tripé in mind. The government wanted the multinationals in to provide technology, and science and technology policy (institutions and regulations) was merely to ensure that technology transfer would take place. Private investment was to be encouraged under the shield of governmental protection. But alas, the multinationals were not interested in tripé. As Saur said, and Velloso later confirmed, "The multinationals here, including the biggest, IBM, declared their lack of interest in this effort."[59]

CAPRE acquired new power at the end of 1975, mainly through the pressure of computer imports on the balance of payments, which had taken a turn for the worse in 1974, and through the efforts of nationalist institutions to implement a market reserve and start domestic production of minicomputers. At the same time, the programs devised to improve the capabilities of Brazilian computer technology were bearing fruit, and the development of Cobra's computers and Serpro's terminals was proceeding as expected. The process thus had a life of its own, which affected the political process.

According to an ex-CAPRE member, CAPRE evolved an "inside policy" of support for maintenance projects but rejection of new ones, and it started to play a "very subtle game." It presented the enterprise involved in a transaction with a situation it was not prepared for, forcing the enterprise to provide new plans, data, specific actions, and objectives in very short order; many failed and so could not pass through "CAPRE's gate." The fact that this was policy developed at the third level (first—president; second—minister/secretary; third—CAPRE/Saur) is important: CAPRE's power derived from the very fact that the third level was able to set guidelines and policies without much interference from above, thereby presenting the higher echelons with *faits accomplis*.

CAPRE's Decisions 01 and 02 of July 1976 creating the bases for reserving the mini/micro market to Brazilian enterprises reflected the government's pragmatism in attempting to protect a weak national industry without giving the multinationals the impression that the policy was still protectionist. Also, the high levels of government continued to hope that IBM and other foreign companies would accept joint ventures with domestic companies. A decision to protect the mini/micro market *for the sole benefit of domestic companies* had not yet been made, and ministers were still divided on this issue.

CAPRE's executive secretariat's strategy had two lines of "containment": the first was to choose 100-percent-Brazilian companies to produce domestic computers; the second was to accept joint ventures or tripé with the multinationals. It must be emphasized that this was *not* a CAPRE council decision, and the high levels of government knew nothing about it. It was strictly guerrilla strategy.

As the time for a final decision approached, Velloso was under heavy fire from both camps. The guerrillas gained an advantage when the media took the case to the front pages. The multinationals, traditionally a subject of hot debate in Brazil, had generated considerable outrage with the tough line they—especially IBM—took, and once the matter went public it became more difficult for the government to accept IBM's position without appearing to be bending under their pressure. CAPRE was further aided by the fact that the powerful banking consortiums, with large investments in Cobra, were pushing for the domestic alternative. Velloso was also under pressure from several key military figures who favored domestic companies and market closure. One such individual, representing a large group of the military, reportedly suggested that CAPRE's council meet in Velloso's home with the ministers involved, directly and indirectly, in the data-processing sector.

The discussions at this meeting concentrated on the question of how to say no to IBM without actually saying it. The ministers, fearing that the movement toward a national computer industry was based on enthusiasm alone, tended to prefer joint ventures but decided nevertheless to invite interested companies to present a bid for CAPRE's council decision according to the conditions specified by the Economic Development Council (CDE) in January (see p. 251). Although all the ministers supported the action, some had misgivings: Finance Minister Simonsen came out of the meeting saying, "Eu no hacho que ou négôcio va dar certo [I don't believe this is going to work]."

The ministers' decision was subtle. They told CAPRE's president that if the national bids were good the nationals should be preferred; if not, then IBM's project would be approved. But since one of the CDE conditions for investment in computers was joint venture with national capital, *the rejection of IBM's bid was almost assured.* Therefore, CAPRE's council decision of June 1977 calling for minicomputer bids from domestic and foreign firms was in fact a cover for a high-level policy decision that had already been reached. For CAPRE and Saur this was a strategic victory, opening the door for the domestic market reserve. For the high policymaker it was a way of saying, "We went according to the rules—we asked for bids, and we let the best win."

Velloso played his cards very diplomatically. He explained to IBM that the joint venture condition was neither mandatory nor the only criterion by which proposals would be judged. The multinationals took Velloso's words as a genuine indication that the door was effectively open to them, interpreting the government's policy to mean that although Brazil would prefer to have local equity—even control—it was prepared to waive this condition if other factors assumed greater weight.[60]

However, while Velloso was saying to IBM, "Everything is OK," CAPRE was telling IBM a different story. At one point when Velloso asked CAPRE for the IBM proposal, "the group" thought that Velloso had yielded to the multinationals' pressure and was going to approve it, a belief reinforced by the fact that an IBM vice-president had met with Velloso and Geisel. But no approval was granted. And fortunately for Saur and his pragmatic dependentistas, the second containment line was not needed, because domestic companies were among the bidders. Some of them were not yet in operation, but the ministers' decision together with CAPRE's own strong determination to leave the multinationals out encouraged CAPRE to go with the domestic bids.

A fundamental landmark in the evolution of Brazil's computer policy was CAPRE's replacement by SEI shortly after CAPRE's responsibilities had been increased to meet the needs of a growing domestic computer market and manufacturing potential. Playing a major role in this change was the military, which except for the navy had so far shown no particular interest in computer policy. However, when the country successfully challenged IBM and set up its own computer companies and developed its own technology, the military took notice. By the end of 1978 they realized that the data-processing sector was too strategically important to be left in the hands of a planning secretariat that might, after the 1979 elections, be led by "internationalists" (as actually happened under first Simonsen and then Delfim Netto) who might retreat from the antidependency policy and again "fall prey" to the multinationals.

Heading the military's interests was the National Intelligence Service, from whose ranks came Figueiredo, Brazil's president since March 1979. In January 1979 it initiated an inquiry commission, known as the Cotrim Commission (headed by Ambassador Paulo Cotrim), which comprised members from the Service, the foreign ministry, and CNPq. The commission findings were critical of CAPRE and its performance, especially for not having placed enough emphasis on software and microelectronics, which were thought to be crucial to computer autonomy. Autonomy would never be achieved, the argument went, if the chips were still being purchased abroad.

But the truth behind the commission's findings was that, with the new Figueiredo government about to take office, CAPRE was losing its power base. This was partly because of the National Intelligence Service's unexpectedly strong interest in the computer sector and its mistrust of the technocrats of the "left" who were managing CAPRE. Also important were changes in the planning secretariat and the science and technology institutions. With Velloso, Pelúcio, and most of the architects of the technological autonomy policy on their way out, CAPRE had become significantly weakened, and the cooperation between government technocrats and the scientific community that had characterized the mid-1970s was eroded.

At the initiative of one minister, the ad hoc Cotrim Commission was turned into a presidential committee to decide the fate of the Brazilian computer industry and informatics policy.[61] On May 10, 1979, a working group composed of the National Security Council, the National Intelligence Service, the foreign ministry, the planning secretariat (SEPLAN), and the Armed Forces Chiefs of Staff was set up to study and suggest policy for the sector within 120 days.

This committee followed the Cotrim Commission's guidelines closely and began work on the basis of its conclusions. One month later the president, by Decree 117, invested in it the authority (and a name: GTE-1 Special Working Group) to study and propose a global informatics policy. The move immediately drew fire from CAPRE and Digibrás as well as some members of the armed forces chiefs of staff who feared the move might result in giving up the market reserve and all the other gains achieved by CAPRE. CNPq, CAPRE, the ministry of industry and commerce (MIC), and Digibrás asked to become members of the GTE-1 Special Working Group but were rebuffed. The show was being run by the National Intelligence Service. Nevertheless, SEPLAN had an official representative: Moacyr A. Fioravante, an ex-president of Serpro. And Saur was his assistant—the pragmatic antidependency guerrillas had penetrated the National Intelligence Service's show.

The GTE-1 worked from May 17 to September 14, 1979. Its main decisions were to abolish CAPRE and to place data-processing policy under the jurisdiction of the National Security Council. It recommended the creation of SEI. Implicit in its recommendations and directions were the following principles: data processing could not be linked only to industry; the sector required incentives for technological development; and policy needed to be coordinated, with special attention given to semiconductors.

The 1979 institutional and political changes were not intended to reverse the achievements of CAPRE's autonomy policy. On the contrary, SEI had adopted de facto "the group's" model and wanted to place the policy on firmer ground. The changes were nevertheless drastic, for they were informed by nationalist strategic considerations rather than by the goals of economic industrial development and equality that had guided CAPRE in the past decade.

This ideological difference led to mistrust and uncertainty among the guerrillas and pragmatic dependentistas, particularly ex-CAPRE members, who feared that the military might not want, or be able, to withstand the multinationals' pressure. Furthermore, they realized that with data processing in the hands of the National Security Council, the danger of its being used to restrict individual liberties for security reasons would increase dramatically. Some guerrilla members even felt guilty for having helped produce this frightening situation. Therefore, the guerrillas decided to keep a watchful eye on SEI's moves and to prepare themselves, if necessary, for a fight. Some of the guerrilla members went back to the universities and to private companies, but others chose to work for SEI, hoping to affect events from within. Still others went to institutions with power to defend "the model." CAPRE was gone, but not its legacy: an ideology and institutions that would ensure that past achievements would serve as the springboard for future success.

Whereas CAPRE had begun its policy formulations and actions in a political environment largely devoid of interest groups and efforts at organization (besides the universities), SEI started off with politically strong organizations watching, controlling, and often opposing its policies. Although some associations in the data-processing sector dated back to the 1960s and early 1970s, such as the Brazilian Association for the Electric and Electronic Industry, created in 1963, and the

Society of Computer and Subsidiary Equipment Consumers, created in 1964, new and important groups emerged between 1976 and 1978, including ABICOMP (Brazilian Association for Computer and Peripheral Equipment Industries), 1978; APPD (Association of Data Processing Professionals), 1978; the Association of Data Processing Service Enterprises, 1976; the National Laboratory of Computer Networks, 1978; and SBC (Brazilian Computation Society), 1978.

SBC became one of the watchdogs of the sector's technological autonomy policy, with Claudio Mammana, one of the most active members of "the group," as its president since 1980. SBC's main target has been the multinationals. Significantly, SBC achieved representation in the National Informatics Commission (CNI), which is where the sector's policy is made, and Mammana was also invited to assist Gennari; he accepted because, as he put it, SEI's objectives were those of the SBC—to contribute to technological autonomy in the data-processing field.[62] Thus the ideas of BNDE, CAPRE, Digibrás, and "the group" achieved continuity through, among other ways, membership at the center of political power on informatics—the CNI.

Another stronghold of pragmatic antidependency is ABICOMP, which was created to defend the domestic industry and thus excludes foreign companies from membership. The organization favors private industry rather than state enterprises and has as its principal aim "the defense of the market reserve . . . fundamental for the survival of the majority of the enterprises."[63] By virtue of its membership, ABICOMP has considerable power. For instance, Cobra, one of its members, is invested in by such government agencies as BNDE, Banco do Brasil, and Serpro, and it has the support of the big Brazilian banks, such as Bradesco, Itaú, and Real. ABICOMP and SEI have found a modus vivendi based on common interests.

The Association of Data Processing Service Enterprises is highly supportive of private enterprise and has called for a more coherent policy for national software and for increased government financial aid and protection. While both it and ABICOMP avoid taking radical positions, not so the SBC, which with its large contingent of university and research institute professionals is very egalitarian nationalist and openly critical of foreign investment.

Domestic producers, scientists, and the pragmatic antidependency guerrillas reacted with anger and concern to SEI's August 1980 decision allowing IBM to produce its 4331 in Brazil. They feared that it would prevent the development of domestic medium-sized and large computers. Giovani Farina, then ABICOMP's president, remarked that the "local industry would be suffocated."[64]

In reaction, ABICOMP, SBC, the Rio de Janeiro Engineering Union, and other institutions created a permanent commission called CEDINI (Coordination of Entities for the Defense of an Informatics National Industry) in 1980 to coordinate actions "to consolidate the national computer industry." CEDINI, "conscious that the market reserve is an indispensable instrument for technological development," decided to watch every single government act having to do with data processing; analyze and make public its impacts; and take all necessary measures at the legislative, judicial, and executive levels to prevent violations of agreements and strategies that might hurt the national industry. It also set itself the task of producing a plan of action for preserving the market reserve and promoting discussions on the subject.

CEDINI decided to look on SEI's approval of the IBM 4331 as nondefinitive. It also proposed that SEI decisions that do not promote the protection of national industry be reversed, that SEI's waiver on microcomputers and desk calculators for scientific use be abolished, and that any new project must be approved by the CNI (in which the national industry and the SBC are represented).[65] CEDINI and its supporters saw the main problem to be that the IBM computer could operate at lower megabyte levels and price ranges and thus would affect the market reserve.[66]

SEI's Advisory Council, created in 1981, was a major gain for market reserve supporters because it provided them with an additional forum in which to air their views and plug their ideology. In mid-1981 Gennari announced that the market reserve for computers would be maintained only for another three years,[67] which elicited the reaction of the nationalists, some fifty enterprises, scientists, technicians, dozens of politically powerful institutions, government technocrats, and the military, all of whom constituted powerful pressure groups that would not allow the market reserve to be terminated. Fioravanti's words reflected the general opinion: "If the country had not adopted the market reserve, IBM's marketing policy would have become our data-processing policy, determining our equipment choices and even the training of our data-processing professionals."[68]

Gennari's apparent lack of strong commitment to the market reserve may have been one reason for his replacement by Joubert de Oliveira Brízida, who, together with Edison Dytz, has strengthened SEI's support for the market reserve, for the Brazilian microelectronics industry, and for the superminis.[69]

When the possibility of manufacturing superminis in Brazil was first raised, SEI had to choose among several alternatives: (1) local production with local technology, (2) local production with foreign technology, and (3) joint ventures with multinational corporations. Some prominent senators, congressmen, and industrialists preferred joint ventures. SEI policymakers were in favor of acquiring foreign technology but rejected joint ventures. Supporters of the market reserve and domestic technological development held out for a completely local industry.

SEI's decision to take road number one pleased the domestic market reserve supporters, but when it approved all five supermini projects using foreign technology they reacted swiftly. ABICOMP, APPD, and SBC, together with the Brazilian Society for the Progress of Science (SBPC), issued a communiqué stating that the SEI decision represented a retreat from the quest for technological autonomy in the computer area and called on SEI to take all possible measures to minimize the negative effects of the decision.[70] Cobra, however, needed the project, and it had little choice but to look for foreign technology. Failure to develop a supermini system would mean the end for the financially ailing firm. With funds supplied by President Figueiredo and the foreign technology supermini project, then, Cobra could look at the future with some hope of shaking IBM's mainframe leadership in the Brazilian market.

SEI took what turns out to have been a tough yet pragmatic position, because it understood that foreign technology was necessary if the domestic industry was to keep pace with development abroad. And despite the fact that some foreign technology was allowed, "the model" has actually been maintained in that only 100-percent-domestic companies were chosen to develop the superminis and the foreign technology being used is ultimately to be transferred.

The supermini battle was not the only one fought by supporters of the market reserve during 1983 and 1984. Even more crucial in their eyes was the struggle to enact a protectionist computer law, in reaction to bills calling for an end to the market reserve. The most threatening bill, proposed by Sen. Roberto Campos of the Social Democratic Party, called for abolishing the market reserve and dismantling SEI, replacing them with a tariff system and joint ventures, and placing informatics policy under the MIC. This proposal had the blessing of internationally oriented business circles, multinational corporations, and the U.S. government, which has always been openly critical of Brazil's computer market reserve and which has used Brazil's financial dependence to pressure the Brazilian government into changing its policy.

On the other side of the political spectrum was an array of bills aimed at protecting the market reserve and import controls and nurturing domestic computer companies. Probably most supportive of the market reserve was the bill of Cristina Tavares, a representative of the Brazilian Democratic Movement party. The guerrillas, several computer associations, the scientific community, and many domestic computer companies organized a propaganda campaign to pressure the congress into passing a law favorable to the market reserve. ABICOMP and APPD issued "In Defense of Brazilian Technology," a document signed by two hundred institutions, which accused the U.S. Commerce Department of interfering with Brazil's computer policy and called for rejection of the antinationalist proposals.[71] As part of the effort there were public meetings at universities, a new journal, entitled *Brazil Informatics,* was founded, and an annual National Informatics Day was declared.

The campaign bore fruit: on September 20, 1984, the military government introduced a bill recognizing these nationalist aspirations, and it was enacted as the National Informatics Law in October, at the same time the National Council on Informatics and Automation (CONIN) was established. Saur reacted to the vote favorably, saying that CONIN represented a refinement of the CAPRE informatics model and adding, "We have returned to what it was."[72] However, the survival of the antidependency policy will depend on how quickly its supporters can move as the technological gap between multinationals and Brazilian companies widens.

Brazil has shown that a developing country can reduce the technological gap with foreign technology but that it must always be ready to take political, industrial, and technological action should the gap widen again, fueling consumer and political unrest.[73] The supermini has recently caused such opposition to the domestic computer policy, and SEI, CONIN, and the industry have had to retreat to the drawing boards. The Brazilian computer case effectively illustrates that overcoming dependency in a high-technology sector is a long evolutionary process with interspersed periods of rapid reductions and then renewed increases in technological dependency.

The Multinational Corporations in an Ideologically Charged Context

The multinationals, led by IBM, resolved to put up a fight to try to kill the domestic industry project before it had a chance to grow. Data General was one of the first to confront the Brazilian government. As the second-largest minicomputer company in the world, it was a likely candidate for association with Cobra and for technology transfer. But Data General was not willing to accept the Brazilian government's requirement that patents and blueprints be transferred to Cobra at the end of the license period.

To enhance its bargaining position, Data General decided to involve the American government. In June 1977 it applied pressure through the president's Special Representative for Trade Negotiations and Congress, trying to convince the U.S. government that if Brazil succeeded in developing its computer industry other countries would be encouraged to go a similar route in the future. Thus Data General suggested bilateral negotiations with Brazil to lower tariffs for minicomputer imports from the United States in exchange for a decrease in U.S. reprisals against Brazilian exports; bilateral negotiations to make Brazil drop its technology transfer requirements; regulations prohibiting U.S. companies from transferring minicomputer technology to foreign companies having no American capital; and regulations prohibiting U.S. companies from providing import and licensing benefits to foreign governments (as Sycor did).[74] The U.S. government did not get directly involved in the negotiations, however, and Data General remained without a share of the Brazilian minicomputer market.

IBM do Brasil, convinced that the Brazilian government would not leave it outside the market, decided to play tough and apply pressure. For one thing, IBM had never been involved in a manufacturing joint venture and as a matter of policy did not intend to do so now. For another, technology transfer was unacceptable, and the low import levels imposed by the Brazilian government were a threat to IBM's international operations. Third, IBM was sure that the Brazilian venture into minicomputers would not be successful. In their lobbying effort, IBM kept emphasizing the "obscurity" of Sycor's technology and the lack of software. Fourth, IBM thought the large number of System 32 computers they had already placed on the market would generate a strong protest against the halting of their production. (In fact, IBM's actions were decisive in bringing about the opposite effect in the end.) Fifth, IBM thought that given the balance of payments difficulties, the government could be lured into accepting an IBM manufacturing venture involving a strong export potential—and they made it clear that they intended to export.

IBM do Brasil's president, José Bonifacio de Abreu Amorim, expressed surprise at the government insistence on full technology transfer and partnership with nationals, saying, "We don't need to ask the government in advance for permission to build System 32. Does Ford ask the government for permission every time it wants to introduce a new model automobile? . . . The government, after all, wants us to export." [75] Amorim's attitude reflected a lack of understanding of what the Brazilian government and its supporters were after. The government did not want to repeat with computers its import substitution industrialization (ISI) experience with automobiles. Instead of 100-percent-foreign companies manufacturing 100-percent-Brazilian products, it wanted at least some 100-percent-domestic companies to manufacture 100-percent-Brazilian computers, after technology transfer had been effected. In other words, IBM confused pragmatic antidependency with ISI and failed to see that Brazil was after much more than exports.

Pressuring the Brazilian government was certainly the wrong strategy, as it generated nationalist sentiments directly supportive of CAPRE objectives. Velloso was reportedly undecided at the time; had IBM been more flexible and accepted a few of the government's conditions, it might still have fit into "the group's" second containment line and so not lost the minis market. However, the framers of the pressuring strategy at IBM World Trade believed that to comply with the government's requests would have meant abandoning its policy of nonjoint ventures, which, given IBM investments in other countries— France, for example—was out of the question. For his part Amorim, who knew the Brazilian military first-hand, did not agree with the strategy, feeling that it would backfire.

There are few more vivid illustrations of the conflict between domestic industry and IBM than their advertisement war in the pages of Brazilian magazines and newspapers. Cobra's ads made ideological statements: "The computer is like oil: it is dangerous to depend on others." "A country that attempts to be big and powerful has to develop its own informatics policy." "Cobra 530 is the first truly national computer. . . . It is Cobra's answer to the market reserve. It is the development and fixation of national technology, where there are no middle terms: independence or death."

IBM decided to run ads of the same caliber, aimed at calming down the nationalist concerns. One read: "For evolution and import substitution, the Brazilian data-processing industry can always count on IBM's support and incentive. IBM is always collaborating with the constant evolution of the national data-processing industry"; then it described how IBM had helped train and educate technicians and had provided technological assistance to clients. It also claimed exports of $150 million in one year, good for Brazil's balance of payment problems. "This is IBM's difference," the ad declared: "incentive."

Other multinationals, such as Burroughs and H-P, watched the gathering storm. They too, although to a lesser degree than IBM, added to the pressure on the government. For example, Burroughs's marketing manager remarked that if Brazil protected the market, the multinationals would set up factories somewhere else, in neighboring countries in Latin America.[76]

As time went by, however, and Brazil showed that its domestic minicomputer industry was there to stay, foreign companies began to adapt to the new reality. Burroughs stated that it would continue to market all its other products in Brazil, and it and other companies indicated that association with Brazilian enterprises was a distinct possibility. For IBM, Brazil's decision to leave it out of the lucrative micro/mini markets provided a learning experience, one that required adaptation, ingenuity, and acceptance. "IBM and Burroughs seem to have made the best of the situation, manufacturing large systems in Brazil since the mid-1970s. Both corporations have gained advantages from the informatics policy because the products produced locally by them benefit from the preference rules regarding imported goods and services."[77]

Although IBM, Burroughs, and H-P did get some of their projects approved through intensive lobbying, and IBM found ways to circumvent domestic manufacturing restrictions, in the end even IBM had to accept the domestic industry's development as a fact and begin to work with rather than against it. Thus IBM has signed an agreement involving nine joint software projects with the Association of Data Processing Service Enterprises, and other IBM projects include a plan to link its system to locally developed terminals (yet to be approved because of SEI's fear that the company is attempting to penetrate the banking computer-automation market) and a proposal to provide the Informatics Technological Center with technological assistance in software, microelectronics, teleinformatics, biomedical applications, technical training, and industrial robot development.[78]

Foreign firms are also working with their Brazilian counterparts through sales agreements, such as that in which Olivetti has joined Scopus to develop a data-entry system with software produced by Olivetti, hardware by Scopus, and peripherals by other local Brazilian firms. The attitude of the multinationals with regard to licensing has also been changing. While in the mid-1970s Cobra had difficulty finding a foreign company willing to sell it technology, in 1980 "18 agreements had been signed involving 16 foreign and 14 local firms."[79]

Brazil's strong commitment to the goal of domestic technological and industrial development is quite apparent in the severely restricted involvement of computer multinationals in Brazil since 1977 and in the fact that the average price of computers in Brazil is approximately two and a half times that in the international market. This determination on the part of the Brazilian government and its technocrats to pay a high economic price to make this goal a reality is what has allowed the policy of autonomy to survive for almost a decade. And now the Brazilian computer policy has started to spread to other Latin American countries. In January 1985, for example, the Mexican government refused an IBM request to set up in Mexico a 100-percent-IBM-owned subsidiary to manufacture 125,000 microcomputers a year, most of them for export. Mexico's stand seemed to work, for, several months later, Mexico reversed its decision when IBM "agreed to set up a semiconductor development center for local industry, purchase a variety of high-technology components from Mexican companies, and produce software for Latin America in Mexico."[80]

The lesson for foreign firms is that they will succeed if they demonstrate sensitivity to the host country's prevalent set of beliefs, expectations, and objectives and perceive that it is to their advantage to change along with these societies. Multinationals can choose between an unyielding position and a pragmatic, flexible position. IBM tried it both ways and learned that, when facing an ideologically determined developing country, rigidity only made things worse. To pragmatic antidependency, there seems to be no better response than consideration and accommodation.

Conclusions

In the fifteen years since the formation of the GTE/FUNTEC 111, Brazil has developed a major industrial venture, one that would never have come into being had it been based on efficiency and other economic considerations alone. By mobilizing its material and ideological resources against IBM and other multinationals, Brazil succeeded in demonstrating that it could determine its own computer policy. The technical know-how of the pragmatic antidependency guerrillas infused their ideology with norms and policy directions. This helped the computer industry maintain its momentum despite, or maybe even because of, the removal or circumvention of some of the individuals and institutions that had been instrumental in the early stages of goal formulation and policy making.

Crucial to the understanding of the pragmatic antidependency guerrillas' success in making "the model" operational is the fact that technological and political factors reinforced each other as much as ideological and institutional actions. For example, moves toward a domestic computer industry were accepted largely because earlier programs to improve Brazilian computer technology were bearing fruit: Cobra's development of computers was proceeding as expected, as was the development of terminals by other institutions. These outcomes were not determined beforehand by the market, the political system, or international forces. Instead, the "actuality" resulted from purposive actions, ad hoc choices and coalitions, reactions by the multinationals, and the reinforcing effects each of these factors had on the others.

Thus, the Brazilian computer case substantiates the hypothesis that ideologies and institutions by their interaction reinforce one another. Institutions such as CAPRE, Cobra, the Federal Data Processing Service, and even *Dados e Idéias* were crucial not only for their formal actions in the computer field but also for giving the guerrillas a base from which to launch "attacks." Specific achievements may have been based on power or influence, but the definition of goals, means, and policy agendas stemmed from the collective understanding that united individuals within or among these institutions.

When CAPRE was eliminated and SEI established to continue its normative policy role, SEI ceased to be dependent on its original constellation of consciousness, that is, Pelúcio, BNDE, FINEP, CAPRE,

and "the group." Nonetheless, "the model" produced by this constellation was kept alive well after CAPRE's demise because it had already succeeded in generating its own institutions, domestic companies, and pressure groups and it had shown nationalists within the
National Security Council how the infant computer industry could
continue growing.

To consider the policy and the industry an unqualified success
would be more than premature. Certain enterprises may still fail and
die, and even the government's commitment to the computer industry
may falter because of political and economic pressures, both international and domestic. Furthermore, dependency on foreign sources
of technology has been only partially reduced—the Brazilian model
actually led to major reliance on foreign semiconductors, against
which Brazil has only in the last couple of years started to look for
countermeasures.

However, Brazil has made some progress in these fifteen years in
developing its scientific and technological capabilities with applications to industry, and some irreversible changes have occurred. Brazil
now enjoys a much more developed electronics industrial structure
than before. It possesses a more developed R & D base, with capacity
for high-technology production and testing, and is producing a "critical mass" of scientists, engineers, and technicians in R & D. There
exists a solid base of political groups that support the ideas behind the
industry and have a personal and institutional commitment to it. Domestic private entrepreneurship is strong, and Brazil has awakened
the general public to the benefits of the "computerization of society,"
something not evident in other developing countries. It has substantially increased its negotiation and bargaining knowledge, which often
proves extremely advantageous when dealing with multinationals, especially in the case of joint ventures. Brazil has allowed national enterprises to assume control of technology management, production, engineering, process engineering, and central modules where technology
is created and has led multinationals not only to adapt to and accept
this situation but also to offer their technology, which never happened
in the past. The computer industry now serves as an example for
other sectors and for other developing countries and, most important,
has provided Brazil with technological know-how that promises much
for future technological development.

In other words, the creative (and for some improbable) aspects of
the Brazilian computer industry have been catalytic in changing the

asymmetrical interdependent situation between Brazil and the multinationals in this sector. Brazil is still very much dependent on foreign
technology and capital for producing certain types of data-processing
equipment, but less so; it has more knowledge, can decide for itself
what it wants, how it wants it, and, to some extent, under what conditions. Some multinationals have adapted to this situation and now rely
on the growing Brazilian computer market. These changes are transforming relative capabilities, Brazil's attitudes about the multinationals
and vice versa, and Brazil's understanding of the limits and possibilities of assertive action.

When comparing computer development in Argentina and Brazil,
we can see that the Argentine case was one of private initiative based
on a vision of technological self-reliance that failed for lack of support
from Argentina's elites, especially the military, amid a chaotic political
and economic environment. The Brazilian case was one of state initiative based on an explicit science and technology policy of self-
reliance that succeeded in rallying the support of broad government
sectors, the military, and private industry, in a stable political and economic environment that supported the domestic computer industry
cause throughout.

Argentina certainly possessed the scientific capacity to build a computer, but it did not have a science and technology policy and financial
institutions oriented expressly toward such technological development, such as Brazil had, influencing and aiding the sector's technological and industrial development. Whereas the Argentine case
was one of private initiative that failed to convince the government of
its merits, the Brazilian case was one of public initiative that nourished
the private. The contest was not between a weak state and a strong
one: both states were strong and interventionist. The difference lay in
the realm of perceptions and an ideology of development. The Brazilians all along perceived the relevance of technological autonomy for
development in general and for the development of a computer industry in particular; the Argentine leaders for the most part did not.

Brazil focused on dependency reduction and management; Argentina chose to focus on efficiency and the market. In Brazil the ideological group that sold the computer idea not only had access to political
power but also for a while was itself part of the political power. Brazil
was bold in its determination to prevent the multinationals from producing minicomputers, but the sparse evidence available on Argentina suggests that there was nobody to withstand the pressure of

Olivetti—certainly FATE could not do it alone. Also, while Brazil concluded that technological change in the sector meant that the project was feasible, those who made the final decision in Argentina arrived at the opposite conclusion.

Thus these two cases show how ideological and political choices with regard to technological development, production, and government intervention in a productive endeavor have led to different capabilities in domestic technological adaptation and innovation within similar dependency settings, how two neighboring countries with the scientific potential to develop computers have taken different roads in their "journeys toward progress."

The Quest for
Nuclear Autonomy
in Argentina and Brazil 11

Immediately after World War II both Argentina and Brazil began an attempt to dominate the nuclear genie. They set up institutions to train nuclear scientists, founded atomic energy commissions, and initiated programs based on natural uranium technology. Their competition to be number one in this crucial area led to an action-reaction pattern. But the similar and slightly parallel paths that characterized their early nuclear development later went in different technological directions and resulted in quite different outcomes.

Argentina's success in developing a near-autonomous nuclear development capacity can be credited mainly to one institution, the National Atomic Energy Commission (CNEA), which centralized all matters of nuclear development, from mining to reprocessing and from technical training to nuclear-plant building. In its infancy the CNEA became a nonpartisan organization, enjoying continuity of leadership and political autonomy, consensus and insulation that allowed it to impose a program and an ideology onto the political elites. Generated by pragmatic antidependency guerrillas within the CNEA, the ideology was aimed at achieving nuclear technological autonomy and reducing industrial dependence. So inspired, the CNEA instituted a program for the gradual development of nuclear autonomy, starting with human resources and metallurgy. This appealed to Peronists and non-Peronists, left and right, alike: the political and military elites tolerated the ideology and accepted the policies for strategic, energy, and prestige reasons; the left backed it for its egalitarian-nationalist orientation, namely, the reduction of foreign dependence. Thus, at a time when other political institutions were being shattered by ideological conflict and political and economic instability, the CNEA was becoming an island of stability. With each choice the CNEA made in the direction of autonomy, and with each ensuing technological success, domestic support for the policy was broadened and strengthened, as was the commitment to withstand external opposition to the autonomy program, especially from the United States.

Nuclear energy has not escaped Argentina's deepening economic crisis and budget cuts. Some projects are behind schedule, and others have even been phased out. But the CNEA's successes in the development of a semi-autonomous nuclear capacity have created a propensity for the country to continue its forward stride—perhaps at a slower pace—toward nuclear technological progress.

Brazil, endowed with a more advanced industrial infrastructure and better physics programs than Argentina, was by the mid-1950s firmly set to develop an independent nuclear program. But subsequent ideological and political conflicts, institutional decentralization and lack of autonomy, and domestic and international opposition to independent nuclear development led to a long period of political stalemate and nuclear inaction. The National Nuclear Energy Commission (CNEN), created in 1956, was no equal to the CNEA and its undertakings.

The conflict intensified when powerful political and economic organizations decided to initiate a nuclear program to generate cheap, efficient electricity—at the expense of nuclear technological autonomy. Political fragmentation regarding a course of action and lack of any early successes hampered not only the determination to pursue an independent program but also the ability to withstand American pressure against such a program. In 1975 Brazil signed an agreement with West Germany for the largest technology package ever to be transferred from a developed to a developing country. Although nuclear self-sufficiency was one of the deal's objectives, it was overshadowed by energy and commercial considerations and by the desire to acquire a "quick fix" rather than to develop indigenous technology incrementally and steadily, as Argentina had chosen to do.

The Brazilian–West German Agreement was for four nuclear 1350 megawatt (Mw) pressurized water reactors, with the option for another four by 1990; "the development of uranium enrichment facilities; a uranium prospecting venture; . . . the construction of a plant to produce fuel elements and pilot plant for reprocessing nuclear fuel; the establishment of an engineering firm to handle key sections in the construction of the plants and a plant to manufacture large components."[1] The program, with an estimated cost of $10 billion, was expected to produce 10,000 Mw of nuclear power by 1990.

This policy was imposed from above, and the scientific community was kept outside the policy-making process. The government set up a holding company called Nuclebrás (Brazilian Nuclear Enterprises) to implement the agreement and be the partner in the joint venture with the Germans; this led to further internal divisions between Nuclebrás and the scientific community. After 35 years of nuclear development, Brazil still does not have a nuclear power plant working at full capacity (Argentina has two and another under construction), has not achieved self-sufficiency in the nuclear fuel cycle, and its military is dismayed by Argentina's progress. Furthermore, the critical economic condition and the foreign debt are waking Brazil from its nuclear dream. Brazil aimed at too much, too soon, and ended up with too little, too late.

Given the military implications of the technology and the competition between Argentina and Brazil, research on this issue faced obstacles and suspicion not encountered in other fields of this study. Nevertheless, a large number of primary sources were assembled and many interviews undertaken. A curious phenomenon was observed: in Argentina the officials were not only eager to talk, but they went out of their way to show off one of their few successes of the last 40 years. In Brazil, however, the issue was enveloped in secrecy—relative failure had made officials reluctant to talk. But former government officials and scientists who opposed the nuclear policy were extremely cooperative, and the richer, larger literature on Brazil's nuclear development, as well as the availability of documentary sources of information, balanced the difficulties involved.

Argentina: Success

The Nuclear Fuel Cycle

Uranium. According to an account by the CNEA's then president, Rear Adm. Carlos Castro Madero, reasonably assured resources of uranium concentrate (yellow-cake), after allowing for the uranium to be used in 1982 and 1983 (643 metric tons of yellow-cake), totaled 30,363 metric tons, enough to fuel nine nuclear power stations for their 30 years of active life. Production of yellow-cake between 1980 and 1982 averaged approximately 200 metric tons annually.[2] There are uranium mines at Sierra Pintada, Los Gigantes, Don Otto, and Pichiñas. One of them, Los Gigantes, with an exploration area of 100 square kilometers and resources estimated at 1,000 tons,[3] is run by a private firm (Sanchez Granel) that produces uranium fuel for the CNEA. Conceived by Argentine professionals using mostly Argentine technology, this is the first such production project outside the developed countries.

Advancing along the nuclear fuel cycle's front end,[4] Argentina produces its own uranium dioxide (UO_2), the basic raw material for the compact fuel pellets that go inside the fuel elements within the reactor's core. The Complejo Fabril Córdoba, in charge of making UO_2, achieves nuclear purification by use of a West German furnace with an annual capacity of 150 tons. A second plant, entirely domestic, is being built by a company in the state of Mendoza and the CNEA, which also were in charge of basic and detailed engineering and industrial architecture.[5] It was estimated that by the end of 1982 production would satisfy the annual consumption of UO_2.[6]

Fuel elements. Two hundred fifty fuel elements built in Argentina with zircalloy developed by the CNEA were used for Atucha I, Argentina's first nuclear power plant. In April 1982 an industrial plant for the production of nuclear fuel elements, located not far from the Buenos Aires international airport, became operational, under the responsibility of Conuar SA, a joint venture of the CNEA (25 percent) and private industry. In 1980 Argentina had agreed to export to Brazil large quantities of zircalloy tubing, developed and manufactured entirely in Argentina.

Heavy water.   In the past, Argentina relied on foreign sources of heavy water for its line of natural uranium reactors, including some 11 metric tons bought from the Soviet Union. But in 1980 it purchased for $300 million a heavy water plant from the Swiss firm Sulzer Brothers, which will produce 250 metric tons a year. First scheduled to be in operation by 1983, the plant's starting date has been postponed several times, and in 1985 construction was proceeding slowly. In addition, a pilot heavy water production plant is being built near Atucha I, with an approximate annual capacity of 3 metric tons.

Uranium enrichment.   Argentina needs enriched uranium only for its research reactors (no military use of enriched uranium is known of at present). Until recently it was imported from the United States, but when its sale was restricted for nuclear antiproliferation reasons Argentina turned to the Soviet Union for supplies. In November 1983 Castro Madero announced that Argentina had developed, without foreign help, the technology and an industrial plant to produce enriched uranium, thus joining a very small and exclusive club of developed countries. Construction of the plant, near Bariloche, started in 1978; it uses the gaseous diffusion method and was planned to yield 500 kg of uranium enriched at 20 percent, beginning in 1985. Future industrial stages to produce annually 150 and 250 metric tons of uranium enriched at 1 percent are planned. The plant was built with an investment of $62.5 million.[7]

Nuclear power stations.   Argentina has two working nuclear power stations: Atucha I, inaugurated in 1974 and located 100 km northeast of Buenos Aires, providing electricity to the Greater Buenos Aires Littoral; and Embalse Rio Tercero, inaugurated in 1983 and located in the Córdoba province. Atucha II, under construction near Atucha I, was expected to become operational by 1989 or 1990. According to Argentina's 1979 nuclear plan, three additional nuclear plants were to be operational by the year 2000, with a total investment of $8 billion. The economic crisis has led to the cancellation of one project, bringing to five the number of nuclear plants to be ready by the end of the century and to $4 billion the estimated amount to be invested.[8]

Atucha I was bought as a turnkey station from the West German firm Siemens AG; it has 335 Mw net power and is of the heavy water, natural uranium type. With an original price of approximately $70 million, 33 percent of the total cost involved local participation. Nuclear Engineering International, in a 1981 survey of the nuclear power stations of the capitalist world, found that Atucha I was third in terms of efficiency worldwide and had, compared to other energy sources, the lowest power generation price in Argentina.[9]

Embalse was purchased from Atomic Energy of Canada, which manufactures and sells the CANDU heavy water reactor. The Italian electrical manufacturer Italpianti also participated in the construction. The 600 Mw plant was originally estimated to cost $250 million, but after many delays it reached $1 billion. Domestic participation was 58 percent, and 50 domestic enterprises were involved, some of which specialized in the engineering and construction of nuclear power stations, such as Nuclar SA and Argatom SA.

Atucha II is similar to Atucha I but larger (750 Mw). It was bought from the Siemens subsidiary, Kraftwerk Union (KWU), for $1.579 billion,[10] not on a turnkey basis as before, but with engineering by an Argentine firm, Enace SA, in which KWU has a 25 percent investment. The reactor design was by KWU, and the base for the metallic vessel holding the reactor is being built in Brazil by Nuclep (Nuclebrás Heavy Equipment Company), a Brazilian–West German joint venture. Domestic integration is expected to be around 65 percent,[11] including components, such as steam generators, moderator/coolers, pressurizer, and so forth.[12] The goal for the fourth station is 100 percent domestic integration of engineering and civil works, 95 percent in the reactor assembly, and 68 percent in components.[13]

Reprocessing.   Argentina has developed domestic reprocessing technology that allows it to reuse the burned fuel in its reactors and to produce plutonium, which can be used to manufacture nuclear explosives or to fuel breeder reactors. A pilot reprocessing facility, built in-house and in operation during the 1960s, yielded the first plutonium manufactured in Latin America.[14] A larger plant for reprocessing uranium under construction at the Ezeiza Atomic Center is expected to be ready by 1986 and to produce 12 kg of plutonium a year.[15] It has been reported that Atucha I has already produced the equivalent of several hundred kilograms of plutonium.[16]

Waste disposal.   Low-level radioactive material is being eliminated annually, but permanent solutions for high-level radioactive material are still to be found. Studies are being made of possible sites for disposing of the dangerous materials; in the meantime they are stored in water tanks 20 meters deep.

Human Resource Development and Autonomous Research

The CNEA has been called the "National Commission of Atomic Education" for its consistent policy of promoting specialized technological training. Physicists, chemists, biologists, metallurgists, geologists, mathematicians, and engineers as well as machine operators, electronics and chemistry technicians, cartographers, and surveyors—over a thousand scientists and technicians in all—have received training through the CNEA.[17] It has organized symposiums, courses, and conferences, sent many scientists to study abroad, and arranged for more than three hundred foreign scientists to come to Argentina to share their expertise.

The CNEA has been in close contact with the universities and was involved in the creation of nuclear medicine and cosmic radiation centers, as well as one of Latin America's most sophisticated and advanced physics institutes, J. A. Balseiro, located at Bariloche. In 1977 the CNEA instituted the nuclear engineer career, which gave a new boost to physics training in that institute, as did a nuclear research reactor, which became critical in 1982. A high percentage of the graduates of J. A. Balseiro have also joined the CNEA's ranks.

The CNEA's Metallurgy Division, established in 1955 and since 1960 under the jurisdiction of the Technology Division, has become the backbone for autonomous nuclear technology development, the transfer of technology to industry, and the training of scientific and technical personnel. Its increasing importance over the years is reflected by its growing share of the CNEA budget: from 1 percent in 1955 (supplemented by funds from the National Council of Scientific and Technical Research [CONICET]) to 5–6 percent in the mid-1970s.[18] According to Jorge Sabato, director of the Metallurgy Division for many years, it

> has solved all the nuclear metallurgy problems identified for it by the CNEA atomic energy programme . . . developed and manufactured all the fuel elements for the five nuclear research reac-

tors installed by the CNEA . . . published more than 250 papers, mainly in well-established international journals . . . solved nearly 500 problems presented by the electro-mechanical-metallurgical industry . . . helped to create other metallurgic research centres and to incorporate metallurgy as a regular subject in the curriculum of several universities, in Argentina as well as in other Latin American countries. . . . Nearly 500 university graduates have received metallurgy training at CNEA, some 130 of them at present working at CNEA, some with industry, and the rest at universities and at other research centres. . . . One hundred and thirty Argentine metallurgists went to Europe and the USA for postgraduate training and research. . . . Thirty Ph.D. theses were prepared at the Metallurgy Division.[19]

Argentina's nuclear scientific and technological capabilities are further demonstrated by the seven nuclear research reactors already built in-house. Table 15 contains information on six of them, plus RA-4, the only one built abroad. None of the reactors was more crucial to Argentina's nuclear development than RA-1, operational since 1958. Though designed and engineered by Argonne National Laboratory in the United States, the Argonaut, as it was known, was totally built and technologically adapted in Argentina. The approach was so innovative that the know-how it generated was sold to a German firm, a milestone for a developing country at the beginning of its "nuclear career." Not included in Table 15 is a 10 megawatts electric (Mwe) enriched uranium reactor sold to Peru, for which Argentina enriches and provides the uranium.[20] Argentina is also said to be manufacturing a new nuclear research reactor, RA-7, the largest of them all with a 70 Mw capacity. Other research projects include Tandar, a heavy ions accelerator now being completed; a high-pressure-and-temperature circuit to test nuclear reactors, completed in 1982; the development of fuel rods; a nuclear fusion program that studies the heating of plasma by the Theta Pinch system; and research on fast breeder reactors and the thorium cycle.

The Domestic Nuclear Power Industry

The Argentine nuclear power industry has developed by design rather than through market forces. Domestic industry helps to mine uranium, produce yellow-cake and fuel elements, provide sophisticated

Table 15. Argentina. Nuclear Research Reactors, 1958–1982

Name	Location	Year	Maximum Thermal Power	Fuel	Moderator	Objective	Builder
RA-0	Constituyentes Atomic Center (later University of Córdoba)	1958	1 w	U-235 20%	H_2O	Education	CNEA
RA-1	Constituyentes	1958	150 kw	U-235 20%	H_2O	Research, production	CNEA
RA-2	Constituyentes	1966	10 w	U-235 90%	H_2O	Research	CNEA
RA-3	Ezeiza Atomic Center	1967	8 Mw	U-235 90%	H_2O	Research, production	CNEA
RA-4	University of Rosario	1971	0.1 w	U-235 19.9%	Polyethylene	Education	Siemens AG
RA-5	Ezeiza	[a]	100 w	Thermic rapid	[a]	Research	CNEA
RA-6	Balseiro Institute	1982	500 kw	U-235 20%	H_2O	Research, education	CNEA

Source: CNEA.

[a]Information unavailable.

inputs for the construction of nuclear power plants, and make nuclear instrumentation and components.

Basic metal has been one of the most active industries; its gross production between 1950 and 1980 expanded at an annual average rate of 7.3 percent, well over the manufacturing industry's 4 percent average and the GNP's 3.3 percent. During the same years, its share of the industrial GDP grew from 2.3 percent to 6 percent, and of the GNP from 0.6 percent to 2 percent.[21]

This dynamism was due partly to support from the CNEA's Technical Assistance Service for Industry (SATI), established in 1962 as a department of the Metallurgy Division to provide industry with technical assistance and technology transfer and consulting services and to study the national potential to build nuclear reactors. It designed R & D projects to help solve some of the problems it found inherent in domestic industry; it acted as a sort of clearing house for information, advising customers where and by whom their problems could be best solved; it developed new products and processes; and it trained personnel. Sabato called it a "window toward reality."[22] SATI actively participated in CNEA's 1965 feasibility study that led to the construction of Atucha I. The relatively high domestic integration achieved in Atucha I and Embalse and planned for Atucha II is proof of the industry's positive response to CNEA activities and to its related institutions.

Important local companies have been established by or with the participation of CNEA. Enace SA, a joint venture of the CNEA (75 percent) and KWU, was created in 1980 as an engineering and nuclear plant production enterprise, with Atucha II its first project. It also provides a mechanism for the transfer of technology from KWU to Argentina.[23] Nuclar Mendoza SA, a joint venture of the CNEA (51 percent) and the state of Mendoza, is a consortium of five domestic enterprises that has built for Embalse steam generators, pressurizers, reactor cooling systems, turbogenerators, and more. It also took part in the construction of Atucha I and Argentina's reprocessing plant.[24] Investigaciones Aplicadas (INVAP) was heavily involved in Argentina's uranium enrichment project and has technical know-how for building nuclear electrical reactors. Other major nuclear companies are the previously mentioned Conuar SA and Sanchez Granel; Industrias Metálicas Pescamona, which produces capital goods for nuclear power plants and a fuel elements plant; and Argatom, which performs electromechanical work for nuclear power plants. Over 20 private companies are directly involved in the engineering, architecture, and construction of power plants and nuclear components. In all, about 60 domestic firms provide products and services to the nuclear field.[25]

History, Policy, and Objectives

Argentina's objectives for developing nuclear power have been remarkably consistent during the last 35 years, most notably the desire to achieve nuclear technological autonomy. Also, in spite of denials to the contrary, one may assume that the military has had a strategic option in mind. During the 1950s another nationalist objective was added to that of domestic nuclear technological development: domestic industrial development, in particular, metalworking. Once the nuclear program showed signs of success it was turned into an energy program, and construction of nuclear power plants began.

After the bombing of Hiroshima and Nagasaki, Argentina stated its intentions with regard to nuclear energy by proclaiming uranium resources to be a national interest. On May 31, 1950, the CNEA was created and authorized to coordinate and promote atomic research and to propose to the executive power measures by which to pursue the use of atomic energy in various economic fields and to protect the population against radiation. The subsequent history of nuclear power development has been strongly linked to that of the atomic energy authority.[26]

The early years of atomic development in Argentina were dominated by odd but politically relevant events. As World War II came to an end, many Nazis escaping Europe found refuge in Argentina, among them scientists who had worked on nuclear energy and related technologies. One of these scientists was Ronald Richter, an Austrian nuclear physicist who turned Argentina's nuclear program into a world affair. In 1949 Perón appointed him to establish a laboratory to research and develop nuclear power. At a press conference in March 1951 Perón told the world that the laboratory had successfully developed nuclear fusion, promising that this would "put a sun in every Argentine home."[27] Richter, who was present, said, "I control the explosion, I make it increase or diminish at my desire."[28]

The whole affair turned out to be a fiasco, and a government investigating committee described it as a fraud. In 1952 Richter was dismissed, the laboratory dismantled, and atomic energy became totally centralized in the CNEA. That same year navy captain Pedro E. Iraolagoitia succeeded the CNEA's first president, army colonel E. P. González, establishing a political tradition, maintained until December 1983, of CNEA rule by the navy.

Iraolagoitia concentrated on the long-range goal of developing human resources. Laboratories of nuclear physics, radiochemistry, analytical and uranium chemistry, electronics, biology nuclear medicine, cosmic radiation, isotope separation, and heavy water were created, and in 1953 the first theoretical course on nuclear reactors was initiated. Iraolagoitia's tenure ended along with Perón's in 1955; the new CNEA president, Adm. Dr. Oscar A. Quihillalt, ruled for over eighteen years and saw eight Argentine presidents come and go. Under his guidance the CNEA and the atomic program grew into the most successful institution and national program Argentina ever had: objectives were formulated and partially achieved, the science and technology infrastructure was built, human resources were nourished, the road toward autonomy was traced, research reactors and other atomic facilities were built, and Atucha I was almost completed.

Two decrees issued in 1956 set the tone for what was to come. The first reorganized the CNEA and gave it broader authority, naming as objectives the promotion of nuclear power, to be applied to scientific and industrial fields in accordance with the public interest. The second decree declared that all raw nuclear material, such as uranium, would become state property and that the CNEA would control its prospecting, production, and commercialization.[29] Regulation of radioisotopes and ionizing radiation became the domain of the CNEA in 1958.[30] A 1960 decree declared the CNEA to be "high national interest" and established it as an independent agency linked directly to the president.

With the help of the U.S. Atoms for Peace program instituted by President Eisenhower in 1953 (designed to control the spread of nuclear weapons technology by providing nuclear know-how to be used for peaceful purposes only), Argentina was able to train two hundred scientists and build its first nuclear research reactor, RA-1. By the beginning of the 1960s the CNEA had developed a science and technology infrastructure, had two working nuclear research reactors, and was ready for larger enterprises. In 1964 it was decided that a nuclear power plant should be built for the Greater Buenos Aires Littoral, and one year later the CNEA resolved that Argentine technicians would conduct the feasibility study for the plant in-house. This decision broke new ground because until then it was common practice to turn to independent foreign firms for such studies, and their recommendations typically required large investment and advanced technological capacity.

The feasibility study concluded that the Greater Buenos Aires Littoral was ready for a nuclear power plant, that the grid was large enough to accommodate a 500 Mwe plant, and that the plant could be operational by 1971. It estimated 40–50 percent domestic industrial participation and decided that the plant would use Argentine uranium reserves and be located at Atucha. As a result of heated discussion within the CNEA on whether to buy a low water reactor (LWR) or a heavy water reactor (HWR), bidding was opened for both types. This strategy, it was thought, would strengthen the CNEA's bargaining position vis-à-vis the competing companies and leave the options open. Bids, 17 in all, were received through July 1967. The most serious contenders were Canada and West Germany offering HWRs, and Great Britain, France, and the United States with LWRs. The final choice was between Canada, with a CANDU, and Siemens AG, with a 335 Mw HWR. The Germans offered a relatively high percentage of national integration (about $25 million worth) and excellent financial conditions, which turned the tide in their favor.

After the Atucha I decision the CNEA issued a nuclear plan for 1967–1977, later extended to 1980, that clearly stated the goal of developing electric power, nuclear fuel, and nuclear technology[31] and contemplated the building of three nuclear power plants, the installation of a nuclear fuel manufacturing facility, and the development of reprocessing, waste disposal, and breeder reactor technologies.

Atucha I was not yet ready when the feasibility study, also in-house, for the second nuclear power station (Embalse) was commissioned. Bids were taken from West Germany, Canada, the United States, Italy, and Japan, and the contract, signed in 1973 after Perón had returned to Argentina, went to the Canadian CANDU. This time the discussions around Embalse and what reactor type to buy transcended the CNEA, and the issue became public and heated. In the corridors of parliament, in newspapers and on television, in the cafés of Buenos Aires, people were taking sides with unusual fervor. LWRs were more efficient and cheaper, but HWRs would avoid dependence on enriched uranium. Even the CNEA was divided on the matter, but enough of its scientists wanted the more independent line of HWRs, as did the Joint Armed Forces Command, to tilt the decision in favor of the CANDU HWR.

With the return of the Peronists to power in 1973, Quihillalt was replaced and Iraolagoitia restored to the CNEA presidency. Although the CNEA was not totally immune from the massive political and economic upheaval of the Peronist years (scientists abandoned the institution in rather large numbers), it proceeded with its work, and in 1974 Atucha I was inaugurated and the Embalse deal ratified. In 1976 a military coup once again produced a change as Rear Adm. Carlos Castro Madero became the CNEA's fourth president. He ratified its commitment to self-sufficiency and, in Decree 3.183 of October 1977, announced new objectives: autonomy in the production of heavy water by 1984, and a third nuclear power plant.

An interministerial commission was established to study the building of the third plant and to set the guidelines for a new nuclear plan. The plan, approved in May 1979, called for the construction of four new HWRs of 600 Mw each, scheduled to start commercial operation in 1987, 1991, 1994–1995, and 1997, respectively; an industrial plant for the production of heavy water; and other necessary installations to close the nuclear fuel cycle. "By laying down long-term policies and commitments," wrote Castro Madero, "the government sought to promote the active participation of Argentine industry in the country's nuclear programme, thereby, it was hoped, creating the desired national capabilities in all aspects of the design, manufacture, and construction of nuclear power plants and associated fuel cycle facilities." [32] The plan also indicated that the decree governing uranium should be changed to allow private companies to participate with the CNEA in mining uranium and manufacturing yellow-cake. The plan authorized the CNEA to close the Atucha II deal, which it did in 1980, choosing KWU over the Canadian CANDU because of the delays and cost increases in the Embalse project and because of the lenient West German conditions in the matter of safeguards.

Three years later Argentines and the world at large had hardly recovered from the news that Argentina had developed an independent nuclear enrichment facility when Raúl Alfonsín became its new president, bringing to an end seven and a half years of military dictatorship. He announced that the CNEA would be even more tightly controlled by the presidency, its reorganization would be studied, and the navy monopoly on its leadership would be ended. Alberto Constantini, a civilian, was appointed to replace Castro Madero. The new administration stressed the peaceful character of the Argentine nuclear program and sent signals suggesting that Argentina might place a greater emphasis on hydroelectric plants and might soften its opposition to the 1968 nonproliferation treaty and the 1967 Tlatelolco Treaty, which established a nuclear-weapon–free zone in Latin America.

When confronted with one of the gravest economic crises in Argentina's history, Alfonsín had no choice but to reduce public expenditures, including the CNEA's budget. Burdened by an external debt of $800 million (1.7 percent of Argentina's total debt in early 1985), the CNEA was forced to cut its 1985 budget by $130 million (23.6 percent); it also had to submit its budget for approval by the energy secretariat, traditionally a hydroelectric power stronghold. Nevertheless, most of the autonomy project's goals remain intact.[33] The project has proven to be too advanced, the technological, industrial, and strategic achievements too numerous, and the national pride payoff too important, for even a deep economic crisis to halt the project entirely. Argentina's nuclear project may be able to traverse the crisis, despite a reduction in its scope and some delays, because it has shown to Argentines and to the world at large that Argentina can produce its own sophisticated technology.

The Autonomy Policy and the CNEA

CNEA's productivity and its success in achieving a substantial degree of nuclear technological autonomy must be partially attributed to its leaders, such as former presidents Quihillalt and Castro Madero, who enjoyed political clout and provided dynamism and—most important of all—continuity. "Quihillalt transformed the Argentine program from a modest research effort into a viable power program,"[34] and Castro Madero turned it into scientific, technological, and industrial reality, achieving partial self-sufficiency in the nuclear fuel cycle besides. Continuity is also reflected in the fact that the CNEA had only four presidents in 33 years.

But the ideology of autonomous technological and industrial development, which Sabato and his colleagues in the Metallurgy Division—the pragmatic antidependency guerrillas—nurtured and transmitted to the entire CNEA, is equally important to an understanding of CNEA's success, having provided permanent guidance for policy choices and implementation. In the CNEA "kingdom," Quihillalt was the king and Sabato, with his guerrillas, was the weathermaker. When Sabato, Wortman, and Gargiulo wrote in the acknowledgments to *Energía atómica e industria nacional,* "Our thanks to . . . Admiral Oscar A. Quihillalt . . . who not only understood our ideas and supported our work, but turned them into reality,"[35] it was not a cursory acknowledgment formula but a statement of crucial, causal fact.

Sabato's pragmatic antidependency guerrillas began to link nuclear energy with technological and industrial dependency in 1955. Assigned by Quihillalt to create the Metallurgy Division, Sabato took a few electromechanical engineers—Mazza, Libanati, Nelly, Ambrosis, and Kittle; a civil engineer—Leyt; a few chemists—Boschi, Coll, Carrea, and Di Primio; and a chemistry student—Aráoz. They defined their objectives as the achievement of nuclear technological and industrial autonomy; the development of a science and technology infrastructure; and the creation of a demonstration effect, showing that indigenous R & D is possible in a dependent (and politically and economically troubled) country before structural economic changes have taken place.[36] Aráoz said, "To work on metallurgy was a way to fight dependency, not in a declamatory, but a concrete way. . . . To expect to be able to make technology only after dependency has been overcome is one of the most dangerous forms of escapism."[37] They decided that Argentina needed a laboratory that would work on all types of metallurgy, not just nuclear, for the benefit of industry in general. Only such a laboratory, Sabato thought, would then be able to take care of all the problems related to nuclear metallurgy.

The guerrillas said to Quihillalt: "We will take it upon ourselves to meet all nuclear metallurgy problems the CNEA entrusts us with. In exchange, we want the most complete liberty of action to select the method we believe will be most convenient to us, even though it might look heterodox to outside observers." The CNEA authorities accepted the proposal, and it was kept throughout metallurgy's history. "Furthermore, they backed it, and it was thanks to this understanding and help that we did what we did."[38]

In order to understand why Quihillalt and other CNEA leaders accepted and tolerated Sabato's *murga* ("group" of popular musicians), as they were called, we must look back at the Richter affair. This fiasco put nuclear energy in the national consciousness and created expectations that, because they were not fulfilled, led to frustration and disappointment. Argentina had to demonstrate that it could and would develop nuclear energy relying on its own human resources.

> [Perón] wanted to appear powerful, not oafish. To show that Argentina was indeed a leader in nuclear technology, Perón ordered the CNEA to hire qualified personnel, regardless of their political creed. The timing was fortuitous, as his purges of other institutions created a large pool of highly qualified unemployed. Largely due to the Richter affair, the Commission became a haven to anti-Peronists, providing a nonpartisan legacy which set the tone for much subsequent development of nuclear energy in Argentina."[39]

Quihillalt thus inherited and strengthened two tendencies: the development of nuclear power as a national project, and the CNEA's nonpartisan character, which attracted the best scientists. The metallurgy "deal" was among these developments. But there was more. Brazil was trying to develop nuclear power itself, and Argentina was not going to let Brazil win the race. The fact that in 1953 Brazil was about to get three centrifuges for uranium enrichment from West Germany was not lost on Argentina.

The first major crossroad was reached with the decision to build Argentina's first nuclear research reactor in-house. Quihillalt had gone to the United States to see about buying a reactor to launch a program of nuclear development when he was approached by an Argentine scientist who told him about the Argonne Laboratory reactor. A consensus developed within the CNEA that it could be built in Argentina. Quihillalt then turned for advice to the Metallurgy Division, which enthusiastically embraced the idea, stating that it could build the fuel elements in spite of the lack of know-how and equipment. Many people, according to Sabato's account, called the decision crazy and irresponsible; but Quihillalt and the guerrillas were not after efficiency; instead, they wanted to learn how to build a reactor by themselves. Quihillalt assumed full responsibility for this enterprise.[40]

The creation of SATI in 1962 was not just another bureaucratic appendix; it came straight from the "triangle" in Sabato's mind. State intervention in the nuclear field was already intense, and the science and technology infrastructure was undergoing rapid development, but no technology was being transferred to domestic industry: the third vertex of Sabato's Triangle was missing. The Metallurgic Industrial Association was not interested in the autonomy idea, however, and the CNEA leadership was hesitant to pay for technology transfer. Fortunately, Sabato found an ally in the industrialist Manuel de Miguel, who offered to finance SATI if the CNEA would also contribute funds. This settled the question. Thus SATI became a key element in the achievement of relatively high, and steadily increasing, levels of domestic integration in the construction of commercial nuclear power plants. Also, by transferring technical knowledge to industry, SATI increased Argentina's bargaining power vis-à-vis industrial giants such as Siemens AG. In other words, SATI turned the guerrillas' goal of opening the technological package into reality.

The decision to carry out Atucha I's feasibility study in-house was another important crossroad. "We did not even know what a feasibility study was," said Sabato, but there was an understanding that it should be done if the CNEA was to learn how to make technology. In fact, very few of the proposals were finally accepted, but this did not bother the guerrillas—they wanted to learn how to do a feasibility study, what domestic industry could contribute, and how to negotiate with potential sellers. As expected, the study called for the active participation of domestic industry, correctly foreseeing that this would be the starting point for a nuclear industry in Argentina.

Local industrial participation was an important factor in the decision to buy Siemens AG's reactor. To implement the agreement with Siemens, SATI organized a committee called National Industries Group (GIN), which was empowered to analyze and evaluate the agreement, examine the capacity of domestic industry to contribute to production of the nuclear plant, and ensure that it would be adequately represented. Thereafter a close relationship between GIN and the metalworking industry developed.[41]

GIN emphasized those items that required domestic industry to make significant improvements, thus stimulating it to become suppliers of nuclear power plant materials and components. The CNEA won an important bargaining battle when it got Siemens to accept responsibility for the items built in Argentina, which at the same time imposed on local industry the highest of standards. It was agreed that there would be three supplier lists: positive, probable, and negative. The positive list included 88 items, all to be manufactured by local industry. There was to be no bargaining about this list and exceptions could be made only with CNEA approval. The probable list included all those items the CNEA could buy from domestic sources if they met the desired standards and prices. This list was to be bargained on item by item, and the guerrillas wanted local industry to take a big chunk of it. The negative list included those items to be supplied by foreign sources. The positive and the probable list items together made up the 33-percent-domestic share of Atucha I's value.

When Quihillalt and Sabato left the CNEA (Quihillalt went to Nuclar, Sabato went first to SEGBA, later to Bariloche, and then was ostracized because of his opposition to the military government), the argument about whether to buy LWRs or HWRs had already been settled in favor of the latter, and the autonomy ideology and policy were strongly rooted in the consciousness of the CNEA's personnel, the political leadership, and the Argentine people.

Not even the ideological polarization under the Peronist government or the subsequent military dictatorship was able to stop the CNEA's work, let alone its ideology of autonomy. There were some casualties under the two regimes: a slow-down under Iraolagoitia, and persecution of scientists under the military regime. But work continued, and Castro Madero maintained the policy almost unaltered. (We must recall that Castro Madero served in the cabinet of a government that decided to free the economy and to throw out the window all ideas of import substitution and self-reliance. As one actor put it, "Outside the CNEA Madero might have been a liberal, but inside he was a strong supporter of the autonomy policy.")

As important as this ideology and the guerrillas were to CNEA's success, the story does not end with them. We still need to know how and why the guerrillas were successful in gaining political support to implement their plans; how and why the policy continued almost unchanged after Quihillalt and Sabato left the scene; and what other factors have been crucial in the development of Argentina's nuclear power and industry.

The foremost factor was institutional. Through its educational, research, technological, and industrial successes, the CNEA acquired clout and legitimacy, an almost free hand in nuclear affairs, larger budgets for its enterprises, and political advantages vis-à-vis the hydroelectric lobby, which opposed nuclear energy development and wanted hydroelectric plants to be built instead.

The ideology's broad appeal, then, tied in with the CNEA's success. Whereas for the pragmatic antidependency guerrillas nuclear autonomy meant the reduction of economic dependency and the development of domestic industry, for at least some policymakers it meant beating Brazil in the race for nuclear power development and showing the United States that Argentina could and would express its self-determination regarding its nuclear development policy, including nuclear weapons.

Almost all of the ruling elites, except the hydroelectric lobby, saw in the development of an additional source of electricity a bonus attached to higher national goals. Argentina had already flexed its muscle against foreign oil companies in the 1960s when it determined to become energy self-sufficient, which two decades later it has almost done. But amid concern about the need to diversify and create new energy sources (in December 1978 oil amounted to 13.2 percent of total domestic energy resources but 62.7 percent of energy consumed), with an average increase in electricity consumption of 8.3 percent between 1962 and 1977 and abundant uranium resources (assessed in 1978 at approximately 11 percent of total energy resources), and the motivation to pursue a nuclear program for technological, industrial, and strategic reasons, nuclear energy was seen as the most natural resource to be developed. It was estimated that by 1995 nuclear energy would represent 10 percent of total installed megawatts.[42]

The nuclear program also appealed to the Argentine masses, looking for one success among so many political and economic failures. Vindicating the saying that "nothing succeeds like success," when the program began to yield results it was embraced as a national cause, a source of pride. It is symptomatic that in Argentina there have been no "greens," anti-nuke movements, Pugwash-oriented physicists, and no fears of Three-Mile Islands or the "day after."

This consensus allowed the CNEA to insulate itself from political and economic ideological contests between right and left, Peronism and liberalism, the civilians and the military: the nuclear autonomy ideology and policy made some sense to everyone. In an embattled and tempestuous sea, the CNEA became a sheltered island. "The continued progress of nuclear power in Argentina," wrote Douglas Tweedale,

> has never been interrupted by bureaucratic infighting and indecision or by intragovernmental rivalries, which seem to be so common in nations struggling with the problem of how best to allocate scarce resources. Despite having over fourteen changes in government since 1950, and despite numerous economic crises that have at times threatened to bankrupt the nation, the desirability of nuclear energy has never been seriously challenged. In fact nuclear energy is one of the few issues on which most impor-

tant political power groups agree . . . popular discussion of nuclear energy has been limited by public ignorance of the subject
and, in recent years, by the harsh restrictions on political expression imposed by an authoritarian government.[43]

Growing government support of the nuclear program can be seen
in the CNEA budget between 1950 and 1982 (Figure 10) and the
percentage of Finalidad 8 allocated to it between 1972 and 1981 (Figure 11). Large sums of money—hundreds of millions of dollars—
began to be channeled into the CNEA after 1976, at a time when the
Argentine economic situation was worsening, and by 1982 the budget
was approximately $750 million. Only twice after 1972 did the CNEA's
share of Finalidad 8 fall below 5 percent, and after 1977 it rose steadily, to 23 percent in 1981.[44] However, the CNEA's 1985 budget shows a
significant reduction (see p. 294), largely attributable to the continued
gravity of the economic crisis. An outcome of this crisis may indeed be
a decrease in CNEA's insulation and in its ability to withstand political
pressure.

Another key institutional factor that helps to explain the CNEA's
success is centralization. In direct contrast to Brazil, where the proliferation of nuclear political and research institutions caused much

Figure 10.   Argentina. CNEA Budget, 1950–1982

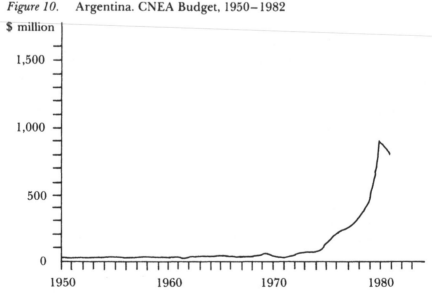

Source: CNEA, Transcript of Rear Adm. Carlos Castro Madero's meeting with the
press, 1982.

conflict, in Argentina the CNEA was in charge of policy making, R & D, human resource training, the building of nuclear power stations and other facilities, and the management of industrial-scientific relations. This centralization allowed the guerrillas to increase their impact on all aspects of policy and action.

Also significant was the fact that policy was not imposed on the CNEA from above but was developed at the CNEA and then sold to the highest political levels. This process, which remained intact throughout military and civilian governments alike, allowed the CNEA to develop a consistent policy without being much affected by the changing moods and ideologies of the many presidents and their advisers, the pressure of interest groups, or the caprice of military officers and factions. This contributed further to the ideological cohesiveness and consensus within the CNEA, enabling it to stick to its technological, industrial, and energy objectives and maintain the civilian nature of the nuclear program.

All this should not, however, be taken as proof that there has not

Figure 11.   Argentina. Finalidad 8 Allocations to the CNEA, 1972–1981

Sources: CONICET, *Programa de centros regionales de investigación científica y tecnológica* (Buenos Aires, Sept. 1977), vol. 2, chap. 2, 48; and Lic. María Lujan Marcon (CONICET).

been or will not be a parallel military program of nuclear development. It is well known that Argentina's nuclear potential has made the atomic bomb possible, as Castro Madero himself acknowledged,[45] that Argentina's strategic policy is one of leaving options open, and that there are political and military groups that would welcome the military option.[46] They see the civilian program as having provided a cover as well as the know-how and facilities necessary to build the bomb. Furthermore, it is clear that Argentina would not sit idle were Brazil to announce its intention to produce nuclear weapons. Finally, it is troubling that Castro Madero announced the uranium enrichment facility only three days after the death of Sabato, the moral force behind the civilian industrial program.

Clearly, the success of Argentina's nuclear program was due to its civilian character, which fostered a sense of solidarity and purpose among the CNEA scientists. Whoever planned the sensitive reprocessing and fuel fabrication facilities only a few thousand feet from where Pan Am and American Airlines planes daily take off and land at Ezeiza International Airport probably did not have a strategic operation in mind (or else showed astonishing naiveté).[47] Therefore, if Argentina did decide to make the bomb, it would kill the goose with the golden eggs—the ideological glue that holds the CNEA, one of its most successful institutions, together would dissolve. As scientists deserted the CNEA it would start to look like most other Argentine institutions: afflicted or even paralyzed by ideological disputes, it would no longer be nonpartisan, and policy would be imposed upon it rather than developed by it.

The Sabato Legacy

Any attempt to use the Sabato Triangle to analyze the interaction between the Argentine state, the CNEA, the Metallurgy Division, SATI, and local industry would be fruitless. The Triangle was a product not only of theory and imagination but also of personal experience acquired within the CNEA. Thus Sabato's ideology of autonomous development generated a model that worked for nuclear power in Argentina, and its success generated theories and models that their authors declared could be applied to all of Latin America. Thus when Sabato, Aráoz, Vidal, and others left the CNEA, they set themselves the goal of steering science and technology policy in all Latin America.

Indeed, the Sabato Triangle intellectually affected the Andean Pact foreign investment and science and technology legislation, Pelúcio's technological development policy, and Mexico's first science and technology plan. The CNEA's story, then, is not just another example of pragmatic antidependency and one of its intellectual guerrillas; it is the story of *the* pragmatic antidependency guerrillas par excellence, and a reason for the existence of such guerrillas in other countries, other sectors, other times. It was to the disappointment of Sabato, and the detriment of his own Argentina, that the message was never systematically applied beyond the gates of the CNEA.

Brazil: Less Than Success

The Nuclear Fuel Cycle

Uranium and thorium.   During the 1950s and 1960s Brazil, France, and the United States cooperated in the mapping and discovery of a limited amount of radioactive resources, but the most significant discoveries were made after 1977. Uranium oxide reserves grew from 11,040 metric tons in 1974 to 236,300 in 1980 and 301,490 in 1982, an increase of nearly 3,000 percent.[48] However, opponents to Brazil's nuclear policy have argued that these figures are exaggerated.

Uranium has been discovered in the states of Paraná, Goias, Ceará, and Bahia, the latter two reportedly possessing 75 percent of the 1981 known reserves. Although Nuclebrás controls a large majority of the uranium reserves (which until 1974 were under CNEN jurisdiction), Nuclam (Nuclebrás Auxiliar de Mineração SA), a joint venture of Nuclebrás (51 percent) with Urangesellschaft of West Germany, has prospected about 10,000 metric tons, which is transferred to and then commercialized by Nuclebrás. (Under Brazilian law all minerals with radioactive content mined in Brazil must be handed over to the government.) West Germany has the option to buy up to 20 percent of the uranium prospected by Nuclam.[49] Brazil's only working uranium mine is in the state of Minas Gerais at Poços de Caldas, where an estimated 10 percent of the overall 1981 reserves seem to be located. Five hundred fifty tons of yellow-cake have been produced there, using an open-pit mine mill and concentrator that cost an estimated $255.6 million. The project was designed and undertaken by Uranium Pechiney Ugine Kuhlmann of France, but some Brazilian firms have been involved in the civil engineering processes.[50]

Thorium reserves, located mainly near Rio de Janeiro, Espíritu Santo, and Bahia, were estimated by Nuclebrás at 43,883 tons in 1979.[51] Monazite, the mineral from which the thorium is extracted, used to be produced and exported without restrictions—the United States bought 18,000 tons between 1941 and 1949—but in the early 1950s its export began to be restricted, and early in the 1960s its mining and research were put under state control. In later years mining was opened up somewhat, with the condition that the radioactive content be surrendered to the government, but only one private company, Orquima, began production of monazite in Brazil. Then, after several changes of ownership, name, and structure, it became Nuclemón (Nuclebrás de Monazita e Associadas Ltda.), a state company holding the thorium production monopoly. In 1980 Nuclemón extracted 2,009 tons of monazite.

Conversion, enrichment, and fuel elements.   In its quest for self-sufficiency in the nuclear fuel cycle, Brazil has acquired capabilities in yellow-cake manufacture and fuel element fabrication, but not in the crucial stages of yellow-cake to uranium hexafluoride (UF_6) conversion and uranium enrichment. These processes are all planned to be performed at a new industrial complex at Rezende, west of Rio de Janeiro, however.

Conversion entails costly and highly sophisticated technology. A plant to produce UF_6 was bought from Uranium Pechiney Ugine Kuhlmann in 1980 at an estimated cost of $69 million; when ready, it will be able to produce 2,000 tons a year.[52] Because of tight safeguards requested by the French, Brazil also decided to develop its own conversion plant at the São Paulo Energy and Nuclear Research Institute (IPEN).[53]

The United States had signed an agreement with Brazil to supply, up to the year 2000, *enriched uranium* for its nuclear research reactors and for the power reactor (Angra I) purchased from Westinghouse, but when Congress in 1978 prohibited the sale of enrichment services to countries having nuclear programs but not applying full-scope safeguards, the supply to Brazil was cut off. Since then Brazil has bought enrichment services from Urenco, the British–West German–Netherland uranium enrichment joint venture. The United States opposed this deal but withdrew its opposition in 1981.

As part of a 1975 agreement with West Germany, Brazil also purchased its own uranium enrichment technology. First choice was the centrifuge system used by Urenco, but West Germany's partners opposed the transfer of this sensitive technology. So Brazil had to settle for "jet-nozzle," or Becker, technology (named for its inventor), similar to the centrifuge but requiring larger amounts of electricity, and at that time still untested at the industrial level. Gaseous diffusion, the third enrichment technology, which the Americans use, was unavailable for purchase by Brazil.

To develop jet-nozzle technology in Brazil, Nuclebrás set up two companies: Nuclei, a joint venture of Nuclebrás (75 percent) and Steag and Interatom of West Germany (15 percent and 10 percent, respectively), responsible for the design, construction, and operation of the pilot plant facility; and Nustep, a joint venture of Nuclebrás and Steag (50 percent each) and located in West Germany, responsible for the commercial development, industrialization, and exploitation of the jet-nozzle technology. Thirty Interatom technicians have worked on the assembly of the first 24 modules of the gas-separation process. A larger plant of 288 modules, scheduled to be already in operation, will not be ready until 1989 at the earliest; its cost is estimated to be $404 million, although some sources have argued that it will end up being much higher.[54] A tripling of capital investment, together with delays and doubts about the commercial potential of the jet-nozzle technology, has led Brazil to also try to develop a domestic reprocessing capability. IPEN, with the help of the University of Campinas, the Aerospace Technical Center, and the CNEN, has been working on laser nuclear enrichment technology. A research team at the Federal University of Rio de Janeiro (UFRJ) is also conducting research on the centrifuge system.

A *fuel element* fabrication facility, completed at the end of 1982, is standing almost idle, waiting for the other front nuclear-fuel-cycle stages to be ready. The plant has an initial capacity of 140 fuel elements, each carrying 1,100 pounds of uranium a year. Nuclebrás contracted a Kraftwerk Union (KWU) subsidiary, RBU, to design and partially build the plant, at an estimated cost of $63.8 million, although a few Brazilian engineering firms are also participating in the project. Technology transfer was scheduled to last until 1984, but technical assistance will be provided until 1991.[55]

Nuclear power plants and equipment.   As of April 1984, Brazil still had no nuclear power plant working at full capacity. Its first plant, bought in 1971 and located at Angra dos Reis, midway between Rio de Ja-

neiro and São Paulo, was scheduled to start operation in 1977 but became critical only in 1982, costing $1.4 billion, five times more than
originally planned.[56] Shortly after it went on-line, Angra I had to be
shut down for repairs when vibrations were discovered in the generator, and since then it has been working at less than full capacity. Subsequent problems such as corrosion have further impaired an already
troubled plant.

As part of the 1975 agreement with West Germany, Brazil bought
two pressurized water reactors, also to be placed at Angra dos Reis
(now called the Álvaro Alberto Complex), and six more nuclear power
plants were planned for completion by 1990, the first two destined for
the São Paulo area, at Iguapé. Angra II and III, whose reactors and
generators were supplied by KWU, were supposed to be ready in 1982
and 1984, respectively. But serious drainage and stability problems,
already a cause of delays at Angra I, also slowed down work at II and
III. At the end of 1981, after three years spent sinking extra foundation pilings, construction of Angra II was only 10 percent complete,
and Angra III only 1 percent;[57] they are not expected to be ready before 1990 and 1991, respectively. In 1983 President Figueiredo decided to delay the construction of Angra II and III for one year and
to suspend for an undetermined time the construction of Iguapé I
and II. Although the original cost estimates for Angra II and III were
$304 and $288 million, recent estimates were as high as $3.5 billion
each. If the original formulation to build eight plants had been followed, cost of the reactors would have amounted to $28 billion.[58]

Three companies were established to construct nuclear plants and
components. Nuclen (Nuclebrás Engenharía SA), a joint venture of
Nuclebrás (75 percent) and KWU, was created to provide engineering
services and to absorb engineering technology for the plants. Technical responsibility was placed in German hands until 1985, and a
small part of its personnel is German. Nuclep (Nuclebrás Equipamen-
tos Pesados SA), a joint venture of Nuclebrás (75 percent) and a European consortium, was set up to manufacture nuclear heavy equipment, mainly steel vessels and turbo generators. With an investment
of $256 million up to 1983, Nuclep built at Itaguí the largest plant of
its kind in the world. It was intended to manufacture one set of reactor components per year, but production uncertainties, economic
costs, and the program's relative failure kept the factory working at 40
percent capacity on such orders as Atucha II's pressure vessel. Nucon,
a Nuclebrás subsidiary, was created in 1980 to carry out the civil works

in nuclear power plant construction. Until then, construction had been in the hands of Furnas, the electrical utility company.

Reprocessing.   Success has also eluded the reprocessing stage of the nuclear fuel cycle. As part of the agreement with West Germany, Nuclebrás signed a deal with KEWA and UHDE, two West German companies, for the transfer of know-how, training, consulting, and technical assistance on nuclear fuel-reprocessing technology and the building of a reprocessing plant. Scheduled to be completed no sooner than 1989, the plant will reprocess plutonium using a procedure called Purex,[59] in which Brazilians are now receiving technical training in Germany. It was also reported that IPEN has built a reprocessing pilot plant that can handle 5 kg of plutonium a year; supposedly ready to operate "cold," after the transfer from West Germany of a radioactive measuring laboratory it could also perform "hot" tests.[60]

Waste.   In accord with a waste disposal training and cooperation agreement with West Germany, Brazilian technicians were trained at the Karlsruhe nuclear center.

Physics, Human Resources, and Research and Development

"Despite a long backward tradition in general education and a slow start in science, Brazil produced in a generation Latin America's largest national grouping of research physicists, including several who attained world renown in atomic fields."[61] The origins of this achievement can be traced to the creation in 1939 of São Paulo University's School of Philosophy, Sciences, and Letters. Gleb Wataghin set up a physics department and recruited foreign scientists to train Brazilian physicists. Some of Brazil's most accomplished scientists were trained in this program. The school branched into theoretical physics, in particular quantum mechanics, and into experimental physics, which was the domain of two groups: one under the direction of Marcelo Dami de Souza Santos, who worked with a Betatron (electron accelerator), and another under Oscar Sala, who used an electrostatic generator.[62]

Rio de Janeiro did not remain behind. A physics laboratory, created in 1939 at the University of Brazil (now UFRJ), became the source of many esteemed studies on the theory of nuclear forces. In 1949 a Center for Physics Research was established under the direction of

César Lattes, who became known worldwide for his work on the méson-pi. Some of Brazil's top physicists worked at the center, such as J. Leite Lopes, and the most important foreign physicists of the time made frequent visits. Other physics institutes were established at Rio Grande do Sul, Pernambuco, and the Pontifical Catholic University in Rio de Janeiro.[63]

While the National Council of Scientific and Technological Development (CNPq) has helped universities and institutes with grants and fellowships to develop theoretical and experimental physics, the National Nuclear Energy Commission (CNEN) has been at the top of nuclear R & D since 1956, in partnership with Nuclebrás after its creation in 1975. Since then the CNEN has been in charge of basic and applied research, reactor technology, fuel, instruments and control, information, and the environment. It is also the parent institution for IPEN, the Radioprotection and Dosimetry Institute (IRD), and the Nuclear Information Center (CIN). Nuclebrás has overseen technological training and development related to the implementation of the 1975 agreement with West Germany, which includes the production of reactors, nuclear fuel cycle technologies, and plants and the training of human resources to handle the technology. Nuclebrás also manages the Center for the Development of Nuclear Technology (CDTN), formerly the Radioactive Research Institute, and the Nuclear Engineering Institute (IEN).

IPEN was known for many years as the Atomic Energy Institute. Created in 1956 as a São Paulo autarky, it became one of Brazil's most advanced nuclear research centers. With over one thousand personnel, IPEN is currently upgrading Brazil's first nuclear research reactor and working on fuel element technology and nuclear metallurgy, uranium and thorium purification, uranium tetrafluoride and hexafluoride, and on projects aimed at enriching and reprocessing nuclear fuel. It is also involved in nuclear training, granting both M.S. and Ph.D. degrees.[64]

The IEN, created in 1963 within UFRJ, employs 290 individuals in R & D on fast breeders and on reactor chemistry and engineering. The institute has a sodium thermal loop that simulates a fast-breeder reactor system, an Argonaut research reactor, a neutron generator, and several laboratories. With the help of France, Brazil is trying to develop a fast-breeder research reactor called Cobra, similar to the French Rhapsode and Phoenix breeders, and Italy is cooperating with the IEN in the field of liquid sodium research.[65]

The CDTN, created in 1953 at the Engineering School of Belo Horizonte, employs five hundred individuals. It has a research reactor, an experimental circuit, pilot plants for enriching uranium and treating minerals, and several laboratories and is currently working on reactor and fuel technology.[66]

Nuclear research also takes place at the Aerospace Technical Center, which works on classified programs of a military nature as well as on centrifuge and laser uranium-enrichment and other nuclear-fuel-cycle technologies; the Military Engineering Institute (IME), which set up the first graduate program on nuclear engineering in the late 1950s and created a research group in heavy water technology; and COPPE (Coordination of Graduate Programs in Engineering; see Chapters 7 and 8), which actively works on nuclear fuel R & D and fast breeders.

Research reactors are a major tool of nuclear research and training. Brazil's first nuclear research reactor, acquired as part of the U.S. Atoms for Peace program initiated in 1953, was designed by Babcock and Wilcox and is located at IPEN. It became operational in 1957, has a power output of 2 Mw, and is upgradable to 10 Mw. The CDTN has both a Triga-MK-1, designed and built by General Dynamics with a power output of 140 kw, and a subcritical CAPITO, designed and built in Brazil. IEN has an Argonaut zero-power reactor, redesigned and built almost in its entirety in Brazil. Two other subcritical reactors are located at São José dos Campos (donated by the U.S. Atomic Energy Commission) and at the University of Pernambuco.

According to Hervásio Guimarães de Carvalho, former CNEN president, the educational system was not ready to take the challenge of the 1975 Brazilian–West German agreement,[67] which required more than 10,000 professionals to absorb and adapt the technology transferred to Brazil. To develop human resources for the nuclear program, Nuclebrás established the Human Resources Program for the Nuclear Sector (PRONUCLEAR), as a result of which over 2,400 engineers and technicians have been trained—1,929 in Brazil and 530 abroad (see Table 16).

The Domestic Nuclear Power Industry

Angra I was built with a minimum of domestic industrial input. Purchased as a turnkey plant, it had a nationalization index of 8 percent,

Table 16. Brazil. Education and Training of Personnel for the Implementation of the Brazilian–West German Agreement, by Area of Specialization

	1974	1975	1976	1977	1978	1979	1980	1981	Total
A. *Brazil*									
Specialization course on nuclear technology	—	71	76	73	107	71	66	49	513
Quality assurance	135	57	—	95	—	—	19	64	370
Training in industry	14	8	54	33	14	128	341	269	861
Prospection and mineral research	—	—	28	33	42	51	28	3	185
Total	149	136	158	234	163	250	454	385	1,929
B. *Abroad* (Missions)									
Reactor engineering	3	10	17	26	51	23	26	15	171
Heavy components	—	—	—	—	41	5	4	1	51
Fuel element fabrication	4	10	3	3	10	28	21	1	80
Enrichment	—	4	—	3	11	9	7	8	42
Reprocessing	3	2	2	2	9	—	1	1	20
Radiological protection and safety	1	1	4	3	3	1	—	3	16
Research and development	15	9	4	3	9	9	4	3	56
Prospection and mineral research	—	1	7	18	49	15	2	2	94
Total	26	37	37	58	183	90	65	34	530

Source: Ronaldo Arthur Cruz Fabrico, "Brazil's General Experience in the Transfer of Nuclear Technology," *Transactions,* Second International Conference on Nuclear Technology Transfer (ICONTT-II), Buenos Aires, Argentina, Nov. 1–5, 1982 (La Grange Park, Ill.: American Nuclear Society, 1982), 8.

and only in civil works.[68] In 1973 Brazil asked the Bechtel Overseas Corporation to make a study of the potential contributions domestic industry could make to the nuclear program. Bechtel, in association with a local firm, reached the conclusion that between 51 and 54 percent of the material, equipment, and services could be domestically supplied, and that with special improvements this index could reach 61–64 percent in 1975 and 1977 and 66–70 percent in 1980 and 1982. The report also listed 34 major enterprises that could come to dominate the domestic nuclear power industry. This information came too late to affect Angra I (studies conducted prior to the acquisition of Angra I had reached similar conclusions but were disregarded), but it helped set the targets for the domestic integration indices of the eight nuclear power stations planned under the Brazilian–West German treaty.

The domestic integration index of Angra II and III (the only nuclear plants that today have some realistic chance of becoming operative in the foreseeable future) was set at 30 percent. However, high integration levels were projected only for "low-technology" items, such as electrical equipment, air conditioning, and steel structures; "high-technology" items, such as turbogenerators, instrumentation and controls, heavy components, and special reactor components, had very low nationalization indices or none at all.[69]

With the creation of Nuclep and its factory, national industry— headed by the Brazilian Association of Capital Goods Industries (ABDIB)—complained that Nuclep was competing with domestic industry and that the government was not using it to its full capabilities. The *1982 Senate Nuclear Inquiry Report* concluded that "the policy adopted was not, in fact, the best for the country."[70]

History, Policy, and Objectives

Nuclear power in Brazil started as a nationalist response to the Brazilian Foreign Office's (Itamaraty) open policy of radioactive-mineral exports during the 1940s. It was fueled by the deep impression the enormous power manifested at Hiroshima and Nagasaki made on Brazilian politicians and scientists and by the fact that Argentina was beginning to move toward nuclear power. It was enabled by early Brazilian advances in nuclear physics, justified by the international prestige of Brazilian physicists like César Lattes, and turned into reality through the tenacity of Álvaro Alberto. Setting up CNPq with nuclear development in mind, Alberto instituted the policy of "specific compensation," demanding from the United States technological compensation for the export of radioactive materials, and as early as 1947 decided to develop a nuclear program aimed at self-sufficiency. Itamaraty was fiercely opposed to the nuclear program, but Alberto found support for it within the armed forces, especially the National Security Council, and in 1949 President Dutra instructed him to implement the plan. The outcome was the creation of CNPq and a policy oriented to the achievement of nuclear technological autonomy.

Brazil's nuclear policy history can be broken into five periods: Romantic Nationalism (1950–1955), Indecision (1956–1968), Crossroads (1969–1975), Nuclear Dream (1975–1981), and Return from Illusion (1981–).

Romantic nationalism.   The main objectives of this period were strategic and nuclear technological development, trailed by science and technology development. Law 1310, passed in January 1951, became the most important nuclear policy instrument of this period. It placed all nuclear activities in the hands of the state, entrusted the president with nuclear policy, established CNPq to exercise his authority on nuclear issues, and gave it the task of taking all research and industrial measures necessary to develop and use nuclear power. CNPq thus became responsible for prospecting and researching uranium, developing and producing nuclear power, training technical and scientific personnel, and carrying out R & D. The law also gave the state a monopoly over the export of radioactive minerals, and another legal measure classified radioactive mineral deposits as national reserves. A year later CNPq issued Memorandum 32 stating its nationalist policy, which the National Security Council and President Vargas approved. Alberto also applied his specific compensation principle: in the exportation of nuclear minerals Brazil should not be guided solely by commercial considerations but should ask in return for training, technology, and hardware aimed at developing nuclear power.[71]

To implement this policy Alberto went to France and Germany looking for partners. In Germany he met some scientists who had developed a centrifuge method of uranium enrichment, and a secret deal was made to have three centrifuges built and shipped to Brazil. Somehow this became known to the Americans, who prevented their shipment. When the centrifuges finally reached Brazil in 1957 they were considered of little value.

Meanwhile, Itamaraty strongly opposed CNPq's and Alberto's policies. In 1952 it co-opted CNPq's control of monazite exports and approved a new deal with the United States, over the opposition of CNPq and the National Security Council. In response, the council issued a secret memorandum (771/772) at the end of 1953 calling for application of the specific compensation principle and for collaboration with *all* friendly nations (an anti-American statement). The memo also asked President Vargas to approve the German centrifuge deal and cooperation with Germany.[72]

Itamaraty and the United States answered the challenge by signing in 1954 still another treaty by which Brazilian monazite was traded for American wheat. Also, four secret documents revealing the extent of American pressure were sent by the American embassy in Rio de Janeiro to Gen. Juarez Távora, head of the military cabinet and National Security Council secretary at the time of João Café Filho's presidency. The first and second documents concerned an American proposal for treaties on atomic minerals and the purchase by Brazil of an American nuclear research reactor. The third and fourth were an explicit critique of Alberto's policies and his attempt to buy the centrifuges. These four documents were attached to a new directive issued by the National Security Council that temporarily reversed its nationalist and pro-autonomy policy.[73] The two agreements proposed by the United States were signed in 1955, and Távora forced Alberto to resign, allegedly because of mismanagement of funds.[74]

Indecision.   The nuclear policy objectives did not substantially change during the second period, but Alberto and the motivation to pursue autonomy were gone. Instead, the major driving forces during this period were Argentina's nuclear development, relations with the United States, and domestic conflicts. When Café Filho was replaced by Kubitschek, the National Security Council returned to its traditional pronuclear development policy. Then, at the instigation of a congressional inquiry commission, Kubitschek named a committee to reassess nuclear policy. It advised creation of the CNEN under the president's control, creation of a National Nuclear Energy Fund, and pursuit of a nationalist nuclear policy. It called for tight state control over all nuclear activities and revision of the agreements reached with the United States. A majority of these proposals were accepted and turned into policy.

The CNEN took over the normative and R & D functions of CNPq. Relatively autonomous while under the presidency, it did not remain an independent institution throughout its 30 years of existence. In 1960 it was attached to the newly created ministry of mines and energy (MME), and in 1962 it was made into a federal autarky, a status it held until 1967, when once again it was placed under MME jurisdiction.[75]

When the government began in 1958 to consider building a 150–200 Mw nuclear power plant, using enriched uranium,[76] it created the Superintendency of the Mambucaba Project to implement the plan. A site for the plant was chosen not far from Angra dos Reis, but the project did not survive. In fact, the only concrete achievement during these years was the creation of the CNEN and the Atomic Energy Institute (IEA). President Jânio Quadros, in power for only a few months, expressed a strong interest in building a natural-uranium atomic plant, with a domestic participation index of 80 percent,[77] but the project was curtailed when he left the political scene.

When Goulart became president, new attempts were made to revive the nuclear power station program. Marcelo Dami, a reputed physicist and strong supporter of an independent "Argentina" type of program, became president of the CNEN. A Three-Year Plan issued in 1963 called for gas-cooled and graphite-moderated natural-uranium reactors to be built and for the resulting plutonium to be used for a new second line of reactors working on the plutonium-thorium and uranium 233–thorium cycles. The first plant was to have 300 Mw and start operating in 1969. Another 600 Mw reactor was planned for 1970, and two more were to be ready by 1975. In addition, the plan envisioned the establishment by 1970 of a plant for the manufacture of 150 fuel elements per year and advised building new nuclear research reactors and developing technologies such as heavy water and enrichment.[78] However, Goulart was overthrown soon after he announced the first nuclear power station, and the plan was not pursued by the new government.

The next four years were characterized by little interest or attention to nuclear policy. The military forces who came to power in 1964 were busy gaining legitimacy—overcoming inflation and setting Brazil on the path of sustained economic growth. Nevertheless, one of the most significant chapters in Brazil's nuclear development history did occur during these years. In 1965 a couple of Radioactive Research

Institute engineers created the reputed "Thorium Group," which set for itself the goal of using the abundant thorium reserves for Brazilian-made nuclear reactors working on the thorium fuel cycle. During the next four years the group established two subprojects: Instinto, to assess the viability of the overall project; and Toruna, to develop a prototype of a 520 Mw natural uranium–fueled power station. The Thorium Group developed sophisticated, competitive techniques, designed in-house a reactor prototype, and published many books and articles on the subject. France became greatly interested in the project and signed an agreement with the CNEN for technical cooperation on thorium technology, which boosted the program. The most appealing aspect from Brazil's point of view was that the project could be adapted to both enriched and natural uranium, as well as to plutonium. Because the reactors were to be built in Brazil they would be adapted to domestic industry, which was to participate heavily in the projects implementation. But by 1968 policy was heading in a different direction; the Toruna project was abandoned, the 520 Mw reactor was never built, and the group was finally disbanded.[79]

Crossroads.   During 1967 and 1968 the military government of President Costa e Silva, having achieved a good measure of economic stabilization and expecting rapid economic growth and an increase in electricity consumption (averaging in those years an estimated 14.3 percent),[80] decided to change the content, direction, and pace of nuclear policy. The CNEN was placed under MME's jurisdiction to emphasize the new objective of cheap and efficient energy, and Eletrobrás, the state-owned electric holding company, was given the responsibility of building and operating nuclear power plants. An understanding was reached that a reactor should be bought as soon as possible, and in April 1968 the CNEN, Eletrobrás, and Furnas, the electric power stations company, signed a ten-year agreement with a goal of producing 50,000 Mw of nuclear energy up to the year 2005. It was at this time that LWRs gained favor over Thorium Group initiatives and proposals to buy natural uranium reactors. The second National Development Plan (II PND) explained that the choice was made because more than 85 percent of the existing nuclear power stations in the world were LWRs.[81] During 1970 the site for the 600 Mw LWR to be bought abroad was finally chosen. West German, Swedish, British, and American companies presented bids, and in May 1971 a joint committee of Furnas, Eletrobrás, and the CNEN decided on the

Westinghouse plant. That same year, to fulfill the I PND directive that a national nuclear program be developed, the government established the Brazilian Nuclear Technology Company (CBTN), the forerunner of Nuclebrás.

Nuclear dream.   Energy and speed remained important objectives in the fourth period, plus an economic commercial objective was added and a middle ground was found between dependency and self-reliance: Brazil would rely on West German technology to acquire nuclear independence. But in the background and parallel to these objectives was the military's dream of developing an independent strategic nuclear option. Nevertheless, no matter how dominant the military objective was, with the signing of the Brazil–West Germany agreement (see pp. 281–82) and the beginnings of implementation, Brazil's nuclear program acquired a life of its own.

The seeds for the agreement had been sown during 1969, when Brazil and West Germany signed a science and technology cooperation agreement, setting the stage for the formal collaboration between the CNEN and the German Nuclear Research Center at Jülich that would crystallize in the 1975 nuclear agreement. When the 1974 Brasília Protocol was signed between these two countries, opening the door for the agreement, Brazil's energy situation had become critical because of rapid economic growth and the 1973 oil crisis. Between 1970 and 1974 electricity consumption grew at an average of 12.9 percent, exceeding the 12.3 percent average GDP growth during that period, and oil imports jumped between 1973 and 1975 from 11.5 to 23.5 percent of total imports.[82] Furthermore, in 1974 India exploded its first atomic bomb, and Argentina was on its way to building its second nuclear power plant, while Brazil was still living in the euphoria of the economic miracle, expecting to become a major international political and economic power in the near future. Something had to be done, and it had to be big. The ten-billion-dollar nuclear program that resulted from the agreement with West Germany, it was thought, would solve Brazil's energy problems, bring about nuclear fuel autonomy, create a nuclear plants industry, develop science and technology, provide the strategic option the military wanted, and give a slap in the face to all American governments that had opposed the development of independent nuclear power in Brazil. Too good to be true for "just" $10 billion.

Return from illusion.   By 1981 the foreign debt was pressing hard, technology transfer was occurring slowly, plants were idle, the completion of Angra II and III was not in sight, the jet-nozzle technology had not yielded an industrial method for enriching uranium, and if the nuclear program had been implemented as planned, it would have cost between $30 and $40 billion. This cost would have been prohibitive for Brazil, whose foreign debt had reached the $102 billion mark in 1985, of which $2.1 billion were accounted for by Nuclebrás.[83] Furthermore, hydroelectric reserves had been upgraded by that time to 213,000 Mw,[84] and the technical problem of transporting hydroelectric power from the northeast to the southeast of Brazil had practically been solved, which meant that Brazil could rely on that cheaper and more efficient power to get it to the year 2000 and beyond. However, the program had to continue: Argentina was ahead, the United States still opposed an independent Brazilian program, and the goal of nuclear autonomy remained of prime importance. Once again there had to be a policy change to bring expectations closer to reality.

Since 1981 policy began to move toward a more limited and indigenous program. Work on Angra II and III was delayed, the fate of the other nuclear stations was left for future administrations to decide, and more emphasis was placed on and financial resources funneled to IPEN and the Aerospace Technical Center. Furthermore, in 1983 Paulo Nogueira Batista, Nuclebrás president since its creation and one of the masterminds of the nuclear program and of the Brazil–West Germany agreement, was replaced and Nuclebrás's budget substantially reduced. When Argentina announced the development of an indigenous method for enriching uranium, Brazil doubled its efforts to do likewise. The energy/commercial project having failed, the military objective was, once again, coming out of the closet.

The Road to Autonomy: For and Against

Brazil's road to nuclear technological autonomy had the potential of being quicker and more fruitful than Argentina's. But Brazil's remarkable lack of ideological and political consensus and continuity regarding nuclear policy and the resulting institutional decentralization and conflict led to only sporadic application of policy to the development of domestic technology and industry. Along the way, some choices opened options for achieving autonomy, while others closed them. Political stalemate, inaction, and even mutually negating actions caused precious time and resources to be wasted. Lack of consensus led to inaction, which bred impatience; impatience led to haste and consequently to lack of success; and lack of success brought about more of the same.

In contrast to what happened in Argentina, Brazil's nuclear policy was pushed and pulled by various groups with clashing ideologies. There was no central institution with the political autonomy or leadership necessary to sell a nuclear plan to the ruling elites and ensure that it would be implemented. Thus the nuclear issue became entangled with other political, economic, and social issues that divided Brazil, and American pressure succeeded in manipulating these ideological and political differences to ensure that Brazil would not acquire an independent nuclear capability. When Brazil finally decided to implement a mammoth independent nuclear program, it could not free itself from its political fragmentation; its expectations were grossly exaggerated; and it lacked a solid science and technology basis, attainable only through incremental, systematic, and consistent furtherance of science, technology, and education.

"Brazil's atomic policy," a Brazilian newspaper concluded, "has been characterized by a fundamental mistake that is nobody's because it is everybody's: the existence of many opposing groups that for a long time have prevented a broad and general agreement." These groups and subgroups, "that of the army, of the navy, from southern Brazil, from Minas Gerais, the various São Paulo groups, the leftist group, the sell-out (*entreguista*) group, the northeast group, etc. . . . [are] at the root of all Brazil's basic atomic policy problems."[85] This same concern about divisiveness and its consequences is evident in the words of a prominent industrialist, Alfredo Marques Viana: "The atomic problem is, for Brazil, a political problem . . . [and] a national problem . . . the choice in Brazil between left and right has only helped those who opposed Brazil's nuclear development."[86]

In the period of "Romantic Nationalism," 1950–1955, Álvaro Alberto embraced an egalitarian-nationalist ideology that not only sought technological autonomy but also was strongly anti-American and anti-Argentinian. "CNPq was created by the Brazilian government," said Mário Schemberg, a distinguished Brazilian physicist, "after the scare and consternation [caused by the information that] Argentina had already developed its atomic bomb project."[87] Therefore, two types of nuclear program were pursued simultaneously: one geared toward autonomous nuclear technological development, the other strategic.[88] Never before had the strategic nuclear objective been so apparent and so explicitly stated. For example, the 1956 Congressional Inquiry Commission, looking back at the program's origin, stated: "It is necessary to admit frankly another truth: it is impossible, at the present time, to dissociate economic development from a military capacity, at least a latent one. . . . To exclude a military option, in this field [nuclear] would amount to neither more nor less than inescapably compromising the essence of any nuclear program."[89]

But the National Security Council, CNPq, and the scientists who favored Alberto's nationalist policies found themselves under heavy fire from Itamaraty, which believed that nuclear development should by no means interfere with the attainment of other "more important" political and economic goals. And when Café Filho and Flavio Távora weighted the domestic balance of forces on the side of those opposing Alberto's policies, continuity of the nationalist nuclear program was destroyed. The effects of this blow were felt, as Leite Lopes said, for at least twenty years.[90]

After the mid-1950s, when Vargas and Alberto had left the political scene, the belief that pursuit of nuclear technological development was at the expense of higher economic goals became even more pervasive, and the strategic and nuclear technological autonomy programs became, for all practical purposes, political orphans. The Kubitschek government feared that provocation of the United States, which opposed these programs, could jeopardize the developmentalist drive, which was fueled in part by American investments; the military probably saw the development of nuclear power as less urgent, since Argentina's first nuclear adventure had resulted in a fiasco; the economic elites and technocrats were forcefully pushing industrialization; and in any case, no institution, guerrillas, or other cohesive political force was able to develop a viable, independent nuclear program and mobilize resources for nuclear technological development. "Economic development as a national goal had been far more legitimized and accepted by the mid-1950s than scientific-nuclear development. . . . The economic-export élite appears to have been more unified as to basic goals and means than the scientific-military élite on what could be done and how with nuclear energy."[91]

Thus an early proposal by Leite Lopes to create a national nuclear energy laboratory was rejected, and it was not until the 1960s that the

National Economic Development Bank (BNDE) goals of linking nuclear power to industrial development materialized, when Pelúcio successfully coordinated the technologists' and the economists' ideologies of dependency reduction. Today we take this linkage for granted, but that was not the case in Brazil in the 1950s or in Argentina before Sabato decided to develop domestic industry on the coattails of the nuclear program.

It would be a mistake to conclude that Brazil's early nuclear policy was the victim only of domestic and foreign constraints. Rather, it was the victim also of the ideological conflict inherent in the nuclear development issue itself and of the lack of awareness that nuclear power and industrialization can be linked (as the CNEA had succeeded in doing in Argentina). An independent nuclear program need not have competed with other economic sectors for financial resources; it could have started as a human resources development program, with relatively low budgetary requirements. Likewise, while American opposition, both implicit and explicit, to Brazil's independent nuclear program did generate obstacles, it also fueled the nationalist ideology and the determination to pursue such a program; once that political determination existed, American pressure need not have mattered. In other words, there was a choice—but choices must first be recognized at the level of policy, and then individual and institutional actors have to implement them.

Why were the scientists not more forceful in bringing about nuclear technological autonomy? Because in contrast to Argentina, where scientists were the core of the CNEA, Brazilian scientists were *outside* the policy-making process. A group of pragmatic antidependency guerrillas could therefore not grow and prosper. In Argentina policy was developed within the CNEA and then sold to the political leaders; the scientists thus had a great deal of leverage over policy decisions. In Brazil the National Security Council dominated the implicit policy-making process; policy was developed through unofficial channels behind the facade of institutions such as the CNEN, which advised the National Security Council but did not really dictate policy.

The chances of a successful group of guerrillas developing in Brazil were further reduced by the fact that R & D, pilot plants, nuclear reactor construction, and nuclear policy were scattered among many research institutes in several cities, separated by large distances. Such fragmentation resulted as much from the size and demographic conditions of the country as from internal ideological divisions, and it remained just as relevant during Goulart's administration. By then, though, the increasingly nationalist leanings of the political forces made a technological autonomy type of program more acceptable.

The Brazilian adaptation and construction of the Argonaut research reactor demonstrated some capacity to develop an independent program. In 1962 a Metallurgy Department was created within the CNEN, which itself was enjoying more autonomy. But the 1964 revolution interrupted this process of domestic nuclear development, and its chances of success today are difficult to gauge.

We recall that in 1967 the nuclear issue was taken from the bureaucratic shelves and placed on the highest desk of policy, primarily to hasten achievement of the energy objective. One of the most important choices in the history of Brazil's nuclear development had to be made: to stick to the traditional goal of nuclear technological self-sufficiency by following a pragmatic antidependency program based on natural-uranium reactors and the incremental mastering of the nuclear fuel cycle, or to choose a new goal of cheap and efficient energy production by following a technological laissez-faire program, best served by proven technologies, such as enriched uranium and LWRs, by imported technology, and by inputs from foreign industry.

The entrance of the MME and Eletrobrás into the political game proved very important in the final decision to buy the Westinghouse reactor. These institutions were informed by an ideology of energy and economic efficiency, and the prospects of technological dependency did not concern them. But the same was true for Electrical Services of Greater Buenos Aires (SEGBA) in Argentina, and nevertheless the CNEA got its way. It is again clear that the CNEN, lacking the power to stand firm against ideologies and interests opposed to nuclear independence, was no match to the CNEA.

Opposition to dependence on American reactors did arise from one very unexpected source: Itamaraty. It is likely that the hard bargaining with the United States over the Tlatelolco Treaty in 1967 and the on-going Geneva nonproliferation talks turned Itamaraty away from supporting the United States, as it had from the 1940s to the 1960s, and toward the anti-American position of backing a nuclear independence program. Perhaps Itamaraty saw U.S. attempts to impose a nuclear-weapons-development freeze on those countries having none, while showing no willingness to give up its own nuclear weapons arsenal, as nuclear imperialism. In any case, when Ambassador Sérgio Corrêa da Costa, who served Brazil in the negotiations of the

Tlatelolco Treaty, became Itamaraty's secretary general in 1967, he brought with him a "nuclear orientation."

> Under Ambassador Corrêa da Costa, a task force of approximately twenty persons drawn from Itamaraty's departments of political affairs and international organizations matters was assembled, and in subsequent months had dealt with matters as various as the gathering of scientific information abroad, improved relations with Brazilian scientists abroad, structuring increased science collaboration with regional and international agencies, and domestic political considerations. Of course the most important task confronting the group was to develop and defend the Brazilian position on nuclear nonproliferation at Mexico City and Geneva.[92]

The new ideology that Itamaraty introduced into the political game was not pragmatic antidependency and did not advocate implementing a slow and incremental indigenous nuclear technological program. Instead, it was based on strong nationalism—or, more accurately, anti-Americanism—and a belief in Brazil's ability to develop one of the largest nuclear programs in the Third World.

Although the decision to buy the American reactor hardly affected Itamaraty's political clout, it was deadly for pragmatic antidependency and the indigenous nuclear technology program. Two circumstantial factors directly affected the decision: presidential succession politics and the man at the head of CNEN. A bitter dispute had developed between rival presidential hopefuls after Costa e Silva's unexpected death. His minister of the interior, Gen. Afonso Albuquerque Lima, representing the nationalist wing of the armed forces, was backed by a great number of officers but opposed by elites interested in continuing Brazil's economic internationalization. Lima lost and Emílio Médici won. An internationalist, Médici favored the Westinghouse deal and the electrical efficiency program, abandoning the various nationalist alternatives. He also named Hervásio de Carvalho, a prominent nuclear scientist, as president of the CNEN, and it has been strongly and widely suggested, by politicians and scientists alike, that de Carvalho played a major personal role in the decision to sign with Westinghouse.

Why didn't Pelúcio, the planning ministry, and those institutions beginning to work for a science and technology autonomy policy take up the banner of indigenous nuclear technological development? According to Pelúcio, BNDE did consider the issue. By then a national computer industry was already being contemplated, the economic miracle was producing resources for technological development, and a few large industrial sectors were being planned with the aim of transforming Brazil into an industrial power. BNDE, looking to support one of these sectors, was placed in the uneasy situation of having to choose between an autonomous nuclear program and a domestic computer industry. Technical experts were consulted as to whether Brazil had sufficient trained scientists and technicians or nuclear technical know-how to reproduce the technology of a nuclear reactor. The answer was negative, and BNDE decided to support the computer option. Past failures and inaction thus crept into the present, affecting it profoundly: when trained human resources were needed to open the path to nuclear technological autonomy, they were not at hand, so a program to produce electricity was instead imposed on scientists and many policymakers—a program that rode on the domestic political conflicts and indecisions and the secretive decision-making process of the time.

These events had two important political consequences: first, they forced Itamaraty to look for an alternative to the program aimed only at producing efficient electricity; second, they placed the scientists outside and definitely against the policy-making process. Now nuclear policy, more than ever before, was decided by informal forces at the presidential and armed forces levels, under a blanket of secrecy and almost completely disregarding the opinion of the scientists who overwhelmingly opposed the Westinghouse deal and were pushing for an "Argentina" type of independent program. Opposition was waged mainly through the Brazilian Society for the Progress of Science (SBPC), whose annual meetings roundly criticized the government's nuclear policy and provided information to fuel a heated public debate.

By the mid-1970s when Ernesto Geisel became president, there was growing evidence that the "crossroads period" policy would yield neither energy nor strategic or technological results. Furthermore, the U.S. government refused to allow Westinghouse to sell enrichment and reprocessing technologies to Brazil. A bolder move was needed, and Itamaraty provided it: a hybrid program of strategic, nuclear technological autonomy and commercial electrical efficiency. The

signing of the Brazil–West Germany agreement signaled that Itamaraty had won the battle; but the overblown nature of the agreement signaled that it was likely to lose the war.

Foreign opposition to the agreement came mainly from the United
States—the *New York Times* called the agreement "nuclear madness."[93]
But rather than constraining the Brazilians, this opposition simply
served to foster ideological support for any policy that would be anti-
American and anti–nonproliferation treaty. When it came to the
United States and nonproliferation, even those domestic forces that so
strongly attacked the agreement with West Germany and the nuclear
program defended Nuclebrás, the Brazilian government, and the
goal of nuclear independence.

The most outspoken groups against the nuclear program were the
industrialists, who complained about being left out, and the scientists,
who argued that the agreement would not result in the effective transfer of German technology to Brazil. This argument was strengthened
when the details of a secret agreement in which Brazil gave West Germany rights to retain control of Nuclen's technology were leaked to a
São Paulo newspaper. The scientists argued further that if independence was really the goal, then the Brazil–West Germany agreement,
in not promoting domestic industry, worked against it. Instead, a
smaller natural-uranium program would be more appropriate. José
Goldemberg, an outstanding Brazilian physicist, former SBPC president, and one of the strongest critics of the agreement, said: "'It is a
great deal for the Germans.' . . . He claimed that the agreement stipu-
lates the Brazilian manufacture of nuclear equipment will be maximized but a clause says that if KWU can produce equipment for a
price 15 percent or more lower than the Brazilians can, the order will
go to the Germans. 'If you knew Brazil, that clause really kills national
industry,' he said. 'Everything costs more in Brazil.'"[94] And Joaquim
Francisco de Carvalho, a former Nuclen director, said: "Brazil, in
order to develop a nuclear program that would really meet its national interests, would have to start, slowly and gradually, with an integrated participation of qualified industrial establishments and of
R & D institutions that provide technical assistance to industry."[95]

The jet-nozzle technology provided another line of attack against
the 1975 agreement and the nuclear program. Goldemberg visited
Germany and talked to the inventor but returned unconvinced of the
technology's commercial viability. The scientists also criticized Nuclebrás for its incompetence to make nuclear technological decisions

and its lack of concern for what the Brazilian scientists had to say. "Why, instead of going to the Germans and the French," they insisted, "didn't Nuclebrás come for answers to IPEN and other R & D institutes?" Still another line of criticism was along the goals for energy production set by the government itself. According to the physicist Luiz Pinguelli Rosa, the nuclear program was based altogether on the wrong premises: an exaggerated assessment of growth in demand for electric energy and an underestimation of the hydroelectric potential, of the viability of bringing hydroelectric power from north to south, and of the program's cost.[96] This last point became one of the stickiest when it appeared that the cost of the entire program would triple, if not quadruple. Public debate on this issue acquired gross dimensions when a former Furnas president attacked Brazil's nuclear agreement with West Germany, and Joaquim de Carvalho resigned his directorship at Nuclen to start a public campaign against the government and expose the program's "real costs."[97] Nogueira Batista responded by setting the program's cost at a somewhat lower level and by suing de Carvalho.

Although dreams sometimes come true, the achievement of technological independence and energy efficiency through a single, quick program (rather than two parallel programs or an Argentina-style program of nuclear technological autonomy later applied to energy) was a dream that would not materialize without political consensus and the support of the scientific community.[98]

When those opposed to the program presented not only arguments but also facts, and when the foreign debt reached unprecedented levels, the National Security Council decided to listen to Goldemberg. He presented in writing to the council a reassessment that called for the completion of Angra II and III, the phasing out of Nuclep, more consistent absorption of domestic technology and human resources, confirmation of uranium deposits that had been in doubt, increased efforts to produce uranium enrichment technology, and reorganization of the nuclear program at the executive and administrative levels to make room for Brazilian scientists.[99] Venturini in turn told Goldemberg that "'in principle' his suggestions had been accepted. The government appears to have already put into practice some of Goldemberg's proposals."[100]

But Goldemberg's dealings with the National Security Council seem to have provoked a new ideological conflict, this time within the scientific community itself, a large proportion of which distrusts the intentions behind the council's move. Thus the SBPC "passed a motion . . . in which it expressed its concern at the support given by the government to nuclear projects with a clear military purpose."[101] And scientists, who were basically reunited in their opposition to the Brazil–West Germany deal, are now divided again. In the meantime, Brazil, in contrast to Argentina, has produced some antinuclear groups that oppose not only this or that policy but nuclear power altogether. In sum, it appears that ideological consensus remains unattainable: power is still divided, nuclear policy continues to be de facto in the hands of the National Security Council, and no institution or leadership has arisen capable of changing the situation. The road to nuclear technological autonomy is still being followed; however, Brazilian elites have yet to agree on how to get there, or even whether it is worth getting there.

Conclusion

This study, most notably through the juxtaposition of the cases it presents, calls into question simplistic deterministic theories of technological dependency that have been used to analyze advanced Latin American countries. The results achieved by Argentina and Brazil in domestic nuclear energy development stand in sharp contrast to those achieved in general technological and computer development. This contrast adds credence to the view that in addition to—and sometimes even in direct opposition to—structural factors, ideologies and institutions can catalytically affect threads of events, thereby leading to outcomes that seem, at first, quite implausible. Natural resources, political power, and economic and technological capabilities, not to mention economic disequilibrium crises, certainly have a profound influence on political economic outcomes; but equally important are institutions, following a historical path set by certain constellations of consciousness: collective understandings of individuals that are as much intellectual, ideological, and political responses to a problem as they are resources for the resolution of that problem.

Each type of institution described in this study, whether a state institution such as BNDE, FINEP, CAPRE, or CNEA or a multinational corporation such as IBM, had some autonomy in determining its own future and that of entities it came in contact with and thus could put its mark on the pattern of interaction, both domestic and international. Even within institutions, certain groups and individuals possessed some degree of autonomy to be creative, to show the way, to call attention to the issues, new and old.

This study has shown that choices at the process level can, if they are perceived, become politically relevant, and that there is nothing automatic or deterministic about these processes. If dependency bred the quest for autonomy and led to the allocation of resources for this purpose, it was not only because resources were available but also because institutions and individuals could recognize dependency as the problem and autonomy as the solution and create new resources and power capabilities to solve the problems. In this connection ideology, while neither sufficient nor necessary as an explanation of international political-economic events, did affect and effect reality and therefore should not be taken for granted.

The power of ideology lies in its ability to make things happen when they might not have otherwise, even though the propensity existed in the form of capabilities or attributes. Ideology tells actors what the goal is, how to seek it, and how important it is compared to other goals. It is dynamic. As ideology changes or is newly created, it affects institutional and policy innovation and creates new factors to deal with both at the domestic and at the international level.

To explain Argentina and Brazil's science and technology, computer, and nuclear energy policies, their evolution and their effect on the appropriate areas and sectors, I have focused on the impact of the ideas and ideologies of subversive elites who, through their influence over political elites, institutions, and processes, created policy and institutional propensities for changing reality—in this case, for reducing dependency, creating a domestic computer industry, or dominating the nuclear fuel cycle. Pragmatic antidependency guerrilla groups—convinced that dependency can be reduced before a world economic structural change takes place, and in the right time and place to influence policy making—helped to inspire, formulate, and implement antidependency policies and to sell to the ruling elites an ideology covered with politically and economically relevant wrapping. I have described several cases in which they, faced initially with considerable hostility, succeeded in affecting the course of progress and in making political systems assimilate the changes and take them for granted.

This was true in Brazil with BNDE and Pelúcio's effort to build a science and technology system by introducing the political and military elites to the idea that development is also a higher capacity of knowing and understanding and that pragmatic measures can reduce technological dependency. It was also true when CAPRE moved the ground under the Brazilian political-economic establishment to start a domestic computer industry that would be impenetrable by IBM. Finally, it was demonstrated by the pragmatic antidependency guerrilla group par excellence—Jorge Sabato's—which helped turn the CNEA into the exception to the rule: the institution that overcame Argentine "fracasomania" and produced one of the most advanced nuclear programs in the Third World.

The guerrilla groups described in this study are but one expression of the role played by ideological and institutional factors; other outstanding individuals and institutions have also influenced the historical process and should not be discounted.

The policy and implementation successes in the general science and technology, computer, and nuclear fields did not happen by themselves, by the invisible hand of the market, or out of the excellence of scientific research and technological innovation and their search for avenues of application. Instead, the computer industry in Brazil and the successful nuclear program in Argentina were the result of design and some measure of planning. The planner, not surprisingly, was the state institution that set the development goals and chose the means to get from here to there. Whereas in the developed countries market efficiency, management, and advanced scientific development and technical training can help travel the distance without state intervention, in the Third World the involvement of guerrillas and institutions and the use of state resources are required.

Ideology and institutions are therefore not meaningless forces. They can drive capabilities; set priorities; open new alternatives; produce political decisions; lead to political and institutional innovations; affect other actors in the international system, such as multinational corporations; catalytically and gradually transform the condition of interdependence; and thus lead to international political-economic change.

I will end this study with a figure that organizes the main thoughts, variables, and linkages together with the domestic and foreign dimensions involved in the processes here described (see Figure 12).

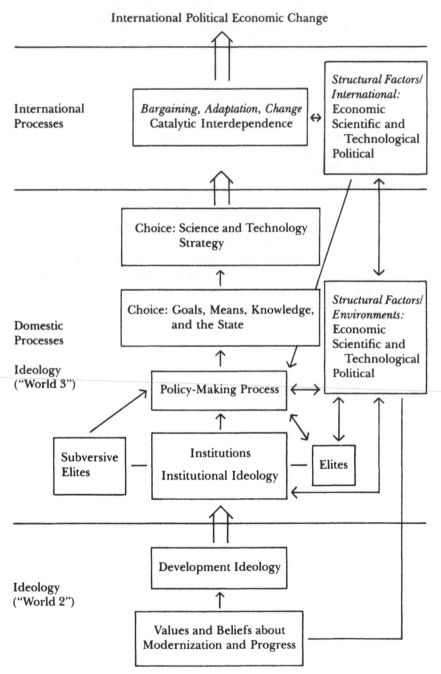

Figure 12. Journeys Toward Scientific and Technological Progress:
A Theoretical Summary

Although international political-economic change may be positively or negatively affected by structural/environmental factors, it is not necessarily mechanistically determined by them. Instead, it may emerge from propensities of change realized through the awareness, values, beliefs, expectations, and political will of people, the institutions they create, and their collective understanding. In many cases the cognitive factors may just make the difference. It is the task of political science to try to understand these cases better by developing theory and empirical studies that actively integrate historical processes into the theory-building effort. My study is but a modest and preliminary effort in that direction.

Notes

Chapter 1

1. Ovid, *Metamorphoses*, trans. M. M. Innes (Harmondsworth, Eng.: Penguin, 1955), 29.

2. Recent studies, however, have added valuable insights and sharpened the debate. See *Change in the International System*, ed. Ole R. Holsti, Randolph M. Siverson, and Alexander L. George (Boulder, Colo.: Westview Press, 1980); and Robert Gilpin, *War and Change in World Politics* (New York: Cambridge University Press, 1981).

3. For a poignant criticism of neorealism and structuralism, see Richard K. Ashley, "The Poverty of Neorealism," *International Organization* 38, no. 2 (Spring 1984): 225–86.

4. International system theory is one of the best examples of a functional-structural, ahistorical approach in international relations. Two of the best-known theorists are Morton A. Kaplan and Kenneth N. Waltz. See Kaplan, *System and Process in International Politics* (New York: Wiley 1962); and Waltz, *Theory of International Politics* (Reading, Mass.: Addison-Wesley, 1979). Marxist dependency arguments are among the best examples of a structural approach in international political economy. We will discuss these arguments and "theories" throughout this study. For a *systemic* Marxist approach, see Immanuel Wallerstein, *The Modern World-System II* (New York: Academic Press, 1980); for an example of non-Marxist structural economic, power-oriented international political economy theory, see Robert O. Keohane and Joseph S. Nye, Jr., *Power and Interdependence* (Boston: Little, Brown, 1977).

5. Waltz, for example, claims to be able to predict the continuity of the current international system, although a critic, John G. Ruggie, has argued that continuity according to Waltz "is a product of premise even before it is hypothesized as outcome" ("Continuity and Transformation in the World Polity: Toward a Neorealist Synthesis," *World Politics* 35, no. 2 [Jan. 1983]: 285). Ruggie, in this sophisticated critique of Waltz's theory of international systems, argues among other things that by banishing unit-level processes from the domain of systemic theory, Waltz has exogenized the ultimate source of systemic change.

6. Robert C. North, *The World That Could Be* (New York: W. W. Norton, 1976), 55.

7. Ibid. Original emphasis omitted.

8. Stephen Toulmin, *Human Understanding: The Collective Use and Evolution of Concepts* (Princeton, N.J.: Princeton University Press, 1972), 21.

9. Karl R. Popper, *Objective Knowledge: An Evolutionary Approach* (London: Oxford University Press, 1979).

10. Thomas S. Kuhn, *The Structure of Scientific Revolutions,* 2d ed. (Chicago: University of Chicago Press, 1970).

11. Henri Bergson, "The Evolution of Life," in *The Philosophers of Science,* ed. Saxe Commins and Robert N. Linscott (New York: Random House, 1947), 293. Original emphasis omitted.

12. Ibid.

13. Stephen Toulmin, *Foresight and Understanding* (New York: Harper & Row, 1961), 57.

14. Karl R. Popper, *The Open Universe: An Argument for Indeterminism,* ed. W. W. Bartley III (Totowa, N.J.: Rowman & Littlefield, 1982), 28.

15. Karl R. Popper and John C. Eccles, *The Self and Its Brain: An Argument for Interactionism* (Berlin: Springer-Verlag, 1977), 127.

16. This catchy phrase is taken from the title of Albert O. Hirschman's *Journeys Toward Progress: Studies of Economic Policy Making in Latin America* (New York: W. W. Norton, 1973).

17. See, for example, Jorge M. Katz, *Importación de tecnología, aprendizaje e industrialización dependiente* (Mexico, D.F.: Fondo de Cultura Económica, 1976); idem, *Oligopolio: Firmas nacionales y empresas multinacionales. La industria farmacéutica argentina* (Buenos Aires: Siglo XXI, 1974); Fabio Stefano Erber, José Tavares de Araújo, Junior, Sergio Francisco Alves, Leonídia Gomes dos Reis, and Myriam Lewin Redinger, *Absorção e criação de tecnologia na indústria de bens de capital* (Rio de Janeiro: FINEP, 1974); José Tavares de Araújo, Junior, ed., *Difusão de inovações na indústria brasileira: Tres estudos de caso* (Rio de Janeiro: IPEA/INPES, 1976); and Simon Teitel, "Tecnología, empresa, e información," *El trimestre económico* 45, no. 2 (1978): 297–324.

18. See, for example, Francisco R. Sagasti, *Technology, Planning, and Self-Reliant Development* (New York: Praeger, 1979); Miguel Wionczek, *Inversión y tecnología extranjera en América Latina* (Mexico: Joaquín Mortiz, 1971); and Jorge Sabato and Natalio Botana, *La ciencia y la tecnología en el desarrollo futuro de América Latina* (Lima: Instituto de Estudios Peruanos, 1970).

19. Guillermo A. O'Donnell, *Modernization and Bureaucratic-Authoritarianism*, Politics of Modernization Series, no. 9 (Berkeley, Calif.: Institute of International Studies, 1973), 11.

20. Interamerican Development Bank (BID), *Progreso económico en América Latina: Informe 1978* (Washington, D.C.: 1978), 33.

21. Ibid., 472.

22. Ibid., 33.

23. Clifford Geertz, "Ideology as a Cultural System," in *Ideology and Discontent*, ed. David E. Apter (New York: Free Press, 1964), 47.

24. Karl Mannheim, *Ideology and Utopia* (New York: Harcourt, Brace, 1936), 2.

25. Burkart Holzner and John H. Marx, *Knowledge Application: The Knowledge System in Society* (Boston: Allyn & Bacon, 1979), 51–52.

26. Geertz, "Ideology as a Cultural System," 52–53.

27. Holzner and Marx, *Knowledge Application*, 48.

28. John S. Odell, *U.S. International Monetary Policy: Markets, Power, and Ideas as Sources of Change* (Princeton, N.J.: Princeton University Press, 1982), 362–63.

29. Lorand B. Szalay and Rita Mae Kelly, "Political Ideology and Subjective Culture: Conceptualization and Empirical Assessment," *American Political Science Review* 76, no. 3 (Sept. 1982): 585.

30. Odell, *U.S. International Monetary Policy*, 62.

31. Yehuda Elkana, "The Myth of Simplicity," in *Albert Einstein: Historical and Cultural Perspectives*, ed. Gerald Holton and Yehuda Elkana (Princeton, N.J.: Princeton University Press, 1982), 205.

32. Ibid., 208.

33. Popper, *Open Universe*, 5.

34. Ibid., 55–56. On these grounds Popper dismisses determinist social science doctrines, such as Marxism, that claim to predict the course of human history in a desired direction. See Popper, *The Poverty of Historicism* (London: Routledge & Kegan Paul, 1957); and idem, *The Open Society and Its Enemies* (Princeton, N.J.: Princeton University Press, 1971), vols. 1 and 2.

A related argument for dismissing determinism in the social sciences is the impact of cultural elements on social change, since ideologies and ideas, policies, and institutions, being historically constructed, are necessarily subject to historically defined standards of judgment (Elkana, "The Myth of Simplicity," 210).

35. A controversy in the fields of physics and philosophy of science arises because some argue that there is little difference between the propensity theory and frequency theory. For a critique of Popper's propensity theory, see Anthony O'Hear, *Karl Popper* (London: Routledge & Kegan Paul, 1980), 132–39.

36. Karl R. Popper, *Quantum Theory and the Schism in Physics*, ed. W. W. Bartley III (Totowa, N.J.: Rowman & Littlefield, 1982), 159. Popper's theory argues that "certain set-ups are random in their outcomes, in the sense that we cannot predict the actual outcome of any particular case where the set-up

is instantiated, but that repeated experiments with or observations of the set-up will show statistical stability. The statistical stability is taken to be the result of the propensities inherent in the set-up. . . . Probabilities are still conjectured statistical frequencies of sequences, but are seen as being the manifestations of actual existing, but indeterministic, forces" (O'Hear, *Popper,* 132).

37. Erich Jantsch, "Unifying Principles of Evolution," in *The Evolutionary Vision,* ed. Erich Jantsch (Boulder, Colo.: Westview Press, 1981), 112.

38. Popper, *Open Universe,* 114.

39. Karl R. Popper, "Replies to My Critics," in *The Philosophy of Karl Popper,* ed. Paul A. Schilpp (La Salle, Ill.: Open Court, 1974), 1050.

40. Popper, *Open Universe,* 116.

41. Mannheim raised the paradox that we cannot distinguish knowledge from ideology because all knowledge is socially determined. Thus, he concluded, social theories cum ideologies are but justifications of reality and far from the objective truth.

42. Holzner and Marx, *Knowledge Application,* 82.

43. Ibid. From this perspective, knowledge can also be defined as the "'mapping' of experienced reality by some observer. It cannot mean the 'grasping' of reality itself. . . . More strictly speaking, we are compelled to define 'knowledge' as the communicable mapping of some aspects of experienced reality by an observer in symbolic terms" (ibid., 93).

44. Ibid., 85. See also Peter L. Berger and Thomas Luckmann, *The Social Construction of Reality* (Garden City, N.Y.: Doubleday Anchor, 1967); and Peter L. Berger, Brigitte Berger, and Hansfried Kellner, *The Homeless Mind: Modernization and Consciousness* (New York: Vintage Books, 1974).

45. Weber also recognized the emergent nature of reality. In his masterpiece on the "protestant ethic," he showed how a constellation of values and attitudes was crucial for the emergence of modern capitalism. See Max Weber, *The Protestant Ethic and the Spirit of Capitalism* (New York: Scribner, 1958).

46. Holzner and Marx, *Knowledge Application,* 58. See also Émile Durkheim, *The Elementary Forms of the Religious Life* (New York: Free Press, 1965).
My use of the collective consciousness concept should also be differentiated from the relatively new fashion in historical studies called "mentalités" (collective states of mind), which deals with popular beliefs, attitudes, and modes of behavior, and from what Marxists in Latin America call "concientización" (consciousness raising). The concept of "mentalités," much more influenced by humanistic anthropology than by the social sciences, rejects the study of elites and prefers to focus on subjective cultural and popular meanings. I, however, study politics, power, and political ideologies that are very much linked to political and intellectual elites directly involved in political-economic processes. For an assessment and critique of "mentalités," see Lawrence Stone, *The Past and the Present* (Boston: Routledge & Kegan Paul, 1981). As for "consciousness raising," Berger et al. (*Homeless Mind,* 135) define it as the "deliberate use of education and political propaganda to make people conscious of the social and political determinants of their situation, particularly their own exploitation, so that they are ready to act politically."

47. Stephen Toulmin, "The Genealogy of 'Consciousness,'" in *Explaining Human Behavior*, ed. Paul F. Secord (Beverly Hills, Calif.: Sage, 1982), 64.

48. Toulmin, *Human Understanding*, 289.

49. See Crawford Young, "Ideas of Progress in the Third World," in *Progress and Its Discontents*, ed. Gabriel A. Almond, Marvin Chodorow, and Roy Harvey Pearce (Berkeley and Los Angeles: University of California Press, 1982).

50. Edward Shils, *Tradition* (Chicago: University of Chicago Press, 1981), 296.

51. Ibid., 80.

52. Leo Hamon (comment) in Alain Birou, Paul-Marc Henry, and John P. Schlegel, *Towards a Redefinition of Development: Essays and Discussion on the Nature of Development in an International Perspective* (Oxford: Pergamon Press, 1977), 61.

53. The inspiration for these thoughts comes from Toulmin, *Human Understanding*, 348–53.

54. I follow the road opened by Ernst B. Haas, who studied the effect of scientific knowledge on political-economic international changes, such as international regime changes. See Ernst B. Haas, "Why Collaborate: Issue-Linkage and International Regimes," *World Politics* 32, no. 3 (April 1980): 357–405; and idem, "Words Can Hurt You or Who Said What to Whom about Regimes," *International Organization* 36, no. 2 (Spring 1982): 207–43.

55. Popper, *Open Universe*, 107.

56. Todd R. La Porte has thus defined what he calls "organized social complexity." See his *Organized Social Complexity* (Princeton, N.J.: Princeton University Press, 1975), 6. See also Harlan G. Wilson, "Complexity as a Theoretical Problem" (Ph.D. diss., University of California, Berkeley, 1978).

57. Raymond Vernon, "Sovereignty at Bay: Ten Years After," *International Organization* 35, no. 3 (Summer 1981): 527.

58. Raymond Vernon, *Storm over the Multinationals: The Real Issues* (Cambridge, Mass.: Harvard University Press, 1977), 146.

59. Ibid., 146–47.

60. Ibid., 203.

61. Isaiah Frank, *Foreign Enterprise in Developing Countries* (Baltimore: Johns Hopkins University Press, 1980), 27, 36.

62. See, for example, Vernon, *Storm over the Multinationals;* and C. Fred Bergsten, "Coming Investment Wars?" *Foreign Affairs* 53 (Oct. 1974): 135–52.

63. This approach argues that at the time its investment is set up, the multinational corporation has the upper hand because it commands the capital and technology that the host country lacks; however, "over time the host country is likely to gain access in varying degrees to the sources of bargaining power which earlier had been controlled by the enterprise. As the country attains greater bargaining power, it forces the balance of benefits to shift in its favor" (Joseph M. Grieco, *Between Dependency and Autonomy: India's Experience with the International Computer Industry* [Berkeley and Los Angeles: University of California Press, 1984], 2–3).

64. Vernon, "Sovereignty at Bay: Ten Years After," 520.
65. Frank, *Foreign Enterprise in Developing Countries,* 145.
66. Vernon, "Sovereignty at Bay: Ten Years After," 524.
67. Frank, *Foreign Enterprise in Developing Countries,* 146.
68. These ideas have been strongly influenced by Toulmin, *Human Understanding,* 349–50.
69. Stephen Toulmin, "Evolution, Adaptation, and Human Understanding," in *Scientific Inquiry and the Social Sciences: A Volume in Honor of Donald T. Campbell,* ed. Marilyn B. Brewer and Barry E. Collins (San Francisco: Jossey-Bass, 1981), 31.

Chapter 2

1. See, for example, Charles W. Anderson, *Politics and Economic Change in Latin America* (Princeton, N.J.: D. Van Nostrand, 1967). See also Economic Commission for Latin America (ECLA) documents on Latin American integration, especially *Influence of the Common Market on Latin American Economic Development,* E/CN 12/C.1/13, April 14, 1959; Roger D. Hansen, *Central America: Regional Integration and Economic Development* (Washington, D.C.: National Planning Association, 1967); Latin American Integration (INTAL), *La integración latinoamericana: Situación y perspectivas* (Buenos Aires, 1965); and Victor L. Urquidi, *Free Trade and Economic Integration in Latin America: Toward a Common Market* (Berkeley and Los Angeles: University of California Press, 1962).
2. Chapter 3 of this study will review the literature on dependency.
3. The term was used by the Organization of American States (OAS), text approved by the Permanent Council on November 10, 1976 (*OAS Chronicle* [Dec. 1976], 4–6).
4. Geoffrey Barraclough, *Turning Points in World History* (London: Thames & Hudson, 1979), chaps. 1 and 2.
5. Francisco R. Sagasti, *Technology, Planning, and Self-Reliant Development: A Latin American View* (New York: Praeger, 1979), 165.
6. Barraclough, *Turning Points,* 26–27.
7. For sources on modernization theory, see Gabriel Almond and James S. Coleman, *The Politics of the Developing Areas* (Princeton, N.J.: Princeton University Press, 1960); Gabriel Almond and Sidney Verba, *The Civic Culture* (Princeton, N.J.: Princeton University Press, 1963); David E. Apter, *The Politics of Modernization* (Chicago: University of Chicago Press, 1965); Cyril E. Black, *The Dynamics of Modernization: A Study on Comparative History* (New York: Harper & Row, 1966); Samuel Huntington, "Political Modernization: America vs. Europe," *World Politics* 18 (April 1966): 378–414; and Marion J. Levy, *Modernization and the Structure of Societies: A Setting for International Affairs* (Princeton, N.J.: Princeton University Press, 1966).
8. On liberalism, see, for example, Alan Bullock and Maurice Slock, eds.,

The Liberal Tradition: From Fox to Keynes (London: Oxford University Press, 1967); and H. J. Lasky, *Political Thought from Locke to Bentham* (London, 1920).

On Marxism, see Shlomo Avineri, *The Social and Political Thought of Karl Marx* (Cambridge: Cambridge University Press, 1968). On nationalism, see Karl Deutsch, *Nationalism and Social Communication* (New York: Wiley, 1953); and Hans Kohn, *The Idea of Nationalism* (New York: Macmillan, 1943).

9. Edward L. Morse, *Modernization and the Transformation of International Relations* (New York: Free Press, 1976), 51.

10. Nazli Choucri, "International Political Economy: A Theoretical Perspective," in *Change in the International System*, ed. Ole R. Holsti, Randolph M. Siverson, and Alexander L. George (Boulder, Colo.: Westview Press, 1980), 105.

11. See V. I. Lenin, *Imperialism, the Highest Stage of Capitalism*, vol. 22 of *The Collected Works* (Moscow, n.d.).

12. Ernst B. Haas, *Beyond the Nation-State* (Stanford, Calif.: Stanford University Press, 1964), 465.

13. Karl W. Deutsch, *The Nerves of Government* (New York: Free Press, 1966), 253–54.

14. I am indebted for some of the ideas of this taxonomy to Douglas C. Basil and Curtis W. Cook, *The Management of Change* (New York: McGraw-Hill, 1974), chap. 2.

15. The term refers primarily to countries of the Third World, in contradistinction to "late industrializers" such as Germany, Italy, and Russia. See Albert O. Hirschman, *A Bias for Hope* (New Haven, Conn.: Yale University Press, 1971), 94.

16. See, for example, Michael Polanyi, *Personal Knowledge* (London: Routledge & Kegan Paul, 1958).

17. See Mancur Olson and Hans J. Landsberg, *The No-Growth Society* (New York: W. W. Norton, 1973).

18. Karl Polanyi argues that evidence from anthropology disputes the idea that men have a "natural tendency" to barter (*The Great Transformation* [Boston: Beacon Press, 1944], 43).

19. See, for example, Charles E. Lindblom, *Politics and Markets* (New York: Basic Books, 1977), esp. chap. 23.

20. The idea of bargaining comprising threats and promises was borrowed from Thomas C. Schelling, *The Strategy of Conflict* (New York: Oxford University Press, 1960), chap. 2.

21. Robert A. Dahl and Charles E. Lindblom, *Politics, Economics, and Welfare* (Chicago: University of Chicago Press, 1976), chaps. 10 and 11.

22. For a study dealing with "postindustrial corrections," see Roger Benjamin, *The Limits of Politics* (Chicago: University of Chicago Press, 1980).

23. Robert O. Keohane and Joseph S. Nye, Jr., *Power and Interdependence* (Boston: Little, Brown, 1977), 8.

24. Morse, *Modernization*, 118.

25. Ernst B. Haas, "Is There a Hole in the Whole? Knowledge, Technology, Interdependence, and the Construction of International Regimes," *Inter-*

national Organization 29 (Summer 1975) : 861; Keohane and Nye, *Power and Interdependence,* 11–19. Keohane and Nye define sensitivity as the liability to suffer costs imposed from outside before policies are altered in an attempt to change the situation, while vulnerability is the liability to suffer such costs even after policies have been altered (13).

26. See Haas, "Is There a Hole in the Whole?" 859–68. See also his "Why Collaborate: Issue-Linkage and International Regimes," *World Politics* 32 (April 1980) : 362–64.

27. See Kenneth N. Waltz, *Theory of International Politics* (Reading, Mass.: Addison-Wesley, 1979), 141–45.

28. For empirical data about the increase of interdependence, see Richard Rosecrance and Arthur Stein, "Interdependence: Myth or Reality?" *World Politics* 26 (Oct. 1973) : 1–27.

29. For some figures about the involvement of multinational corporations in Latin America, see Richard Newfarmer and Willard F. Mueller, "Multinational Corporations in Brazil and Mexico: Structural Sources of Economic and Noneconomic Power," *Report to the Subcommittee on Multinational Corporations of the Committee on Foreign Relations,* U.S. Senate, 94th Cong., 1st sess. (Washington, D.C.: GPO, 1975).

30. Simon Kuznets, as quoted by Edward L. Morse, "Transnational Economic Processes," in *Transnational Relations and World Politics,* ed. Robert O. Keohane and Joseph S. Nye, Jr. (Cambridge, Mass.: Harvard University Press, 1970), 46.

31. Ibid., 29–30. Original emphasis omitted.

32. That is, the effects of transfers of alien ways of life on nationals of a country (habits of consumption, modes of dress, patterns of education, leisure, recreation, etc.).

33. "The Third World Forum: Intellectual Self-Reliance," *International Development Review* 17, no. 1 (1975) : 8–9.

34. Carlos F. Díaz-Alejandro, "Delinking North and South: Unshackled or Unhinged?" in *Rich and Poor Nations in the World Economy,* ed. Albert Fishlow, Carlos F. Díaz-Alejandro, Richard R. Fagen, and Roger D. Hansen (New York: McGraw-Hill, 1978), 140.

35. For a look at the strategy of "basic human needs," see Roger D. Hansen, *Beyond the North-South Stalemate* (New York: McGraw-Hill, 1979), chap. 8.

36. Pugwash Symposium, "The Role of Self-Reliance in Alternative Strategies for Development," *World Development* 5, no. 3 (1977) : 258–59.

37. Ibid., 259.

38. Ibid. A similar idea was also advanced by Díaz-Alejandro ("Delinking North and South"), who called it "selective delinking."

39. Pugwash Symposium, 259.

40. A minority school of thought in the self-reliance movement takes a different approach. It prescribes complete, or almost complete, autarky. One source for this ideological position is dogmatic Marxism, which advocates the strategy of abandoning the ship before it sinks; the ship is, of course, the capitalist international economic system. Those who see in Western values and

technology the "evil of this world" also agree with ideas of autarky. That is, since Western technology is responsible for inequality, exploitation, alienation, apathy, the depletion of the earth's resources, and environmental catastrophes that threaten survival, the only solution for the developing countries is to shield themselves against Western science and technology.

For an example of these approaches, see *Alternatives* 4, no. 3, ed. Ward Morehouse (Jan. 1979).

41. Sagasti, *Technology, Planning, and Self-Reliant Development*, 168.
42. Ibid., 151.
43. Pugwash Symposium, 259–60.
44. Ibid., 261.

Chapter 3

1. Obviously scholars and practitioners do not agree on the best method for classifying technology transfers. Each typology depends on what argument is being advanced. The most complete listing of practices relevant to technology transfers and the devices adopted to control them in the importing country (with examples of national legislation) is in United Nations Conference for Trade and Development (UNCTAD), *The Possibility and Feasibility of an International Code of Conduct on Transfers of Technology*, TD/B/AC. 11/22, June 6, 1974 (reprinted in Karl P. Sauvant and Hajo Hasenpflug, *The New International Economic Order* [Boulder, Colo.: Westview Press, 1977], 283–84). See also Miguel S. Wionczek, "La transferencia de tecnología en el marco de la industrialización mexicana," in *Comercio de tecnología y desarrollo económico*, ed. Miguel S. Wionczek (Mexico, D.F.: Coordinación de Ciencias, 1973), 249.

2. On the implantation of information retrieval systems and the attendant problems in Latin America, see Win Crowther, *The Search for Relevance: Political Ideology, Culture, and Political Choice as Factors of Technological Development and Dependence in Latin America* (Santiago, Chile: ECLA, Jan. 17, 1976, mimeo).

3. Some implications of systematic national development planning for global planning efforts are explored in *RIO: Reshaping the International Order*, ed. J. Tinbergen (New York: Dutton, 1976); and A. Herrera et al., *Catastrophe or New Society?* (Ottawa: International Development Research Centre [IDRC], 1976).

4. Jorge Sabato and Natalio Botana, *La ciencia y la tecnología en el desarrollo futuro de América Latina* (Lima: Instituto de Estudios Peruanos, 1970). Also known as the IGE (Infraestructura, Gobierno, Empresas) Triangle, the scheme presupposes routinized communication and feedback links among the three poles, not a hierarchy. Consensus is expected to result from the cybernetic interaction. At the same time, Sabato spelled out some aspects of phasing and interaction effects among the components of the triangle. At first, the state should simply fund and symbolically encourage R & D centers, allowing the researchers at the centers to proceed according to their own preferences.

Only after the prestige and capability of the centers is established should the state *impose* research priorities according to a strategy of adapting foreign technology to national development.

5. Harlan G. Wilson, "Complexity as a Theoretical Problem" (Ph.D. diss., University of California, Berkeley, 1978), 53–56.

6. Nuno Fidelino de Figueiredo, *A transferência de tecnologia no desenvolvimento industrial do Brasil* (Rio de Janeiro: IPEA/INPES, 1972), 39. My translation.

7. Luis Carbonell, "Interaction Between the Scientific-Technological Infrastructure and the Industrial Sector: Points of View of the Scientific Researcher," in *Science, Government, and Industry for Development. The Texas Forum,* ed. Earl Ingerson and Wayne G. Bragg (Austin: Institute of Latin American Studies, University of Texas, 1975), 77–78 (hereafter, *Texas Forum*).

8. Gabriel Palma, "Dependency: A Formal Theory of Underdevelopment or a Methodology for the Analysis of Concrete Situations of Underdevelopment?" *World Development* 6, no. 7/8 (Aug. 1978):881–924.

9. See, for example, Paul Baran, *La economia política del crecimiento* (Mexico, D.F.: Fondo de Cultura Económica, 1969); and André G. Frank, *Latin America: Underdevelopment or Revolution?* (New York: Monthly Review Press, 1969).

10. Palma, "Dependency," 899.

11. See, for example, Raul Prebisch, *Towards a Dynamic Development Policy for Latin America* (New York: United Nations, 1963). This disenchantment of Prebisch and others must be viewed in the light of ISI's performance in Latin America. The early phase of ISI is normally characterized by a shift toward locally produced consumer goods; the number of people employed in industry and per capita and national income rise. But, in the absence of planning, the importation of capital goods and raw materials also increases as new industry becomes dependent on more sophisticated foreign intermediary technology, fuel, and primary goods not produced locally. Hence, the total national involvement with foreign trade does not necessarily change in this phase; the new types of imports must still be paid for, industrial exports do not enjoy demand abroad, and traditional exports may decline for a variety of reasons or suffer from adverse terms of trade.

12. Latin America's GNP grew by 5.2 percent annually between 1945 and 1955. Despite a population increase of 26 percent during this period, per capita GNP rose by 31 percent and consumption by 40 percent. Total production rose by 38 percent, industrial production by 45 percent, rates that declined to 12 percent and 15 percent, respectively, between 1955 and 1958. By 1960, manufacturing accounted for about 20 percent of the aggregate gross product and employed 15 percent of the labor force (Victor L. Urquidi, *Free Trade and Economic Integration in Latin America: Toward a Common Market* [Berkeley and Los Angeles: University of California Press, 1962], 4–7).

13. Palma, "Dependency," 908.

14. See Maria da Conceição Tavares, *Da substituição de importações ao capitalismo financeiro,* 8th ed. (Rio de Janeiro: Zahar, 1979); and Osvaldo Sunkel, "Big Business and Dependencia," *Foreign Affairs* 50, no. 3 (April 1972): 517–31.

15. Cardoso's arguments, as explained in Palma, "Dependency," 909.

16. Furtado refers to capitalism in developing countries as peripheral: "a capitalism unable to generate innovations and dependent for transformation upon decisions from the outside" (quoted in Ronald Chilcote and Joel Edelstein, *Latin America: The Struggle with Dependency and Beyond* [New York: Wiley, 1974], 44).

17. Amílcar Herrera, *Ciencia y política en América latina* (Mexico, D.F.: Siglo XXI, 1971), 13. My translation.

18. Charles Cooper, "Science, Technology, and Production in the Underdeveloped Countries: An Introduction," in *Science, Technology, and Development: The Political Economy of Technical Advance in Underdeveloped Countries*, ed. Charles Cooper (London: Frank Cass, 1973), 5.

19. Ibid., 7.

20. Amílcar Herrera, "Social Determinants of Science in Latin America: Explicit Science Policy and Implicit Science Policy," in Cooper, ed., *Science, Technology, and Development*, 22.

21. Enrique Leff, "El desarrollo de la ciencia y la tecnología y su integración dentro de un marco de desarrollo económico y social: El caso de México," *Comercio exterior* 23, no. 4 (April 1973):336. My translation.

22. Constantine Vaitsos, "Patents Revisited: Their Function in Developing Countries," in Cooper, ed., *Science, Technology, and Development*, 81, 85.

23. See Celso Furtado, *Development and Underdevelopment* (Berkeley and Los Angeles: University of California Press, 1964).

24. Herrera's idea, as analyzed by Cooper, "Science, Technology, and Production," 4.

25. Sergio Ortíz Hernán and Federico Torres Arroyo, "Necesidad de una política de ciencia y tecnología en México," *Comercio exterior* 23, no. 5 (May 1973):427. My translation.

26. Amílcar Herrera, "Notas sobre la ciencia y la tecnología en el desarrollo de la sociedad latinoamericana," in *América latina. Ciencia y tecnología en el desarrollo de la sociedad*, ed. Amílcar Herrera (Santiago, Chile: Universitaria, 1970), 25. My translation.

27. Amílcar Herrera, "La creación de tecnología como una expresión cultural," *Comercio exterior* 23, no. 10 (Oct. 1973):993.

28. Herrera, *Ciencia y política en América latina*, 91. My translation.

29. "In order to achieve alternative technology there is no need for a highly advanced technological development; it is enough to utilize the technological elements already known in a different way" (Herrera, "La creación de tecnología," 993). My translation.

30. Jorge Sabato, "El cambio tecnológico necesario y posible en América latina," *Comercio exterior* 26, no. 5 (May 1976):544.

31. Carbonell, "Interaction Between the Scientific-Technological Infrastructure," 79–80.

32. Francisco Sagasti, "A Framework for the Formulation and Implementation of Technology Policies: A Case Study of ITINTEC in Peru," *Texas Forum*, 207–10.

33. Sabato argues that "we are fundamentally technology consumers but

not producers . . . thus spectators, and not authors, passive recipients of what others perform in their own interest and not in ours, which leads us inexorably to adopt this Weltanschauung against which there has been so much protest and which some have pretended to defeat with mere rhetoric. Two positions, equally fatal, are reached: the worst of technologies, of imitation and copy, or the frantic denunciation that sterilizes because it does not propose viable options" ("El cambio tecnológico," 544). My translation.

34. Miguel S. Wionczek, "La planificación de la ciencia y la tecnología en México," *Comercio exterior* 26, no. 11 (Nov. 1976):1276. My translation.

35. Luis Soto-Krebs, "Tecnología en el Grupo Andino," *Comercio exterior* 25, no. 1 (Jan. 1975):55. My translation. See also Aldo Ferrer, "Políticas y planes de desarrollo científico y tecnológico," *Ciencia interamericana* 15, no. 3/4 (July/Sept. 1974):2–7.

36. The approach is consistent with theories that accept the possibility of development with both economic growth and income redistribution. See Hollis Chenery et al., *Redistribution with Growth* (London: Oxford University Press, 1974).

37. Charles W. Anderson, *Politics and Economic Change in Latin America* (Princeton, N.J.: D. Van Nostrand, 1967), 41.

38. Jorge Sabato, Roque G. Carranza, and Gerardo Gargiulo, "Ensayo de régimen de tecnología: La fundición ferrosa en Argentina," *Comercio exterior* 26, no. 11 (Nov. 1976):1302. My translation.

39. Francisco Sagasti and Mauricio Guerrero, *El desarrollo científico y tecnológico de América latina* (Buenos Aires: BID/INTAL, 1974). See also Alberto Aráoz, "Compras estatales y desarrollo tecnológico," *Comercio exterior* 27, no. 6 (June 1977):654–70.

40. See, for example, Eduardo White and Jaime Campos, "Elementos para el estudio de las empresas conjuntas latinoamericanas," *Integración latinoamericana* 1, no. 3 (June 1976):11–30; and Luciano Tomassini, "Elementos para un estudio de los procesos de integración y otras formas de cooperación en América latina," *Integración latinoamericana* 2, no. 12 (April 1977):22–42.

41. Alberto Aráoz, "Compras estatales," 654. My translation.

Chapter 4

1. Exceptions are Peter S. Cleaves, *Bureaucratic Politics and Administration in Chile* (Berkeley and Los Angeles: University of California Press, 1974); Merilee S. Grindle, *Bureaucrats, Politicians, and Peasants in Mexico: A Case Study in Public Policy* (Berkeley and Los Angeles: University of California Press, 1977); *Politics and Policy Implementation in the Third World*, ed. Merilee S. Grindle (Princeton, N.J.: Princeton University Press, 1980); Oscar Oszlak, "Critical Approaches to the Study of State Bureaucracy: A Latin American Perspective," *International Social Science Journal* 31, no. 4 (1979):661–81; and Clarence E. Thurber and Lawrence S. Graham, *Development Administration in Latin America* (Durham, N.C.: Duke University Press, 1973). See also Fred W. Riggs,

Administration in Developing Countries: The Theory of Prismatic Society (Boston: Houghton Mifflin, 1964).

2. Graham Allison, *Essence of Decision* (Boston: Little, Brown, 1971). The rational model assumes optimal choices among alternatives, each of which has a set of consequences. The government, nation, state, or nation-state thus acts according to consistent, value-maximizing choices made within specified constraints. In the organizational-process model, government means large organizations, each of which attends to a special set of problems and acts in a quasi-independent way. Policy, then, is the routinized output of organizations coordinated by governmental leaders. The bureaucratic-politics model assumes that power is shared among persons who act according to diverse international, national/organizational, and personal goals. Policy is thus the outcome of bureaucratic give and take, of a political process. (For the rational model, see ibid., 10–66; for the organizational process model, 67–143; and for the bureaucratic politics model, 144–244.)

3. Riggs, *Administration in Developing Countries*, 15.

4. Ibid. Riggs argues here that given high degrees of overlapping in administrative organization, considerable formalism can be expected.

5. "'Politicians' are unable to formulate clear and enforceable policies. Lack of clarity in the 'political' policy-making process means that 'Administrators' have to fill the gap by supplementing the vague goals and objectives given them by the Politicians. This supplementing process means that Administrators themselves have to engage in political action, as well as in the administrative process. In other words, they become part-time or 'quasi' politicians" (ibid., 55).

6. Thurber and Graham, *Development Administration*, 18.

7. Ibid., 24.

8. Grindle, ed., *Politics and Policy Implementation*, 18–19.

9. Robert T. Daland, *Brazilian Planning* (Chapel Hill: University of North Carolina Press, 1967), 4. Science and technology plans are frequently included in the economic development plan. When set up independently, they can be seen as a de facto appendix to the economic development plan.

10. Ibid., 215.

11. Klaus Mehnert, "The Weather Makers," in *The Intellectuals in Politics*, ed. Malcolm MacDonald (Austin: University of Texas, Humanities Research Center, 1966), 93.

12. Thomas Sowell refers to six types of ideas used for the production of knowledge: "ideas systematically prepared for authentication ('theories'), ideas not derived from any systematic process ('visions'), ideas which could not survive any reasonable authentication process ('illusions'), ideas which exempt themselves from the authentication process ('myths'), ideas which have already passed authentication processes ('facts'), as well as ideas known to have failed—or certain to fail—such processes ('falsehoods—both mistakes and lies')" (*Knowledge and Decisions* [New York: Basic Books, 1980], 4–5).

13. Franklin L. Baumer, *Modern European Thought: Continuity and Change in Ideas—1600–1950* (New York: Macmillan, 1977), 7. As Joseph A. Schumpeter remarked, the intellectual's influence is as "the moral code of an epoch

that exalts the cause of some interests and puts the cause of others tacitly out of court" ("The Sociology of Intellectuals," in *The Intellectuals: A Controversial Portrait*, ed. George B. de Huszar [Glencoe, Ill.: Free Press, 1960], 78).

14. Aloysio Biondi, as quoted in João Paulo dos Reis Velloso, *Brasil: A solução positiva* (São Paulo: Abril-Tec, 1978), 51. My translation. "Intellectuals," says Schumpeter, "rarely conquer responsible office. But they staff political bureaus, write party pamphlets and speeches, act as secretaries and advisers, make the individual politician's newspaper reputation which, though it is not everything, few men can afford to neglect. In doing these things they to some extent impress their mentality on almost everything that is being done" ("Sociology of Intellectuals," 78). In so doing they also affect administrative practice.

15. The military are inclined to see the technocrat, or *técnico*, as a substitute for the politician, or *político*. This comes from a distrust of the politician—both for his links with popular sectors and with various clienteles and for his compromising and bargaining-oriented attitude, which becomes a roadblock to the achievement of economic development goals. On the other hand, the *técnico* "knows what is best" and applies it. If the regime is authoritarian, the *técnico* need not worry about popular and clientelistic pressures.

16. For an example of Mexico, see Raymond Vernon, *The Dilemma of Mexico's Development* (Cambridge, Mass.: Harvard University Press, 1963), 136. See also Roderic Ai Camp, *The Role of Economists in Policy Making: A Comparative Case Study of Mexico and the United States* (Tucson: University of Arizona Press, 1977).

17. Guy Benveniste, *Bureaucracy and National Planning* (New York: Praeger, 1970), 87.

18. Quoted in ibid.

19. The term *subversive* is by no means intended to be derogatory. It is used only in a metaphorical sense to convey the process by which people having ideas of a certain ideological flavor purposefully and incrementally influence political action.

20. See Frank Bonilla, "Cultural Elites," in *Elites in Latin America*, ed. Seymour M. Lipset and Aldo Solari (London: Oxford University Press, 1967).

21. Adrien C. Taymans, "Facts and Theory in Entrepreneurial History," *Exploration in Entrepreneurial History* 1 (Jan. 1949): 17–18.

22. Remarks of the Peruvian philosopher Francisco Miro Quesada, at a public lecture at the University of California, Berkeley, Winter 1981.

23. Edward Shils, "The Intellectuals in the Political Development of the New States," *World Politics* 12 (April 1960): 344.

24. Ibid.

25. Edited, with introduction and notes, by Gordon Brotherston (Cambridge: Cambridge University Press, 1967).

26. Miro Quesada, Berkeley lecture.

27. Edward Shils, *The Intellectuals and the Powers* (Chicago: University of Chicago Press, 1972), viii.

28. All references and quotations not otherwise noted are based on inter-

views I conducted in Latin America in the spring and summer of 1980 and the fall of 1982. The,persons interviewed are identified here by their titles at the time of my research, which may not reflect their current (1986) positions.

29. The term is used in a figurative way and is by no means intended to be derogatory.

30. Albert O. Hirschman, "The Turn to Authoritarianism in Latin America and the Search for Its Economic Determinants," in *The New Authoritarianism in Latin America,* ed. David Collier (Princeton, N.J.: Princeton University Press, 1979), 86–87.

31. Torcuato Di Tella, "La función política de la inteligentsia latinoamericana," in *Los intelectuales políticos,* ed. Juan F. Marsal (Buenos Aires: Nueva Visión, 1971), 339, 338. My translation.

32. "The 'client' of the bureaucratic intellectual is a policy maker who is concerned with translating certain vague or well-defined purposes into programs of action" (Robert K. Merton, "Role of the Intellectual in Public Bureaucracy," *Social Forces* 23, no. 4 [May 1945]: 410).

33. Carlos Estevam Martins, in his study of Brazilian technocrats (*Tecnocracia e capitalismo* [São Paulo: Editora Brasiliense, 1974], 89–99), also deals with the question of alliances between politicians and technocrats, but from a Marxist, class-oriented perspective. (My use of the term *alliance* is coincidental and has nothing to do with the literature Martins is discussing.)

34. Vernon, *The Dilemma of Mexico's Development,* 137–38.

35. Benveniste, *Bureaucracy and National Planning,* 104.

36. Di Tella, "La función política," 326.

Chapter 5

1. Albert O. Hirschman, *Journeys Toward Progress: Studies of Economic Policy Making in Latin America* (New York: W. W. Norton, 1973), 229, 231.

2. Albert O. Hirschman, *Essays in Trespassing: Economics to Politics and Beyond* (Cambridge: Cambridge University Press, 1981), 140.

3. Hirschman, *Journeys Toward Progress,* 245.

4. Hirschman, *Essays in Trespassing,* 155–56.

5. Rogelio Frigerio, *Síntesis de la historia crítica de la economía argentina* (Buenos Aires: Hacette, 1979), 72. See also Aldo Ferrer, *Economía internacional contemporánea: Texto para latinoamericanos* (Mexico, D.F.: Fondo de Cultura Económica, 1976), 175.

6. Salvador Treber, *La economía argentina: Análisis, diagnóstico, y alternativas* (Buenos Aires: Macchi, 1977), 54.

7. Laura Randall, *An Economic History of Argentina in the Twentieth Century* (New York: Columbia University Press, 1978), 4.

8. Quoted from National Council of Science and Technology (CONACYT), "La investigación industrial y el INTI," in Oscar Oszlak, "State Policy and Organization of Scientific and Technical Activities in Argentina: A Cri-

tique of Current Models and Prescriptions," paper presented at the Seminar on Implementation in Latin America's Public Sector: Translating Policy into Reality, Austin, Texas, April 29–May 1, 1976, 42.

9. UNESCO, "Política científica y organización de la investigación científica en la Argentina," Estudios y documentos de la política científica 20 (Paris, 1970):29–30.

10. Of the total public expenditures for research and development, 4.3 percent were devoted to the industrial sector in 1961, 3.4 percent in 1963, and 2.9 percent in 1965. UNESCO, "Política científica," 45, 47, 55, 65.

11. The Presidency, Secretariat of Planning, Subsecretariat of Science and Technology (SUBCYT), *Informe de situación* (Buenos Aires, July 1982), 25–26.

12. Message to the Congress attached to Law 20.794 (September 27, 1974), *Derecho de la integración* 8, no. 18/19 (March/July 1975):229. My translation.

13. Ferrer, *Economía internacional contemporánea*, 197. My translation.

14. *Economic Information on Argentina*, no. 104 (Feb. 1980), 6.

15. Robert Mallon and Juan V. Sourrouille, *Economic Policy Making in a Conflict Society: The Argentine Case* (Cambridge, Mass.: Harvard University Press, 1975), 76–77, 87.

16. *Economic Information on Argentina*, no. 111 (Nov. 1980), 11.

17. Elva Roulet, "Análisis de instrumentos de política científica y tecnológica" (Buenos Aires, Jan. 1973, mimeo), 45–46.

18. José-María Dagnino Pastore, "Multinational Corporations and Transfer of Technology: The Case of Argentina," in Organization of Economic Cooperation and Development (OECD), *Transfer of Technology by Multinational Corporations*, ed. Dimitri Germidis (Paris: OECD Development Center, 1977), 163.

19. Ibid., 164–65.

20. Business International Corporation (BIC), *Investing, Licensing, and Trading Conditions Abroad* (New York), May 1978, 10 (hereafter *ILT* Argentina). For the legislation on technology imports and exports since 1977, see Interamerican Development Bank/Latin American Integration (BID/INTAL), *Régimen de la transferencia de tecnología en los países de América latina. Textos legales y procedimientos administrativos* (Buenos Aires, 1977–1981).

21. Approval could have been denied when the contract prohibited or limited exports; obliged the licensee to cede innovations or the licensed product; failed to grant the licensee the right to receive improvements in the licensed technology; required the purchase of raw materials, intermediary goods, or capital goods from a specified source; established a resale price; permitted the licensor to control production or marketing to an extent greater than necessary for protection of its right under the agreement; or prohibited the use of competitive technology without good cause (*ILT* Argentina, May 1978, 11).

22. Ibid.

23. Ibid., June 1981, 10.

24. *Integración latinoamericana*, May 1981, 87–88. A new trademark law (22.362) had been passed on February 2, 1981, one of its main purposes being

to stipulate that ownership of trademark begins at registration (*ILT* Argentina, June 1981, 10).

25. Secretariat of the National Council of Science and Technology (SECONCYT), *El sistema científico-técnico y su relación en el sistema socio-económico* (Buenos Aires: National Planning Directorate, Dec. 1970), 32; United Nations Conference for Trade and Development (UNCTAD), *Legislation and Regulations on Technology Transfer: Empirical Analysis of Their Effects in Selected Countries*, TD/B/C.6/55 (Geneva: United Nations, Aug. 1980), 26.

26. National Institute of Industrial Technology (INTI), *Aspectos económicos de la importación de tecnología en la Argentina en 1972* (Buenos Aires, Nov. 1974), 10; Leopoldo H. Tettamanti, "La brecha tecnológica," *Estudios sobre la economía Argentina* 20 (Jan. 1975):68; and INTI, "Contratos de transferencia de tecnología con el exterior," *Derecho de la integración* 16 (July 1974):186–89.

27. Argentine Republic, National Directorate of Industrial Property, *Publicación complementaria de información estadística* (Buenos Aires), 1975, 12; 1976, 32; 1977, 50; 1978, 31.

28. For a description of Law 18.587 (February 1970) on industrial promotion and Law 19.151 (July 1971) on foreign investment, as well as other laws and regulations, see Roulet, *Análisis de instrumentos*, 45–62.

29. Ferrer, *Economía internacional contemporánea*, 206–9.

30. *ILT* Argentina, Jan. 1977, 6; and Executive Power of the Argentine Republic, Ministry of Economy, *Legal System Governing Foreign Investment* (Law 21.382).

31. *ILT* Argentina, May 1978, 15–16. Law 21.608 of 1977 eliminated the 1973 regime of industrial promotion and replaced it with a liberal regime.

32. Daniel Chudnovsky, "Empresas multinacionales y tecnología en la industria argentina," *Comercio exterior* 25, no. 4 (April 1975):453–54.

33. Mallon and Sourrouille, *Economic Policy Making*, 73.

34. INTI, "Contratos de transferencia," 186; and INTI, *Aspectos económicos*, 49.

35. *ILT* Argentina, February 1974, 3.

36. UNESCO, *Statistical Digest 1981* (Paris, 1981), 144.

37. *Economic Information on Argentina*, no. 108 (June/Aug. 1980), 32.

38. UNESCO, *Statistical Yearbook 1982* (Paris, 1982), V-36 (hereafter *USY*).

39. *USY 1982*, V-36, 48–49.

40. *USY 1974*, 631; *USY 1980*, 873; and *USY 1982*, V-116.

41. *USY 1980*, 825, 852.

42. Argentine Republic, Secretariat of Planning and Government Action, Subsecretariat of Science and Technology (SUBCYT), *Política nacional de ciencia y técnica: Plan operativo 1973* (Buenos Aires, n.d.), 45 (hereafter *1973 Science and Technology Plan*). The 22 priorities were metallurgy, machinery and equipment, food technology, leather technology, cellulose and paper, chemistry, beef production, wool, grains, fruits, industrial farming, energy, electronics, water resources, educational research, environment, nutrition, endemic illnesses, housing, oceanographic research, informatics, and computation and social research (ibid., 53).

43. Francisco Sagasti, *Science and Technology for Development: Main Comparative Report of the Science and Technology Policy Instruments Project (STPI)* (Ottawa: International Development Research Centre [IDRC], 1978), 57.

44. Finalidad 8 has historically been divided into three categories: science and technology training and promotion, R & D, and other expenditures. The amounts devoted to R & D have been the largest, ranging from a low of 58.7 percent in 1973 to a high of 74.5 percent in 1981. The average expenditure for R & D out of Finalidad 8 for the 1972–1981 period was 65 percent while the average expenditure for science and technology training and promotion during those years was 8.7 percent. During the last period expenditures for science and technology training and promotion went up, taking a bite from "other expenditures" (Ministry of Culture and Education, National Council of Scientific and Technical Research [CONICET], *Programa de centros regionales de investigación científica y tecnológica* [Buenos Aires, Sept. 1977], vol. 2, chap. 2, 46; and data furnished by Lic. María Lujan Marcon [CONICET]).

45. This and the information on Finalidad 8 that follows in the text comes from the two sources listed in note 44, above.

46. CONICET, *Primer encuentro de directores de institutos del CONICET. Bahía Blanca, 29 y 30 de junio de 1979* (Buenos Aires, Sept. 1979), 32.

47. Ibid., 52.

48. *Economic Information on Argentina*, no. 124 (Sept./Oct. 1982), 40.

49. *1973 Science and Technology Plan*, 69–79. The UNDP figures total 101 percent in the source.

50. The program was aimed at strengthening scientific and technological development in new areas within Argentina, expanding and building installations and providing scientific and technical personnel for nine regional centers, and improving and extending the research capacity in agriculture, industry, fisheries, infrastructure, navigation, transportation, and natural resources. The state also purchased and installed equipment, contracted about 200 international consultants to provide technical assistance for the centers to be built, gave 275 scholarships for training personnel, and increased research, professional, and technical personnel in seven centers by about 600 (*Economic Information on Argentina*, no. 97 [June 1979], 14–15).

51. *USY 1976*, 701; *USY 1977*, 686; and *USY 1980*, 844.

52. Law 18.527 referred to fiscal incentives to promote R & D in specific areas. It exempted from taxes local private and public industries and institutions that would perform R & D. Roulet, *Análisis de instrumentos*, 33–35. The tax incentives for public enterprises were set up by Decree 466 of 1971.

53. Ibid., 42–43. Law 18.587 for industrial promotion encouraged the creation of "industrial parks" in areas being developed, the participation of the state in the capitalization of productive units, and the creation of R & D and technical assistance organizations (ibid., 53–55).

54. Ibid., 63–67.

55. BID/INTAL, "El marco jurídico de la innovación tecnológica en América Latina," *Estudios* 2, no. 19 (Dec. 1976):37.

56. Interamerican Development Bank (IDB), *Economic and Social Progress in Latin America: The External Sector*, 1982 Report (Washington, D.C., 1982),

142, 143, 152, 155. Although multinationals accounted for 70 percent of infrastructure sales (and 40 percent of contracts), they were dominated by domestic industry in industrial plant sales, accounting for only 28 percent (and 3 percent of contracts) (ibid., 152).

57. The enterprise enters into a contract with INTI to undertake technological innovation, and INTI may finance up to 50 percent of the needs. If the innovation is successful, the enterprise must repay the loan without interest; in case of failure it is not required to repay INTI. The research takes place in the facilities of the enterprise and in temporary research centers. If the final product is patentable, the patent is registered in INTI's name, but after reimbursement of the loan the patent right is transferred to the enterprise (INTI, *Régimen para la promoción de desarrollos tecnológicos*, 1978).

58. Argentine Republic, National Development Council (CONADE), *Plan nacional de desarrollo y seguridad 1971–1975* (Buenos Aires, 1971).

59. *1973 Science and Technology Plan*, 45. My translation.

60. Argentine Republic, National Executive Branch, *Plan trienal para la reconstrucción y la liberación nacional* (Buenos Aires, Dec. 1973).

61. CONICET, *Objetivos, políticas 1982/85* (n.d., internal use).

Chapter 6

1. The best-known explanation of the 1966 coup from an economic perspective is O'Donnell's bureaucratic-authoritarianism theory. See Guillermo A. O'Donnell, *Modernization and Bureaucratic-Authoritarianism*, Politics of Modernization Series, no. 9 (Berkeley, Calif.: Institute of International Studies, 1973), chaps. 2–4.

2. The government imposed a long-term wage freeze, lowered import duties, increased taxes, reduced the deficit, and decreed a "once-and-for-all 40 percent devaluation of the peso" (Robert R. Kaufman, "Industrial Change and Authoritarian Rule in Latin America: A Concrete Review of the Bureaucratic-Authoritarian Model," in *The New Authoritarianism in Latin America*, ed. David Collier [Princeton, N.J.: Princeton University Press, 1979], 176).

3. Ibid., 236.

4. Edward S. Milenky, *Argentina's Foreign Policies* (Boulder, Colo.: Westview Press, 1978), 84.

5. Gary W. Wynia, *Argentina in the Postwar Era: Politics and Economic Policy Making in a Divided Society* (Albuquerque: University of New Mexico Press, 1978), 177, 180.

6. UNESCO, "Política científica y organización de la investigación científica en la Argentina," *Estudios y documentos de la política científica* 20 (Paris, 1970): 129. My translation.

7. "Appendix B: Action of the OAS in Relation to Scientific and Technological Development in Latin America," in *Texas Forum* (see chap. 3, n. 7), 323.

8. Aldo Ferrer, *Economía internacional contemporánea: Texto para latino-*

americanos (Mexico, D.F.: Fondo de Cultura Económica, 1976), 195, 200. My translation.

9. Business International Corporation (BIC), *Argentina in Transition: Evaluating Business Prospects in a New Setting* (New York, 1978), 28. The value of manufactured goods exports grew from $420 million in 1970 to $442 million in 1971 and $588 million in 1972. Kaufman, "Industrial Change," 244.

10. Jorge Katz and Eduardo Ablin, "Tecnología y exportaciones industriales: Un análisis microeconómico de la experiencia argentina reciente," *Desarrollo Económico* 17, no. 65 (April/June 1977):99. In the period between 1969 and 1974 the ratio was approximately 15 percent.

11. José Serra, "Three Mistaken Theses Regarding the Connection Between Industrialization and Authoritarian Regimes," in Collier, ed., *The New Authoritarianism,* 146–47.

12. Aldo Ferrer, *La posguerra* (Buenos Aires: El Cid, 1982), 34–35. My translation.

13. Ferrer, *Economía internacional contemporánea,* 202. My translation.

14. This view, conveyed to me by Ferrer, was also shared by Diamand, but he added that an ideological evolution of some industrialists toward the center allowed CGE in these years to become more influential.

15. Wynia, *Argentina in the Postwar Era,* 223–24.

16. Among the strategies chosen were public-sector austerity, redistribution of public expenditures in favor of social services, progressive income and land tax reform, consolidation of public enterprises, slow growth of the money supply, channeling of credit to domestic enterprises, avoidance of devaluation, reduction and control of consumer goods prices, institution of a "voluntary" two-year wage freeze, selective import controls, gradual improvement in real working-class income, and a shift from foreign to domestic investors (ibid., 215).

17. Economic Commission for Latin America (CEPAL), *Estudio económico de América latina: 1977* (Santiago, Chile: United Nations, 1978), 36, 44.

18. This information was taken from Ministry of Economy, "Ciencia y tecnología en la nueva ley de ministerios: Documentos de trabajo" (Nov. 1973; internal memorandum, mimeo).

19. Wynia, *Argentina in the Postwar Era,* 214.

20. Elements of the system included the National Center for Electronics Research, the National Institute of Physics Research for the Production of Energy, the National Program for Oceanographic Research, the Special Fund for the Support of Young Scientists, and the National System of Science and Technology Research (Dr. Julio H. G. Olivera, *Política científica y desarrollo económico* [Buenos Aires: National Academy of Economic Science, 1974], 10).

21. Adolfo Canitrot, "Discipline as the Central Objective of Economic Policy: An Essay on the Economic Programme of the Argentine Government Since 1976," *World Development* 8 (1980):918.

22. Ibid., 922. Among the provisions of the new economic strategy were the establishment of a new equilibrium level for real wages about 40 percent lower than the previous five-year average; application of a program of progressive tariff reductions; liberalization of financial and foreign-exchange

markets; and reduction of government expenditures, public employment, and the borrowing requirement, together with the return of public enterprises to private ownership (ibid., 917).

In 1978 the government devalued the peso at 30 percent behind the domestic price increase for that year, adjusted to international inflation. In 1979 import duties were slashed across the board, and duties for capital goods projected for 1984 were already implemented, creating in fact a negative protection for several branches of this industrial sector. The lowering of import duties and the overvaluation of the peso became the basic instruments to change relative prices, reallocate resources, and fight inflation (Aldo Ferrer, "The Argentine Economy: 1976–1979," *Journal of Interamerican Studies and World Affairs* 22, no. 2 [May 1980]: 140–44).

23. *Economic Information on Argentina*, nos. 98 (July 1979) and 119 (July/Aug. 1981).

24. Ferrer, "The Argentine Economy," 151. In the first nine months of 1978, for example, value added (the increase in value of a raw material once it becomes a finished product) in the basic metalworking industry dropped 18.5 percent; similar drops occurred in other industries: machinery and equipment, 9.7 percent; textiles, 5.3 percent; chemicals, 2.8 percent; and automotive output plunged 30 percent for automobiles and 76 percent for tractors. Although there was a partial recovery in 1979, especially in the first quarter, 1980 saw further stagnation. For example, a drop in steel production of 15 percent reduced its economies of scale and increased prices relative to imports (*Business Latin America*, Nov. 29, 1978, 381; Feb. 18, 1981, 52).

Chapter 7

1. External demand decreased, exports fell from $455.9 million in 1929 to $180.6 million in 1932, and capacity to import was severely curtailed. "With large stocks of unsalable coffee on their hands, politically powerful coffee growers succeeded in obtaining huge subsidies from the government in the form of government coffee purchases. This increased aggregate demand combined with imports effectively tariffed by the Depression created the opportunity for capital holders to invest in industry" (Richard Newfarmer and Willard F. Mueller, "Multinational Corporations in Brazil and Mexico: Structural Sources of Economic and Noneconomic Power," *Report to the Subcommittee on Multinational Corporations of the Committee on Foreign Relations*, U.S. Senate, 94th Cong., 1st sess. [Washington, D.C.: GPO, 1975], 96).

2. Werner Baer and Carlos Von Doellinger, "Determinants of Brazil's Foreign Economic Policy," in *Latin America and World Economy: A Changing International Order*, ed. Joseph Grunwald (Beverly Hills, Calif.: Sage, 1978), 148.

3. In the 1940s the production output increased sixfold, the number of industrial establishments grew by more than two-thirds, and the industrial labor force expanded by over 60 percent. Changes in the industrial pattern reflected the country's increased maturity. Chemicals and pharmaceuticals,

building materials, metallurgy, and machinery for the transportation, electrical, and communications industries acquired greater importance relative to food and textiles. Between 1948 and 1953 production of capital goods rose by 77 percent, while the output of consumer goods increased about 30 percent (Business International Corporation [BIC], *Operating Successfully in a Changing Brazil* [New York, 1975], 11).

4. Carlos Lessa, "Fifteen Years of Economic Policy in Brazil," *Economic Bulletin for Latin America* 9, no. 2 (Dec. 1964):153.

5. Stephen H. Robock, *Brazil: A Study in Development Progress* (Lexington, Mass.: Lexington Books, 1975), 29.

6. "Although the instruments at BNDE's [National Economic Development Bank] disposal played a vital role in regulating and furthering the development of the basic and capital goods industries, the most important group of instruments for the channeling of private investment comprised the complex exchange system, the customs tariff and the policy applied to attract savings from abroad" (Lessa, "Fifteen Years of Economic Policy," 196).

7. Newfarmer and Mueller, "Multinational Corporations," 97. On foreign investment in Brazil, see also Nathaniel H. Leff, *Economic Policy-Making and Development in Brazil, 1947–1964* (New York: Wiley, 1968), chap. 4; and José María Gouvêa Vieira, *O capital estranjeiro no desenvolvimento do Brasil* (São Paulo: Difel, 1975). See also Peter Evans, *Dependent Development: The Alliance of Multinational, State, and Local Capital in Brazil* (Princeton, N.J.: Princeton University Press, 1979).

8. Simon Schwartzman, "Struggling To Be Born: The Scientific Community in Brazil," *Minerva* 16, no. 4 (Winter 1978):545.

9. IPT flourished during its first years and then entered a period of decline. In the words of one of its directors: "The technical assistance agreements and the intense implantation of foreign industry in our environment took away from IPT the leadership in the introduction of innovations" (José Murilho de Carvalho, "A política científica e tecnológica do Brasil" [Rio de Janeiro: IUPERJ, 1976, mimeo], 52). My translation.

10. See Regina Lúcia de Morâes Morel, *Ciência e estado: A política científica no Brasil* (São Paulo: Queiroz, 1979), 41–42. See also "A SBPC e a função social (e política) do cientista," *Cadernos de tecnologia e ciência* 1, no. 3 (Oct./Nov. 1978):37–49; Simon Schwartzman, *Formação da comunidade científica no Brasil* (Rio de Janeiro: FINEP, 1979); idem, "Struggling To Be Born," 545–80; and idem, "Por uma política científica no Brasil," *O estado de São Paulo*, March 24, 1979, 4–5.

11. Morel, *Ciência e estado*, 43.

12. Presidency of the Republic, Planning Secretariat (SEPLAN), "Ciência, tecnologia, e desenvolvimento," *Relatório de atividades CNPq 1975–1978* (Brasília: National Council of Scientific and Technological Development [CNPq], 1979), 7 (hereafter CNPq, *Relatório de atividades 1975–1978*).

13. Morel, *Ciência e estado*, 49.

14. But the fact that other ministries and banks (Banco Central, Banco do Brasil, and BNDE) can decide on tax exemptions diminishes the effectiveness of CDI's coordinating role.

15. Shing K. Fung and José Cassiolato, "The International Transfer of

Technology to Brazil Through Technology Agreements: Characteristics of the Government Control System and the Commercial Transactions" (Cambridge, Mass.: MIT Center for Policy Alternatives, 1976, mimeo), 110–11.

16. Eduardo Rappel, "PBDCT: Plano ou programão?" *Cadernos de tecnologia e ciência* 1, no. 3 (Oct./Nov. 1978):5–6.

17. In the mid-1970s there was a proposal in Congress that a science and technology secretariat be created, but the motion was considered counterproductive and was not approved. Opposition was based on the fact that science and technology policy is an interagency policy and it was believed preferable to have a coordinating mechanism between the various agencies and sectors rather than a science and technology minister who would unavoidably get into confrontations with other ministers. See Eduardo Rappel, "O retrato de um paradoxo brasileiro," *Revista brasileira de tecnologia* 12, no. 2 (April/June 1981):3–8. Nevertheless, in 1985 the new civilian government created a science and technology ministry, headed by Renato Archer.

18. Presidency of the Republic, *I PBDCT, Basic Plan for Scientific and Technological Development, 1973/74* (Brasília, June 1973), 5; and Federal Republic of Brazil, *II Plano Nacional de Desenvolvimento 1975–1979 (II PND)* (1974), 135. My translation.

19. Julian M. Chacel, "Summary of Brazilian Policies on Science and Technology," in *Business-Government Cooperation in Science and Technology for Development*, ed. Alejandro D. Sans and Harvey W. Wallender III, Seminar Report, Fund for Multinational Management Education, Rio de Janeiro, May 10–11, 1979, 2.

20. BIC, *Investing, Licensing, and Trading Conditions Abroad* (New York), Nov. 1978, 34–35 (hereafter *ILT* Brazil).

21. William Tyler, "Brazilian Industrialization and Industrial Policies: A Survey," *World Development* 4, nos. 10/11 (Oct./Nov. 1976):869. (Figures refer to tariffs' nominal rates.)

22. *ILT* Brazil, Nov. 1978, 35. The Council of Tariff Policy (CPA) could award some duty reductions on imports of capital goods for approved investment in certain industries, but only after recommendation by CDI or any other agency allowed to grant incentives (ibid.).

23. Ibid., Dec. 1980, 31.

24. José Serra, Working Document presented to ECLA (São Paulo, 1981), tables 33 and 54.

25. Eduardo Augusto de Almeida Guimarães and Ecila Mutzenbecher Ford, "Science and Technology in Brazilian Development Plans: 1956–1973," in *Science and Technology for Development: Planning in the STPI Countries*, ed. Francisco R. Sagasti and Alberto Aráoz (Ottawa: International Development Research Centre [IDRC], 1979), 54. See also A. L. Figueira Barbosa, *Propriedade e quase propriedade no comércio de tecnologia* (Brasília: CNPq, 1978); Francisco Almeida Biato, Eduardo Augusto de Almeida Guimarães, and Maria Helena Poppe de Figueiredo, *A transferência de tecnologia no Brasil* (Brasília: IPEA, 1973); and Murillo F. Cruz Filho and Anne-Marie Maculan, *Propriedade industrial e transferência de tecnologia*, Coleção Estudos de Política Científica e Tecnológica (Brasília: CNPq, 1981).

26. Fung and Cassiolato, "International Transfer of Technology," 12–14.

27. *ILT* Brazil, Nov. 1978, 14–15.

28. Richard D. Robinson, *National Control of Foreign Business Entry: A Survey of Fifteen Countries* (New York: Praeger, 1976), 235.

29. *ILT* Brazil, April 1974, 10.

30. As quoted in A. L. Figueira Barbosa, "Considerações sobre categorias tecnológicas e política de desenvolvimento," *Conjuntura econômica* 33, no. 1 (Jan. 1979):85. My translation.

31. Robinson, *National Control*, 238.

32. Ministry of Industry and Commerce (MIC), Industrial Technology Secretariat/National Institute of Industrial Property (STI/INPI), Revista de propriedade industrial, *Normative Act 15 of September 11, 1975*, 3. My translation.

33. In addition, the "contractual duration for licenses was changed from the former maximum of five years to not to exceed the period of validity of the protection granted to industrial property as follows: Fifteen years for patent of invention; and ten years for patent of utility model, industrial model, or design. The contractual duration for the supply of industrial technology and technical-industrial cooperation was to be determined by the time necessary to enable the recipient to master the technology, whereas that for specialized technical services was to be the time needed for the rendering of services or the completion of the project" (Fung and Cassiolato, "International Transfer of Technology," 9).

34. United Nations Conference for Trade and Development (UNCTAD), *Legislation and Regulations on Technology Transfer: Empirical Analysis of Their Effects in Selected Countries,* TD/B/C. 6/55 (Geneva: United Nations, Aug. 1980), 24; and MIC, STI/INPI, *Relatório de atividades 1980,* 10.

35. Fábio Stefano Erber, "Science and Technology Policy in Brazil: A Review of the Literature," *Latin American Research Review* 16, no. 1 (1981):40; and Biato et al., *A transferência de tecnologia,* 213.

36. Erber, "Science and Technology Policy," 40–41.

37. MIC, STI/INPI, *Relatório de atividades 1980,* 9–10.

38. Barbosa, *Propriedade,* 120–22; and MIC, STI/INPI, *Relatório de atividades 1980,* 16.

39. Fung and Cassiolato, "International Transfer of Technology," 4.

40. *Business Latin America,* May 4, 1977, 144.

41. Ibid., Feb. 22, 1978, 57.

42. See ibid., April 6, 1977, 109; and May 9, 1979, 146.

43. *ILT* Brazil, Dec. 1980, 4.

44. Fung and Cassiolato, "International Transfer of Technology," 5.

45. *ILT* Brazil, Dec. 1980, 4, 5; and *Veja,* July 15, 1981, 93.

46. Sylvia Ann Hewlett, *The Cruel Dilemmas of Development: Twentieth-Century Brazil* (New York: Basic Books, 1980), appendix. The concern of Brazilians with what they defined as the negative effects of multinational corporations in Brazil prompted a congressional investigation: see House of Representatives, *Commissão parlamentar de inquérito das multinacionais,* Voto em Separado do MDB, vol. 1, doc. 33.85C733 (n.d.).

47. *ILT* Brazil, Dec. 1980, 4, 14.

48. Gary Gereffi and Peter Evans, "Transnational Corporations, Dependent Development, and State Policy in the Semiperiphery: A Comparison of Brazil and Mexico," *Latin American Research Review* 16, no. 3 (1981):48.

49. Schwartzman, "Struggling To Be Born," 566–67. Between 1960 and 1965 the number of students in higher education went up 67 percent and the number of faculty rose 57 percent; between 1965 and 1970 the former category increased 173 percent, the latter category 64 percent (Morel, *Ciência e estado*, 147).

50. Schwartzman, *Formação da comunidade científica*, 293.

51. Schwartzman, "Struggling To Be Born," 570.

52. Morel, *Ciência e estado*, 148.

53. *Planejamento e desenvolvimento* 9, no. 83 (Jan./Feb. 1982):79–81; and Statistical Appendix, *Revista brasileira de tecnologia* 13, no. 3 (June/July 1982):61.

54. A master's thesis in the engineering field tended to be a lengthy and original piece of work and was often competitive, at least in length, with doctoral dissertations in other countries.

55. See Marcia Bandeira de Melo Nunes, Nadja Volia de Souza, and Simon Schwartzman, "Pós-graduação em engenharia: A experiência da COPPE" (Rio de Janeiro, Oct. 1978, mimeo), 1.

56. Schwartzman, "Struggling To Be Born," 573. The average annual growth rate of student enrollments between 1963 and 1967 was about 140 percent; between 1967 and 1971 it came down to 25 percent and between 1972 and 1976 it decreased even further to 18.2 percent (de Melo Nunes et al., "Pós-graduação em engenharia," 20). From 1964 to 1977 COPPE granted over 900 master's and 25 doctoral degrees. COPPE has an extremely flexible program; it invites foreign professors and sends the best students to study abroad, and it decides on its own curriculum and research plans (ibid.).

57. CNPq, *Relatório de atividades 1975–1978*, 34.

58. See Schwartzman, "Struggling To Be Born," 574.

59. Presidency of the Republic, *II PBDCT, II Plano Básico de Desenvolvimento Científico e Tecnológico* (Brasília, March 1976), 141.

60. Francisco Almeida Biato, Eduardo Augusto de Almeida Guimarães, and Maria Helena Poppe de Figueiredo, *Potencial de pesquisa tecnológica no Brasil* (Brasília: IPEA, 1971), 52.

61. Ibid., 63–64. Thirty-eight percent of the enterprises were installed with domestic know-how (ibid., 77).

62. Fábio Stefano Erber et al., *Reflexões sobre a demanda pelos serviços dos institutos de pesquisa* (Rio de Janeiro: FINEP, 1974), 16.

63. De Melo Nunes et al., "Pós-graduação em engenharia," 31. My translation.

64. De Carvalho, "A política científica," 29.

65. See "Pesquisas e materiais e processos para engenharia," *Conjuntura econômica* 28, no. 1 (Jan. 1974):112–13. For a historical account of INT, see Simon Schwartzman, "The Bureaucratization of Technology," paper presented at the Technological Policy Seminar, Instituto Torcuato Di Tella, Buenos Aires, Oct. 4–6, 1982.

66. [CTA] initially served as a center for the development of personnel and basic aeronautical research. The CTA sought to transfer the developed technology endeavoring to create industries in prime materials, motors, turbines, aviation systems, electronic instruments, etc. In the beginning of the 1970s, the CTA attempted to create appropriate conditions among a small group of enterprises in order to support an aeronautical program. More recently, the Institute for Industrial Growth and Coordination (IFI/CTA) has sought to stimulate the development of national industries in forging and smelting, and factories for the various needs of the aeronautical complex including finishing industries. The action of the IFI, in its attempt to create an industrial park, complements the policies of EMBRAER [the state aeronautical industry] in its internal search for quality goods at reasonable prices. The activities of IFI have been aided financially by FINEP, BADESP [São Paulo State Development Bank], etc., guaranteeing rapprochement with the small and medium industries seen by EMBRAER as forming an industrial base. The institute also participate [sic] in negotiations concerning the transfer of technology dealing with a new product.

EMBRAER employs the services of two private enterprises previously created completing the cycle in this sector which also includes the purchase of parts or components from the small and medium enterprises already mentioned. . . . Since the majority of these enterprises are small, the difficulties in adapting and assimilating ever more sophisticated technology are significant. It is in this area that the role of IFI/CTA has been essential in stimulating industry to develop from rudimentary technological stages to greater complexity (Paulo D'Arrigo Vellinho, "Considerations on the Creation and Transfer of Technology in Brazilian Industry," in Sans and Wallender, eds., *Business-Government Cooperation*, 13–14).

67. *Índice do Brasil: Brazilian Index Yearbook, 1979/80* (Rio de Janeiro: Data Bank Enterprise, n.d.), 136. An in-house study of 67 projects financed by FINEP, assessing their impact on public policy and the expansion of domestic enterprise, found that 34.3 percent of the projects led to the introduction of products, 30 percent improved products, 73.1 percent helped the diffusion of techniques, 70.1 percent saved or produced foreign exchange, 41.8 percent increased sales, 64.1 percent led to diversification of production, 26.9 percent led to the training of specialized human resources, and 65.6 percent consolidated the enterprises' technological capacity (Marcelino José Jorge [Coordinator] et al., *Avaliação do impacto da atuação da FINEP: Uma análise exploratória* [Rio de Janeiro: DEP/FINEP, April 1982], 25–26).

68. CNPq, *Relatório de atividades 1975–1978*, 36. According to one of the best-known Brazilian scientists, José Goldemberg, CNPq lacks "teeth" because the system is still too decentralized and based on the granting of fellowships and grants for research. Instead, he proposed the creation of the "researcher" (*pesquisador*) career, which would match a university degree, and the creation of laboratories associated with CNPq. This is the model the French used to solve their crisis at the universities and to stimulate scien-

tific research, and that Argentina has adopted (*Agenda CNPq,* no. 11 [Jan. 1981], 4).

69. UNESCO, *Statistical Yearbook 1975* (Paris, 1975), 527; *USY 1978–1979,* 845; Studies and Projects Financing Agency (FINEP), "Recursos financeiros destinados a C & T" (n.d.; internal memo); and Statistical Appendix, *Revista brasileira de tecnologia* 13, no. 2 (April/May 1982):61.

70. Statistical Appendix, *Revista brasileira de tecnologia* 13, no. 2 (April/May 1982):62, 64.

71. *USY 1978–1979,* 741; and *USY 1980,* 774, 787, 832, 852–53.

72. For the history of the development of SNICT and the original proposals by the interministerial body in charge of creating the system, see Fernando de Mendonça, "Criação de um sistema nacional de informações tecnológicas," *Revista de administração de empresas* 14, no. 3 (May/June 1974).

73. *II PBDCT,* 28.

74. José Goldemberg, "Uma aliança necessária: Govêrno e ciência," *Revista brasileira de tecnologia* 12, no. 2 (April/June 1981): 16.

75. MIC, STI/INPI, *Relatório de atividades 1980,* 23.

76. The role of the state in the economy has been steadily increasing. For example, public expenditures grew from a 1947–1957 average of 19 percent of the GDP to 24 percent between 1969 and 1978. The public sector represented 38.1 percent of the total gross formation of fixed capital in 1965, but this share had increased to 43 percent by 1978 (Gesner Oliveira Filho, "A 'Estatização' da economia," pt. I, *Folha de São Paulo,* Oct. 4, 1981, 39). In 1980 the state controlled 83 of the 200 largest firms in terms of net worth, accounting for one-third of the GDP and 46 percent of sales (*ILT* Brazil, Dec. 1980, 5; and *Visão,* July 27, 1981, 61). See also Thomas J. Trebat, "Public Enterprises in Brazil and Mexico: A Comparison of Origins and Performance," in *Authoritarian Capitalism: Brazil's Contemporary Economic and Political Development,* ed. Thomas C. Bruneau and Philippe Faucher (Boulder, Colo.: Westview Press, 1981), 48. No manufacturing sector is reserved exclusively for the state, but it has a virtual monopoly in oil refining and nuclear energy and a commanding role in steel, iron-ore mining, aircraft manufacture, petrochemicals, hydroelectric power, and other utilities. It is the largest entrepreneur in the Brazilian market and the largest purchaser of machinery and equipment.

77. Statistical Appendix, *Revista brasileira de tecnologia* 13, no. 2 (April/May 1982):63; Biato et al., *Potencial de pesquisa,* 48; and *USY 1978–1979,* 807.

78. *I PBDCT,* 143. Brazil has set up bilateral agreements with Argentina (1968), Canada (1968), Costa Rica (1976), France (1975), Mexico (1976), the United Kingdom (1968, 1969), the United States (1968, 1971), Venezuela (1978), and West Germany (1971, 1974, 1978) (CNPq, *Relatório de atividades 1975–1978,* 90).

79. SEPLAN/FINEP, *Relatório de atividades 1978* (Rio de Janeiro: FINEP, Feb. 1979), sec. 7.

80. CNPq, *Relatório de atividades 1975–1978,* 91.

81. *Agenda CNPq,* no. 13 (March 1981), 10.

82. FINEP, *Relatório de atividades 1965–junho 1973,* 27, 28, 41.

83. BIC, *Operating Successfully in a Changing Brazil*, 60.

84. *Legislação brasileira para o desenvolvimento industrial e tecnológico*, comp. Eury Pereira Luna Filho (Rio de Janeiro: CNPq, 1979), 65–66 (hereafter CNPq, *Legislação brasileira*).

85. *ILT* Brazil, Dec. 1980, 23.

86. *Business Latin America*, May 4, 1977, 144.

87. *ILT* Brazil, Dec. 1980, 21–24.

88. CNPq, *Legislação brasileira*, 67.

89. Evans, *Dependent Development*, 188–90.

90. IDB, *Economic and Social Progress in Latin America: The External Sector*, 1982 Report (Washington, D.C., 1982), 142–57.

91. Fung and Cassiolato, "International Transfer of Technology," 117, 118, 121, 143. Although this study provides an excellent description of technology transfer policy, we have to be careful with its technology transfer data because its extensive analysis of 4,443 contracts between 1972 and 1975 was based on the *Revista de propriedade industrial*, not a very reliable source. But given that other data are unavailable, I am using their data in this instance only to give a general idea of the scope of the state enterprises.

92. *Planejamento e desenvolvimento* 5, no. 59 (April 1978):68.

93. *Agenda CNPq*, no. 13 (Jan. 1981), 4, and no. 14 (April 1981), 4; *Informe FINEP* 3, no. 31 (May 1982):2; and FINEP/CEP, internal memo CEP024/81 (May 19, 1981).

94. Victor Urquidi, *International Encyclopedia of Higher Education* (San Francisco: Jossey-Bass, 1977), 8:3790.

95. Goldemberg, "Uma aliança necessária," 16.

96. FINEP/CEP, internal memo CEP024/81.

97. Harvey W. Wallender III, *Technology Transfer and Management in the Developing Countries* (Cambridge, Mass.: Ballinger, 1979), 184.

98. D'Arrigo Vellinho, "Considerations," 7.

99. FINEP, *Relatório de atividades 1978*.

100. D'Arrigo Vellinho, "Considerations," 8.

101. *Índice do Brasil*, table 11.36.

102. José Mindlin, "A Perspective of Latin American Technological Development—Interaction Between Industry and Research: The Brazilian Industrial Experience," in *Texas Forum* (see chap. 3, n. 7), 150–70. This description cannot substitute for detailed research on the system.

Chapter 8

1. Nathaniel H. Leff, *Economic Policy-Making and Development in Brazil, 1947–1964* (New York: Wiley, 1968), 139.

2. Ibid., 139–40. The "emphasis on economic infrastructure and the 'basic industries' stems from a perception of the importance of potential external economies and diseconomies for growth. Similarly, many of the long-term investments to which the ideology led reflected an effort to overcome various

capital-market imperfections which were felt to be distorting resource allocation" (140).

3. For planning in this early stage, see Roberto de Oliveira Campos, "A Retrospect Over Brazilian Development Plans," in *The Economy of Brazil*, ed. Howard S. Ellis (Berkeley and Los Angeles: University of California Press, 1969); and Robert T. Daland, *Brazilian Planning* (Chapel Hill: University of North Carolina Press, 1967).

4. Leff, *Policy-Making in Brazil*, 143.

5. According to Leff, in the 1930s and 1940s Roberto Simonsen attempted to formulate a private-sector industrialist ideology of Brazil's economic development. But in his advocacy of government intervention and planning, his views were not very different from those of the "intelligentsia"; where there was disagreement, his views were ignored (ibid., 141).

6. *Ciência, tecnologia e independência*, ed. Severo F. Gomes and Rogério C. de Cerqueira Leite (São Paulo: Livraria duas Cidades, 1978), 225. My translation.

7. Leff cites as an early example of this phenomenon the relations between the intellectual Rómulo de Almeida and President Vargas (*Policy-Making in Brazil*, 145–46).

8. Simon Schwartzman, "Struggling To Be Born: The Scientific Community in Brazil," *Minerva* 16, no. 4 (Winter 1978):548.

9. While the 1964 coup undoubtedly led to a strong mood of cooperation between the military and technocrats, some authors have raised doubts about whether the coup was the result of economic "deepening," as stated by O'Donnell in his theory of the bureaucratic-authoritarian state. See José Serra, "Three Mistaken Theses Regarding the Connection Between Industrialization and Authoritarian Regimes," in *The New Authoritarianism in Latin America*, ed. David Collier (Princeton, N.J.: Princeton University Press, 1979), 99–163.

10. As quoted in Carlos Estevam Martins, *Tecnocracia e capitalismo* (São Paulo: Editora Brasiliense, 1974), 126.

11. Leff, *Policy-Making in Brazil*, 147. "Creation of these new institutions . . . saved the *técnicos* the years of struggle that would have been necessary to take over the older agencies and oust their personnel from key positions in the decision-making process" (ibid.).

12. Mário Schemberg, in an article (debate) called "Ciênçia no Brasil," *Cadernos de tecnologia e ciência* 2, no. 2 (March/April 1980):9–11. My translation.

13. Ibid., 16. My translation.

14. Leff, *Policy-Making in Brazil*, 146.

15. See debate: "Um convite á interação de economistas e tecnólogos," *Cadernos de tecnologia e ciência* 1, no. 2 (Aug./Sept. 1978):55–69.

16. Simon Schwartzman, "Por uma política científica no Brasil," *O estado de São Paulo*, Mar. 24, 1979, 5. My translation.

17. Caio Navarro de Toledo, *ISEB: Fábrica de ideologias* (São Paulo: Ática, 1977).

18. See Hélio Jaguaribe, *O nacionalismo na atualidade brasileira* (Rio de Janeiro: ISEB, 1958); Cândido Mendes, *Nacionalismo e desenvolvimento* (Rio de

Janeiro: IBEAA, 1963); Álvaro Vieira Pinto, *Conciência e realidade nacional* (Rio de Janeiro: ISEB, 1960); and Wanderley Guilherme, *Introdução ao estudo das contradições no Brasil* (Rio de Janeiro: ISEB, 1963).

19. Toledo, *ISEB,* 161. My translation. As an example of the radicalization of ISEB's writings, see Guilherme, *Introdução ao estudo.*

20. Toledo, *ISEB,* 170. My translation.

21. See, for example, Alfred Stepan, *The Military in Politics: Changing Patterns in Brazil* (Princeton, N.J.: Princeton University Press, 1961); and idem, "The New Professionalism of Internal Warfare and Military Role Expansion," in *Authoritarian Brazil,* ed. Alfred Stepan (New Haven, Conn.: Yale University Press, 1973).

22. Stepan, "The New Professionalism," 54.

23. Ibid., 56–57.

24. Waldimir Pirró e Longo, "Ciência e tecnologia e o poder militar" (n.d., mimeo), 7. My translation. See also idem, "Tecnologia e transferência de tecnologia," *Cadernos de tecnologia e ciência* 1, no. 2 (Aug./Sept. 1978):8–28.

25. Daland, *Brazilian Planning,* 35.

26. Carlos Lessa, "Fifteen Years of Economic Policy in Brazil," *Economic Bulletin for Latin America* 9, no. 2 (Dec. 1964):191.

27. *Conjuntura econômica* 2, no. 27 (June 1973):99–100, 101.

28. Daland, *Brazilian Planning,* 38.

29. J. Leite Lopes, *Ciência e desenvolvimento* (Rio de Janeiro: Tempo Brasileiro, 1964).

30. J. Leite Lopes, *Ciência e libertação* (Rio de Janeiro: Paz e Terra, 1978), 12. My translation.

31. The physicists' fascination with science and technology and its power and the subsequent desire to become scientifically and technologically self-sufficient were born, according to Pelúcio, when the United States exploded its first nuclear device.

32. Since 1958, BNDE had been devoting a limited amount of resources for scientific and technical training with the so-called Scientific and Technical Training Quota.

33. Debate: Z. Vaz, "Princípios básicos de administração das ciências," *Seminario FINEP–PROTAP–MEC/CAPES–MEC/SESU* (Rio de Janeiro: FINEP, Sept. 1979), 2. My translation.

34. The "policy measures initially taken," Erber pointed out, "implicitly contained a diagnosis of the causes of technological dependency, a strategy to overcome it, and—in varying degrees of clarity—a vision of an alternative future that would not be merely an amplified reflection of the present situation" (Fábio Stefano Erber, "Um ensaio de evolução conceptual," *Revista brasileira de tecnologia* 12, no. 1 [Jan./March 1981]:48). My translation.

35. According to Lessa, when Castelo Branco's policies failed to lead Brazil into the club of developed countries, his military successors decided to use a *desenvolvimentista* policy of the kind followed by Kubitschek in the 1950s. However, they themselves had abolished those earlier policies as ineffective to meet desired goals, so they had to "discover" science and technology as a new focus of *desenvolvimentismo* and as the way to transform Brazil into a powerful

nation and lead it to sustained economic growth (Carlos Lessa, "A estratégia de desenvolvimento 1974–1976: Sonho e fracasso" [thesis presented to the Faculty of Economics and Administration of the Federal University of Rio de Janeiro (UFRJ), 1978], 57–60). But Lessa's theory sounds too much like a "conspiracy": "Given that 'x' failed, let's do 'y' for our legitimation"; instead, I believe the answer is more subtle, more incremental, and more compartmentalized—more subtle because the change was not a "master plan" of the state but the result of the action of individual groups within the state; incremental because it was a process of trial and error; and compartmentalized because although some policymakers, such as Velloso, held the views implied by Lessa, others, such as Delfim Netto, did not.

36. Jean-Jacques Servan Schreiber, *The American Challenge* (New York: Atheneum, 1968); and Herman Kahn and A. J. Wiener, *Toward the Year 2000* (London: Macmillan, 1967).

37. See João Paulo dos Reis Velloso, in Ministry of Planning and General Coordination, *Desenvolvimento e planejamento: Pronunciamento dos ministros João dos Reis Velloso e o Hélio Beltrão* (Nov. 1969), 16, 36.

38. João Paulo dos Reis Velloso, *A solução positiva* (São Paulo: Abril-Tec, 1978), 133. My translation.

39. Organization of Economic Cooperation and Development (OECD), Gaps in Technology Between Member Countries Series, *General Report* and sector reports: *Scientific Instruments; Electronic Components, Electronic Computers, Plastics; Pharmaceuticals; Non-Ferrous Metals; Analytical Report* (Paris, various dates). See, for example, Organization of American States (OAS), *Estudios sobre el desarrollo científico y tecnológico*, OAS, Regional Program of Scientific and Technological Development, Department of Scientific Affairs, General Secretariat (Washington, D.C.: OAS, various dates).

40. Francisco Almeida Biato, Eduardo Augusto de Almeida Guimarães, and Maria Helena Poppe de Figueiredo, *A transferência de tecnologia no Brasil* (Brasília: IPEA, 1973); and idem, *Potencial de pesquisa tecnológica no Brasil* (Brasília: IPEA, 1971).

41. Fábio Stefano Erber et al., *Reflexões sobre a demanda pelos serviços dos institutos de pesquisa* (Rio de Janeiro: FINEP, 1974); and idem, *Absorção e criação de tecnologia na indústria de bens de capital* (Rio de Janeiro: FINEP, 1974).

42. *Agenda CNPq*, no. 13 (March 1981), 3.

43. Leonídia G. Dos Reis and Myriam L. Redinger, *Pesquisa tecnológica em empresas estatais*, preliminary study (Rio de Janeiro: FINEP, 1975).

44. "Um convite á interação de economistas e tecnólogos," 67.

45. Schwartzman, "Por uma política científica no Brasil," 5. My translation. For a more comprehensive statement of this view, see Simon Schwartzman, *Ciência, universidade e ideologia: A política do conhecimento* (Rio de Janeiro: Zahar, 1980).

46. For more information on the making of the PBDCTs, see Eduardo Rappel, "PBDCT: Plano ou programão?" *Cadernos de tecnologia e ciência* 1, no. 3 (Oct./Nov. 1978): 4–8; and Peter R. Seidl, "Um técnico do governo analisa o PBDCT," *Cadernos de tecnologia e ciência* 1, no. 5 (n.d.): 4–12.

47. Ronald M. Schneider, *Brazil's Foreign Relations: Environment, Institu-*

tions, Outlook, A Study Prepared for the Department of State under Its External Research Program, Jan. 1976, 9–10. For a detailed description of the political system, see Peter Flynn, *Brazil: A Political Analysis* (Boulder, Colo.: Westview Press, 1978); and Stepan, ed., *Authoritarian Brazil.*

48. Pedro S. Malan and Regis Bonelli, "The Brazilian Economy in the Seventies: Old and New Developments," *World Development* 5, nos. 1/2 (Jan./Feb. 1977):36.

49. United Nations Industrial Development Organization (UNIDO), *Industrial Priorities in Developing Countries* (New York: United Nations, 1979), 2, 3.

50. Malan and Bonelli, "The Brazilian Economy," 38.

51. José Serra, Working Document presented to ECLA (São Paulo, 1981), table 5.

52. Malan and Bonelli, "The Brazilian Economy," 26, 36, 38, 39.

53. Ibid., 38; and Serra, Working Document presented to ECLA, tables 1, 4, 5, 9.

54. "Such complacency was bred by the government's wish not to break the cycle of prosperity presided over by their predecessors, by a fear of the political consequences of lowering popular growth expectations and by the absence of immediate external pressures; Brazil had a large cushion of foreign exchange reserves ($6.4 billion as of December 1973) and, although the rate at which the current account was in deficit would have eliminated these reserves in twelve months, Brazil succeeded in obtaining new loans" (John R. Wells, "Brazil and the Post-1973 Crisis in the International Economy," in *Inflation and Stabilization in Latin America,* ed. Rosemary Thorp and Laurence Whitehead [London: Macmillan, 1979], 242).

55. Velloso, *A solução positiva,* 166. Sales of the 13 largest private industries increased by 402.1 percent in three years, with an increase in real earnings of 473.5 percent (167).

56. Federal Republic of Brazil, *II Plano Nacional de Desenvolvimento 1975–1979* (II PND) (1974), 40.

57. Serra, Working Document presented to ECLA, tables 1, 4, 5, 9, 26.

58. The huge agromineral complex of Carajás alone will provide a major boost to the sector, and one single company is going to invest $2.8 billion by 1985. According to a spokesman of the capital goods industry, only 9 percent of the value of Carajás's orders will be placed abroad (*Latin American Regional Report,* July 3, 1981, 6–7).

59. See, for example, Ricardo A. Bielschowsky, "Notas sobre a questão da autonomia tecnológica na indústria brasileira," in *Indústria: Política, instituições e desenvolvimento,* ed. Wilson Suzigan (Rio de Janeiro: IPEC/INPES, 1978); and Erber, "Um ensaio de evolução conceptual," 45–50.

60. The 1973 ratios are from UNIDO, *Industrial Priorities in Developing Countries,* 5. Brazil's 1978 ratio was elaborated with data on manufactured value-added from *Latin America Weekly Report* (August 2, 1981, 8) and with data on manufactures exports from Business International Corporation (BIC) (*Trading in Latin America: The Impact of Changing Policies* [New York, July 1981], 152).

Chapter 9

1. The big jump in semiconductor technology came about when the transistor was integrated with other necessary components into a single silicon base, or "chip." This resulted in a huge reduction in cost, higher efficiency, and a tremendous increase in information storage capacities (Atud Wad, "Microelectronics: Implications and Strategies for the Third World," *Third World Quarterly* 4, no. 4 [Oct. 1982]: 629). But the largest leap to date occurred in 1971 with the invention of the microprocessor, which can be programmed to carry out information-processing and control functions: in essence, a computer-on-a-chip. A very dynamic world semiconductor market evolved, with companies from several nations entering in, able to supply millions of microprocessors every year. At the same time, technological and market changes have partially transformed the highly concentrated and oligopolized international computer industry. In the 1970s this industry grew at an annual rate of between 10 and 15 percent, doubling the number of computers in use worldwide every couple of years (*World Business Weekly*, April 20, 1981, 30).

Sales data are still more startling. The rise of computer sales has outstripped GDP growth in the United States, Japan, and the European Economic Community. Thus, for example, in 1979 the combined average sales growth of these three markets was 16.6 percent, whereas their combined average GDP growth was only 3.9 percent (*Electronics*, Jan. 5, 1978, and Jan. 4, 1979; and *OECD Main Economic Indicators*, Oct. 1978). By 1978 the sales market in the developed countries alone was estimated at $37.6 billion, more than a third of the total for electronic equipment worldwide. The total market for data-processing equipment went up from $11.7 billion in 1970 to $53.5 billion in 1980 (United Nations Conference for Trade and Development [UNCTAD], *Electronics in Developing Countries: Issues in Transfer and Development of Technology*, AD/B/C.6/34 [Geneva: United Nations, 1978], 1; and *Business Week*, June 8, 1981, 84).

2. The worldwide computer industry has lived under the shadow of one giant: International Business Machines (IBM), which started the 1970s with a 60 percent share of the computer market and ended them with a still-impressive but slimmed-down 40 percent (*Time*, July 11, 1983, 47). It is the world's most profitable industrial corporation, with 1984 revenues of $46 billion and a net income in 1983 valued at $5.5 billion (*New York Times*, Jan. 20, 1985, IV-5). Today, U.S. companies hold 80 percent of the computer market—seven of the top ten companies in the industry are American—but Japan has been making large inroads and holds close to 10 percent, or about $9 billion, of that market, and the remaining three of the top ten companies are Japanese. In 1983 Japan exported $3.9 billion worth of computers. Smaller Japanese companies are supplying computer hardware to U.S. firms and, together with Taiwanese and Korean companies, are selling components and personal computers on world markets (*Business Week*, July 16, 1984, 61–62).

3. Joseph M. Grieco, *Between Dependency and Autonomy: India's Experience with the International Computer Industry* (Berkeley and Los Angeles: University of California Press, 1984).

4. These studies include Superintendency of Military Industries, *Estructura de la industria electrónica* (Directorate of Development, Department of Military Industries, 1975); Alberto Petrocolla, Roberto Zubieta, Hector Abrales, and Julio Nogues, *Industria electrónica y proceso técnico en un contexto de industrialización* (Buenos Aires: Instituto Torcuato Di Tella, 1974); Ministry of Research and Technology (BMFT), *Estudio sobre el desarrollo de la industria electrónica argentina: Conclusiones. Fase 2* (Munich, Mar. 1981); and Juan F. Rada, *The Impact of Microelectronics and Information Technology: Case Studies in Latin America* (Paris: UNESCO, 1982).

5. Eugenio Lahera Parada, "FATE y CIFRA: Un estudio de caso en difusión y desarrollo de tecnología digital en Argentina" (1976, mimeo).

6. It should be stressed from the start that Argentina's industrial statistics of the 1960s and early 1970s are well known for their deficiencies, making an accurate representation difficult.

7. Petrocolla et al., *Industria electrónica*, 7.

8. BMFT, *Estudio sobre el desarrollo*, 58, 101; Military Industries, *Estructura*, 136, 143; and UNCTAD, *Electronics*, 26.

9. Military Industries, *Estructura*, 117.

10. UNCTAD, *Electronics*, 31.

11. Military Industries, *Estructura*, 113–19.

12. BMFT, *Estudio sobre el desarrollo*, 66.

13. UNCTAD, *Electronics*, 11, 29.

14. Military Industries, *Estructura*, 52.

15. Petrocolla et al., *Industria electrónica*, 119.

16. "La electrónica y sus razones," *El economista*, December 30, 1976, 6.

17. BMFT, *Estudio sobre el desarrollo*, 101.

18. Ibid.

19. Lahera, "FATE y CIFRA," 24.

20. Lee Wayne, ed., *The International Computer Industry* (Washington, D.C.: Applied Library Resources, 1971), chapter on Argentina, 1–2. Although the Centralized Systems Center of Data and Information, the ministry of social welfare, and the ministry of education had IBM 360s, the army operated an IBM 1401 and a Burroughs B3500, the navy and air force each had a Bull/GE Gamma 30, and the air force had in addition an IBM 1401. Also, at the end of 1969 there were 14 computer systems used in education (IBM 360s, IBM 1440s, and an IBM 1620). The CNEA owned an IBM 1130 and three H-P 2116s, and the Di Tella Institute had an IBM 1130 (ibid., 1–9).

21. Military Industries, *Estructura*, 85; Lahera, "FATE y CIFRA," 12; and *Computación: Revista iberoamericana* 2, no. 23 (Sept. 1982): 24.

22. IBM, internal memo.

23. Lahera, "FATE y CIFRA," 22; and Military Industries, *Estructura*, 85.

24. Lahera, "FATE y CIFRA," 1, 2.

25. Ibid., 12.

26. Oscar Varsavsky, *Estilos tecnológicos* (Buenos Aires: Ediciones Periferia,

1974); "Ciencia y tecnología en la industria," (Fundación Bariloche, April 1972); and *Marco histórico constructivo* (Buenos Aires: Centro Editor de América Latina, 1975).

27. Lahera, "FATE y CIFRA," 9, 14–15.

28. Ibid., 5.

29. Ibid., 5–6, 16.

30. Ibid., 10.

31. As the economic situation crumbled and FATE became involved in the scandal, the military did not want to take any measures to aid the computer idea.

32. Lahera, "FATE y CIFRA," 34. My translation.

Chapter 10

1. Ivan da Costa Marques, *Computadores: Parte de um caso amplo de sobrevivencia e da soverania nacional* (Rio de Janeiro, July 30, 1979, mimeo), 12. My translation.

2. Simon Nora and Alain Minc, *The Computerization of Society: A Report to the President of France* (Cambridge, Mass.: MIT Press, 1981), 3.

3. Wando Pereira Borges, president of Digibrás, *Hearings Before the Parliament* (House of Representatives), (Brasília, Aug. 31, 1977, mimeo); Joseph M. Grieco, *Between Dependency and Autonomy: India's Experience with the International Computer Industry* (Berkeley and Los Angeles: University of California Press, 1984), 158; Paulo Bastos Tigre, "Indústria de computadores e dependência tecnológica no Brasil" (M.A. thesis, UFRJ, 1978), 75; Commission for the Coordination of Electronic Data-Processing Activities (CAPRE), *Boletim técnico* 1, no. 1 (Jan./March 1979): 38–39; and G. B. Levine, "Brazil 1976—Another Japan?" *Datamation* 21 (Dec. 1975): 63–66.

4. SEI, *Boletim informativo*, no. 11 (June/Sept. 1983): 10.

5. *Data News*, May 3, 1983, 9; and *Brazil Trade and Industry*, May 1982, 11.

6. The classifications have been drawn up by SEI according to the mean value of computers: *Class 1*, $20,000; *Class 2*, $90,000; *Class 3*, $180,000; *Class 4*, $670,000; *Class 5*, $1.9 million; and *Class 6*, $3 million (SEI, *Boletim informativo*, no. 8 [July/Aug./Sept. 1982], 5). The six classes represent, roughly, micros, minis, small, medium, large, and very large computers. The micro category includes electronic accounting machines and desk processors.

7. SEI, *Boletim informativo*, no. 11 (June/Sept. 1983): 7, and no. 8 (July/Aug./Sept. 1982): 4.

8. "O novo teto para importações e a experiência de 76," *Dados e Idéias*, no. 5 (April/May 1977): 30.

9. Robert A. Bennett, "IBM in Latin America," in *The Nation-State and Transnational Corporations in Conflict: With Special Reference to Latin America*, ed. Jon P. Gunneman (New York: Praeger, 1975), 225 (app. B).

10. United Nations Center on Transnational Corporations, *Transborder Data Flows and Brazil* (prepared by the Special Secretariat of Informatics of

the National Security Council of the Presidency of the Republic of Brazil in cooperation with the Ministry of Communications of Brazil), ST/CTC/40 (New York: United Nations, 1983), 80; *Brazil Trade and Industry*, May 1982, 12; and IBM brochures.

11. *Data News*, May 15, 1984, 4.

12. SEI, *Boletim informativo*, no. 11 (June/Sept. 1983):11.

13. Ibid., 13, 18.

14. Paulo Bastos Tigre, *Technology and Competition in the Brazilian Computer Industry* (New York: St. Martin's Press, 1983), 94.

15. *Transborder Data Flows and Brazil*, 98.

16. Mário Ripper, "Professor Zezinho," *Dados e Idéias*, no. 1 (Aug./Sept. 1977):59. My translation.

17. Ibid., 61. My translation.

18. Ferranti built a general-purpose, realtime 16-bit word computer designed for use in data communications networks, realtime information systems, and process control (Steve Yolen, "Computer Production Prospects in Brazil Brighten," *Electronic News*, June 7, 1976, 32).

19. Paulo Bastos Tigre, "Brazil: A Future in Homemade Hardware," *South* 16 (Feb. 1982):99.

20. Ibid.

21. For more information on the G-10, see "A difícil afirmação de G-10, o computador nacional," *Dados e Idéias*, no. 4 (Feb./March 1976):35–37.

22. Pereira Borges, *Hearings*, 23.

23. "Como vai o mini nacional," *Dados e Idéias*, no. 2 (Oct./Nov. 1978):43. My translation.

24. See article on Cobra, *Cadernos de tecnologia e ciência* 1, no. 5 (n.d.):n.p. Cobra also signed several contracts with the army for the supply of computer equipment and peripherals. See also *Planejamento e desenvolvimento* 4, no. 46 (March 1977):40.

25. Other banks that invested in Cobra were Bahia de Investimento, Auxiliar Nordeste, NCN, Banespa, and Bamerindus. See Silvia Helena, "Os banqueiros e a Cobra," *Dados e Idéias*, no. 2 (April/May 1977):35.

26. *Planejamento e desenvolvimento* 4, no. 46 (March 1977):40.

27. Grieco, *Between Dependency and Autonomy*, 159.

28. *Planejamento e desenvolvimento* 1, no. 4 (Oct. 1973):64.

29. CAPRE, *Boletim técnico*, 105.

30. "Our main emphasis has been in the area of human resources," said Saur, "because we know how dependent the computer is on the human elements that command it" (Ricardo A. C. Saur, *Hearings Before the Parliament* [*House of Representatives*] [Brasília, 1977, mimeo], 16). My translation.

31. The government approved Cr 13 million for the sector in 1974, of which CAPRE was to get Cr 5.3 million. In 1975 the sector was set to receive Cr 42 million, with Cr 16 million to go to CAPRE (*Diario oficial*, July 17, 1974, 8033; and Oct. 3, 1975, 13254).

32. Grieco, *Between Dependency and Autonomy*, 160.

33. "Um perfil artificial," *Dados e Idéias*, no. 1 (April/May 1980):8.

34. Marília Rosa Millan and João Lizardo Hermes de Araújo, "Na palavra

dos técnicos, um ponto de vista nacional," *Cadernos de tecnologia e ciência* 1, no. 4 (Dec./Jan. 1979):36.

35. CAPRE, *Boletim informativo*, no. 4 (July/Sept. 1976):53. My translation.

36. Ibid., no. 5 (April/June 1977):69.

37. See Silvia Helena, "Minis: A decisão final," *Dados e Idéias*, no. 2 (Oct./Nov. 1977):34–45.

38. See Angeline Pantages, "Cracking Brazil Nuts," *Datamation* 25 (Feb. 1979):78.

39. Steve Yolen, "Brazil Seen as Medium-Scale System Maker in Three Years," *Electronic News*, Dec. 3, 1979, 40.

40. Millan and de Araújo, "Na palavra dos técnicos," 39.

41. *Conjuntura econômica*, Feb. 1979, 95.

42. *Business Latin America*, June 6, 1979, 178.

43. *Transborder Data Flows and Brazil*, 69.

44. Silvia Tavora, "De Capre à SEI, e as grandes promesas," *Dados e Idéias*, no. 6 (Feb./March 1980):16–17.

45. SEI, "Ato Normativo #001/80" (mimeo).

46. SEI, "Ato Normativo #005/80" (mimeo).

47. SEI, *Boletim informativo*, no. 1 (Aug./Sept. 1980):12, and no. 2 (Nov./Dec. 1980):30–32.

48. *Business Latin America*, Oct. 22, 1980, 344.

49. *Diario oficial*, Aug. 21, 1980, sec. 1, 16549.

50. SEI, *Boletim informativo*, no. 5 (Aug./Sept./Oct. 1981):49–50b.

51. *Diario oficial*, Dec. 7, 1982, sec. 1, 22803–4.

52. Digibrás, *Boletim informativo*, no. 5 (Nov./Dec. 1982):4. My translation.

53. *Transnational Data Report on Information Policies and Regulation* 7, no. 8 (Dec. 1984):431–32.

54. CNPq, *Avaliação e perspectivas* 3 (1978):47. My translation.

55. Saur, *Hearings*, 17.

56. CNPq, *Avaliação e perspectivas*, 47; *Transborder Data Flows and Brazil*, 91, 97.

57. Saur, *Hearings*, 16.

58. *Recomendações*, Seminar on Computation at the University, Florianópolis, Sept. 29, 1977, 11.

59. Saur, *Hearings*, 4. My translation.

60. *Business Latin America*, Oct. 19, 1977, 331.

61. *Latin American Economic Report*, July 6, 1979, 207.

62. "A participacão da SBC," *Dados e Idéias*, no. 1 (April/May 1980):39.

63. José Martinez, "Uma comunidade à procura de caminhos," *Dados e Idéias*, no. 6 (Feb./March 1980):28. My translation.

64. *Business Week*, Nov. 3, 1980, 58.

65. CEDINI, "Análise da decisão da SEI de 6 de Agôsto de 1980," 2d draft (Aug. 14, 1980, mimeo, incomplete).

66. According to Gennari, SEI's decision to allow IBM to manufacture its model 4331 did not mean retreat from "the model," because at the time there was no national potential in that computer size. He argued that the permit resulted not from IBM pressure but from an understanding at SEI that the

market would gain (customers were unattended at that size level) and "the model" would not lose.

67. Gennari confided that the statements attributed to him about the end of the market reserve were the result of newspapers' selective editing of his remarks and that the words were then used as part of the "war" between those who wanted and those who opposed the market reserve. SEI used these remarks to scare the national industry into becoming competitive.

68. Moacyr Antoñio Fioravante, quoted in *O globo*, May 3, 1981, 30. My translation.

69. On the one hand Brazil's foreign debt crisis is aiding the pragmatic antidependency strategy by increasing the need for import substitution, but on the other hand Brazil's financial dependence on the United States and the international banking community is making it less able to withstand their pressures.

70. *Data News*, July 24, 1984, 8.

71. Ibid., Oct. 18, 1983, 6.

72. Ibid., Oct. 9, 1984, 2. My translation.

73. I would like to thank Manuel Fernando Lousada Soares (Informatics and Communications Coordinator, CNPq) for calling my attention to this process.

74. Tigre, "Indústria de computadores," 147–48. See also Jack Baranson, *North-South Technology Transfer* (Mt. Airy, Md.: Lomond, 1981), 38–42.

75. Steve Yolen, "Brazil Move May Impact IBM/32 Plans," *Electronic News*, Dec. 13, 1976, 30, 40. See also Levine, "Brazil 1976–Another Japan?" 63–66.

76. Maria de Conceição, "Uma luta desigual," *Dados e Idéias*, no. 3 (Dec. 1976/Jan. 1977): 17.

77. *Transborder Data Flows and Brazil*, 78.

78. *Data News*, Nov. 1, 1983, 6, and July 24, 1984, 4; and *O estado de São Paulo*, Jan. 13, 1984, 23.

79. Tigre, *Technology and Competition*, 144.

80. *The Economist*, Feb. 9, 1985, 49–50; "IBM Concessions to Mexico," *New York Times*, July 25, 1985, 5.

Chapter 11

1. *8 Days*, March 13, 1982, 59.

2. National Atomic Energy Commission (CNEA), Transcript of Rear Adm. Carlos Castro Madero's meeting with the press, 1982.

3. "Energía atómica: El uranio de los privados," *Mercado*, Nov. 11, 1982, 43–44.

4. The fission nuclear fuel cycle consists of those industrial operations necessary for the civilian use of nuclear power. After mining, the uranium is separated and processed into yellow-cake, which has a high quantity of uranium oxide (U_3O_8). The yellow-cake is then "cemented" into either uranium diox-

ide (UO_2) or uranium hexafluoride (UF_6), depending on the type of reactor. Some reactors use enriched uranium; the most common enrichment methods are gaseous diffusion, gas centrifuge, and jet-nozzle. The fissionable material goes next into tubs of an alloy of zirconium (zircalloy), which are arranged in special geometrical configurations to improve the nuclear chain reaction, and then are placed in the reactor. The heat released by the nuclear reaction creates steam for driving a turbine, which in turn creates electricity. After the fuel has been burned it is stored for cooling until it can be disposed of as waste or reprocessed. In the latter event, the fuel goes to a reprocessing plant where it is dissolved and plutonium and uranium are extracted and purified. The final stage is waste disposal, usually within water pools. No entirely safe disposal method has yet been found. See U.S. Congress, Senate Committee on Government Operations, *Facts on Nuclear Proliferation: A Handbook Prepared by the Congressional Research Service, Library of Congress* (Washington, D.C.: GPO, Dec. 1975); and William C. Potter, *Nuclear Power and Non-Proliferation: An Interdisciplinary Perspective* (Cambridge, Mass.: Oelgeschlager, Gunn, & Hain, 1982), chap. 3: "The Technology of Nuclear Power."

5. Argentine Republic, The Presidency, CNEA, Nuclear Supplies Division, *Complejo Fabril Córdoba* (Buenos Aires: CNEA, 1982), 3, 13.

6. Jorge Sabato, "Energía nuclear en Argentina, autonomía tecnológica y desarrollo industrial," Discussion Paper Prepared for the Seminar on Industrial Energy Management, UNIDO, São Paulo, Oct. 18–21, 1982, 6.

7. *La nación*, Nov. 24, 1983, 10.

8. Latin American Newsletters, Special Report: Latin America's Nuclear Development Programmes Today, SR–85–02, April/June 1985, 2. Nuclear reactors are usually classified by coolant and/or moderator. The coolant is the material used to remove the heat and transfer it out of the reactor's core. The moderator (regular water, heavy water, or graphite) slows neutrons to keep the nuclear reaction going. The most common reactors are light water reactors (LWRs) and heavy water reactors (HWRs). LWRs use common water as moderator and coolant, and slightly enriched uranium (U-235) as fuel. (U-235 is the only naturally occurring isotope that can be obtained by enriching natural uranium. A fissionable material is one that can sustain chain reactions. Isotopes are atoms of the same chemical element whose nuclei have different numbers of neutrons.) LWRs come in two versions, the pressurized water reactor, which is the most widely used in the world, and the boiling water reactor. HWRs use heavy water as coolant and moderator. They require no enriched uranium, and thus are preferred by countries pursuing nuclear power independence. But a supply of heavy water must be guaranteed. One of the most common HWRs is the Canadian model CANDU. Other types of reactors are the high-temperature gas-cooled, which use graphite as a moderator and helium as coolant, and breeder reactors, which convert nonfissionable uranium and thorium into fissionable plutonium and uranium at a faster rate than that at which the original fissionable isotopes are consumed, thus breeding a surplus of fuel (John V. Granger, *Technology and International Relations* [San Francisco: W. H. Freeman, 1979], 141).

9. Sabato, "Energía nuclear en Argentina," 3.

10. Daniel Poneman, *Nuclear Power in the Developing World* (London: George Allen & Unwin, 1982), 80.

11. Sabato, "Energía nuclear en Argentina," 5.

12. Carlos Castro Madero, "Planning for Nuclear Self-Sufficiency in Argentina," in Argentine Republic, CNEA, *The Argentine Nuclear Development Plan* (selected reprints from *Nuclear Engineering International,* Sept. 1982) (Buenos Aires, Nov. 1982), 6.

13. *La nación,* Dec. 11, 1982, 14.

14. Poneman, *Nuclear Power in the Developing World,* 74.

15. Sabato, "Energía nuclear en Argentina," 6.

16. John R. Redick, "The Tlatelolco Regime and Nonproliferation in Latin America," *International Organization* 35 (Winter 1981): 118.

17. Douglas L. Tweedale, "Argentina," in *Nuclear Power in Developing Countries,* ed. James Everett Katz and Onkar S. Marwah (Lexington, Mass.: Lexington Books, 1982), 83.

18. Sara Volman de Tanis and Jorge Kittl, "20 años de investigación y desarrollo," (Buenos Aires: The Presidency, CNEA, Department of Metallurgy, mimeo, 1976), 7.

19. Jorge Sabato, "Atomic Energy in Argentina: A Case History," *World Development* 1, no. 8 (Aug. 1973): 26.

20. The deal also included an ample transfer of technology; the technical help to set up a radioisotope production plant; physics, chemistry, and biology laboratories; and a national center for radiological protection and nuclear safety. Argentina also signed nuclear agreements with Bolivia, Brazil, Chile, Ecuador, India, Iraq, Libya, Mexico, Paraguay, Peru, Uruguay, and Venezuela involving research, power, training, and uranium exploration activities.

21. Jorge Sabato, "Desarrollo tecnológico de la metalurgia argentina en las industrias siderúrgicas del aluminio y del uranio" (n.d., mimeo), 5.

22. Sabato, "Atomic Energy in Argentina," 29.

23. For a description of Enace, see H. Leibovich, J. Coll, and K. Backhaus, "Good Experience of Transferring Technology," in CNEA, *Argentine Nuclear Development Plan,* 15.

24. R. Frydman, "Establishing a Nuclear Engineering and Construction Company in Argentina," *Transactions,* Second International Conference on Nuclear Technology Transfer (ICONTT-II), Buenos Aires, Argentina, Nov. 1–5, 1982 (La Grange Park, Ill.: American Nuclear Society, 1982), 194.

25. José Mirabelli, "Argentina: Steady Progress Amid National Turbulence," *International Herald Tribune,* Oct. 30–31, 1982, 9S; and Latin American Newsletters, *Latin America's Nuclear Development,* 3.

26. Sabato, "Atomic Energy in Argentina," 23–24.

27. Ibid., 24.

28. Norman Gall, "Atoms for Brazil, Dangers for All," *Foreign Policy* 23 (Summer 1976): 180–81.

29. CNEA, Legal Documentation and Information, Decree Law 22.498, Dec. 28, 1956.

30. Sabato, "Atomic Energy in Argentina," 24.

31. José Cafasso and Enrique Recchi, *Economía energética argentina* (Buenos Aires: Don Bosco, 1976), 194–201.

32. Castro Madero, "Planning for Nuclear Self-Sufficiency," 5–6.

33. Latin American Newsletters, *Latin America's Nuclear Development*, 2–3.

34. Poneman, *Nuclear Power in the Developing World*, 175.

35. Jorge Sabato, Oscar Wortman, and Gerardo Gargiulo, *Energía atómica e industria nacional*, SG/P.1 PTT/47 (Washington, D.C.: OAS, General Secretariat, 1978), 2. My translation.

36. Sabato, "Atomic Energy in Argentina," 25; and Sabato, "Energía nuclear en Argentina," 4.

37. Alberto Aráoz and Carlos Martínez Vidal, "Ciencia e industria: Un caso argentino," *Estudios sobre el desarrollo científico y tecnológico*, OAS, Regional Program of Scientific and Technological Development, no. 19 (Washington, D.C.: OAS, 1974), 45. My translation. See also Jorge Sabato, *Ensayos en campera* (Buenos Aires: Juárez, 1979), 153.

38. Sabato, *Ensayos en campera*, 154. My translation.

39. Poneman, *Nuclear Power in the Developing World*, 70.

40. Sabato, *Ensayos en campera*, 157.

41. See Aráoz and Martínez Vidal, "Ciencia e industria," 66; and Sabato et al., *Energía atómica e industria nacional*, 67–70.

42. *Economic Information on Argentina*, no. 104 (Feb. 1980): 17, 19, 21. The hydroelectric establishment was not interested in technological autonomy, or in the strategic/military aspects of nuclear power. It preferred a LWR to a HWR as being cheaper and more efficient. The state-owned utility, SEGBA, also complained that nuclear power stations should be placed under its control, rather than under the CNEA (Poneman, *Nuclear Power in the Developing World*, 185).

43. Tweedale, "Argentina," 80.

44. Embalse's increasing costs, the purchase of Atucha II and the heavy water plant, and the development of an enrichment process have led to increasing budget allocations since 1977 in both absolute and relative terms.

45. Poneman, *Nuclear Power in the Developing World*, 132.

46. See, for example, the Argentine magazine *Estrategia*, especially the articles by Juan E. Guglialmelli.

47. The CNEA leadership consistently explained Argentina's failure to ratify the nonproliferation treaty and the Tlatelolco Treaty on the grounds that Argentina could not "accept remaining subordinate to a continuing dependence on the great powers for nuclear technology, especially when our country has laid the foundations for a nuclear technology needed for economic development" (Poneman, *Nuclear Power in the Developing World*, 122). Castro Madero stated that it "is not reasonable to discriminate and expect the victim to make no attempt to free himself. As a result, such 'antiproliferation' measures have an effect opposite to that which was intended" ("Planning for Nuclear Self-Sufficiency," 7).

48. Federal Republic of Brazil, *A questão nuclear: Relatório da Comissão Parlamentar de Inquérito do Senado Federal, Resolução 69/78*, National Congress Record, sec. II, Supplement to no. 104 (Brasília, Aug. 17, 1982), 67 (hereafter

1982 Senate Nuclear Inquiry Report); and Juan de Onis, "Uranium Exports Could Offset Brazilian Oil Import Bill," *International Herald Tribune*, Oct. 30–31, 1982, 6S.

49. *1982 Senate Nuclear Inquiry Report*, 67.

50. Ibid., 67, 238; and de Onis, "Uranium Exports," 6S.

51. Thorium is a radioactive material extracted from monazite, a heavy mineral abundant in sand beaches that can hold as much as 18 percent thorium oxide. When a fertile thorium 232 isotope is bombarded with neutrons, it is transformed into U-233 (*1982 Senate Nuclear Inquiry Report*, 68).

52. Ibid., 239.

53. "The Nuclear Plan Mushrooms," *Latin American Weekly Report*, July 31, 1981, 9.

54. *1982 Senate Nuclear Inquiry Report*, 240; and "A aventura nuclear," *O estado de São Paulo*, Oct. 18, 1983, 4.

55. *1982 Senate Nuclear Inquiry Report*, 238; and Juan de Onis, "Brazil's Crash Program Slows to Realistic Pace," *International Herald Tribune*, Oct. 30–31, 1982, 5S.

56. *Veja*, Jan. 19, 1983, 74.

57. Poneman, *Nuclear Power in the Developing World*, 45.

58. Joaquim Francisco de Carvalho, "A controvérsia dos custos nucleares," in *Os custos do programa nuclear*, ed. Paulo Brossard (speech delivered at the Jan. 11, 1982, session of the Federal Senate, Brasília) (n.d.), 15; and "A aventura nuclear," 4.

59. *1982 Senate Nuclear Inquiry Report*, 135.

60. "The Nuclear Plan Mushrooms," 10.

61. James W. Rowe, "Science and Politics in Brazil: Background of the 1967 Debate on Nuclear Energy Policy," in *The Social Reality of Scientific Myth*, ed. Kalman H. Silvert (New York: American Universities Field Staff, 1969), 115.

62. Ibid., 102–3.

63. Renato de Biasi, *A energia nuclear no Brasil* (Rio de Janeiro: Atlántida, 1979), 18–19. See also J. Leite Lopes, *Ciência e libertação* (Rio de Janeiro: Paz e Terra, 1978); and Ricardo Guedes Ferreira Pinto, "Liliputianos e Lapucianos: Os caminhos da fisica no Brasil (1810 a 1949)" (M.A. thesis, IUPERJ, 1978).

64. *1982 Senate Nuclear Inquiry Report*, 148; and Rui Ribeiro Franco, "Education and Training in Brazil for the Nuclear Technology Transfer Program," *Transactions*, Second International Conference on Nuclear Technology Transfer (ICONTT-II), Buenos Aires, Argentina, Nov. 1–5, 1982 (La Grange Park, Ill.: American Nuclear Society, 1982), 136–37.

65. Juan de Onis, "'Jet Nozzle' Fuel Enrichment System Could Bring Brazil into an Elite Club," *International Herald Tribune*, Oct. 30–31, 1982, 7S.

66. Biasi, *Energia nuclear*, 105, 113.

67. *1982 Senate Nuclear Inquiry Report*, 147.

68. Ibid., 259.

69. Ibid., 140.

70. Ibid., 146. My translation.

71. Pitanguy Romani, *Apoio institucional á ciência e tecnologia no Brasil* (CET/SUP/CNPq, April 1977, preliminary draft); and Maria Cristina Leal, "Caminhos e descaminhos do Brasil nuclear, 1945–1958" (M.A. thesis, IUPERJ, 1982).

72. Therefore, the 1970s conflict between the United States and West Germany over Brazil's nuclear purchases had its origins several decades before.

73. *1982 Senate Nuclear Inquiry Report*, 96.

74. Távora later denied that Alberto's resignation had anything to do with the secret documents and American pressure.

75. For more information on the CNEN's first ten years, see Rowe, "Science and Politics in Brazil," 110–22.

76. Gall, "Atoms for Brazil," 186.

77. Regina Lûcia de Morâes Morel, *Ciência e estado: A política científica no Brasil* (São Paulo: Queiroz, 1979), 106.

78. National Nuclear Energy Commission (CNEN), *Pesquisa e tecnologia nucleares no Brasil* (1977, mimeo), 4–6.

79. Ibid., 8–9; and Biasi, *Energia nuclear*, 26–29.

80. *1982 Senate Nuclear Inquiry Report*, 104.

81. Federal Republic of Brazil, *II Plano Nacional de Desenvolvimento 1975–1979 (II PND)* (1974), 140.

82. *1982 Senate Nuclear Inquiry Report*, 49, 51.

83. "A aventura nuclear," 4; and Latin American Newsletters, *Latin America's Nuclear Development*, 4.

84. Margarete K. Luddemann, "Nuclear Power in Latin America: An Overview of Its Present Status," *Journal of Interamerican Studies and World Affairs* 25, no. 3 (Aug. 1983): 393.

85. "A integração da política atômica brasileira," *Correio da manhã*, July 23, 1967. My translation.

86. "Industrial opina sobre a energia atômica do Brasil," *Correio da manhã*, Aug. 5, 1967. My translation.

87. Mário Schemberg, "Ciência no Brasil," *Cadernos de tecnologia e ciência* 2, no. 2 (March/April 1980): 5–6. My translation.

88. According to Roberto Gomes de Oliveira, a former IEN and INT director, although these programs cannot be mixed, as they are intrinsically different and respond to different markets, they can nevertheless be developed simultaneously. Although the technological autonomy program requires open international cooperation, the development of a domestic nuclear industry, and natural uranium HWRs, the strategic program is characterized by a modest reserve of uranium outside the safeguards system, a strong human resources reservoir, and a modest HWR (40 Mw) ("Politicas de mercado industrial e de ciências e tecnologia: O caso do programa nuclear brasileiro," in David N. Simon et al., *Energia Nuclear em Questão* [Rio de Janeiro: Instituto Euvaldo Lodi, 1981], 64–67).

89. Leal, *Caminhos e descaminhos do Brasil nuclear*, 133. My translation.

90. Leite Lopes, *Ciência e libertação*, 198.

91. Rowe, "Science and Politics in Brazil," 116.

92. Ibid., 117.

93. *New York Times*, June 13, 1975, 36.

94. U.S. Congress. House of Representatives. Committee on Science and Technology, *Oversight of Energy Development in South America*, 96th Cong., 2d sess., 1980, 26.

95. Joaquim Francisco de Carvalho, "Aspectos econômicos e estratégicos

do acordo nuclear Brasil-Alemanha," *Ciência e cultura* 33 (1981):17. My translation.

96. Luiz Pinguelli Rosa, "O papel da energia nuclear na geração de energia elétrica no Brasil," in Simon et al., *Energia Nuclear em Questão,* 35.

97. See de Carvalho, "Aspectos econômicos e estratégicos"; and Brossard, ed., *Os custos do programa nuclear.*

98. This argument is from de Oliveira, "Políticas do mercado industrial," 70–71.

99. Brossard, *Os custos do programa nuclear,* 29.

100. "The Nuclear Plan Mushrooms," 10.

101. Ibid.

List of Interviews

Name	Position or Organization	Date of Interview
Argentina		
Albertoni, Jorge	Former President, INTI	June 9, 1980
Aráoz, Alberto	Enace SA	June 6, 1980; November 10, 1982
Bargagna, R.	Manager, Fábrica Argentina de Telas Engomadas (FATE)	June 5, 1980; November 15, 1982
Bustamante, Jorge	Undersecretary of Industry	May 30, 1980
Bustos, Edgar C.	Vice-President, Video Cable Communication, SA	November 22, 1982
Campos, Jaime	Centrosur	November 16, 1982
Capra, Bruno	President, Servotron	November 17, 1982
Caputo, Dante	CISEA	June 4, 1980
Castex, Roberto R.	Technology Transfer Registry	May 30, 1980
Castro Madero, Carlos	President, CNEA	December 23, 24, 1982
Coll, Jorge	CNEA	June 3, 1980

Name	Position or Organization	Date of Interview
Argentina (continued)		
Comín, Luis	Engineer, formerly of FATE	June 2, 1980
Davie, Alberto G.	Former Minister of Industry	May 28, 1980
Diamand, Marcelo	Economist; industrialist	June 3, 1980
Fernandez, Pio C.	IBM Argentina	November 18, 1982
Fernandez Ocampo, Jorge	Director, Technology Transfer Registry	May 30, 1980
Ferrer, Aldo	Former Minister of Economy	June 10, 1980
Halperin, Marcelo	Integración Latino-Americana (INTAL)	June 6, 1980
Henning, Roberto	INTI	November 19, 1982
Katz, Jorge	Director of S&T Program, BID/ECLA	May 27, 1980; November 10, 1982
Kitroser, Jorge	Director, Servotron	November 12, 1982
Kohanoff, Rafael	Former Adviser to Minister of Economy for Science and Technology	June 2, 1980; November 12, 1982
Lapadula, Eduardo	Project Manager, Standard Electric Argentina; formerly of FATE	November 19, 1982
Leivobich, Harry A.	CNEA	June 3, 1980
Marcon, María Lujan	CONICET	November 11, 1982
Martínez Favini, Jorge	Director of Juridical Affairs, CNEA	November 16, 1982
Monti, Roberto C.	OAS, Argentina	November 22, 1982
Nochteff, Hugo Jorge	General Manager, CADIE	November 17, 1982
Olivera, Julio	Dean, School of Economic Sciences; former Secretary for Science and Technology	June 5, 1980
Oszlak, Oscar	Center for the Study of the State and Society	June 6, 1980
Pietragalla, Inés	INTI	November 19, 1982
Quihillalt, Oscar	Former President, CNEA	November 17, 1982

Name	Position or Organization	Date of Interview
Argentina (continued)		
Remetín, Mario Antonio	Director, SECYT	November 17, 1982
Rodriguez, José A.	President, INTI	June 3, 1980
Sabato, Jorge	Bariloche Foundation; formerly of CNEA	May 30, 1980; November 15, 1982; November 22, 1982
Sanjurjo, Carlos	IBM Argentina	November 18, 1982
Serebrinsky, Horacio	Coacin Computación	November 17, 1982
Suarez, Velasco	SECYT	June 9, 1980
Valeiras, Juan A.	Former Director, Technology Transfer Registry	June 4, 1980
Vercino, Juan Carlos	CONICET	November 11, 1982
Volman, Sara de Tanis	CNEA	November 16, 1982
Wortman, Oscar	ENACE	November 12, 1982
Zadunaisky, Gerardo	Formerly of Olivetti Argentina	November 11, 1982

Brazil

Name	Position or Organization	Date of Interview
Abreu, Marcelo de Paiva	Director, FINEP	July 2, 1980
Albuquerque, Ruy Henrique Pereira Leite de	CNPq	July 15, 1980
Amorim, J. B. de Abreu	Former President, IBM do Brasil	December 1, 1982
Arruda, Mauro Fernando Maria	Director of Technology Transfer, INPI	July 9, 1980
Barboza, Antonio	INPI	June 27, 1980; November 26, 1982
Bethlem, Mario	IBM do Brasil	December 2, 1982
Biato, Francisco A.	IPEA	July 11, 1980
Calabi, Andrea Sandro	Economist, USP	December 13, 1982
Candotti, Enio	Physicist, UFRJ	December 6, 1982
Carneiro, Dionísio Dias	Vice-President, FINEP	July 4, 1980
Caulliraux, Heitor	National Nuclear Energy Commission	December 1, 1982
Coaracy, Gastão Roberto	Director of Information, INPI	June 27, 1980

Name	Position or Organization	Date of Interview

Brazil (continued)

Name	Position or Organization	Date of Interview
Cotrim Rodrigues Pereira, Paulo Augusto	President, Digibrás	December 13, 1982
Dytz, Edison	Executive Secretary, SEI	December 9, 1982
Erber, Fábio S.	UFRJ	June 26, 1980
Façanha, Luis Otavio	FINEP	June 25, 1980
Fernandes, Jorge Monteiro	Joint Chiefs of Staff	December 10, 1982
Gennari Netto, Octávio	Former Secretary General, SEI	December 17, 1982
Goldemberg, José	Physicist, USP	December 11, 1982
Guedes, Hélio Octávio Pinto	FINEP	November 29, 1982
Guillaumón, João André	Physicist, USP	December 16, 1982
Iída, Itiro	CNPq	July 10, 1980
Kahl, Ary Barbosa	FINEP	December 2, 1982
Leal, Alexandre	National Technological Institute	July 4, 1980
Leite, Rogério C. de Cerqueira	Director, Physics Institute, University of Campinas	December 15, 1982
Lessa, Carlos	UFRJ	June 30, 1980
Mammana, Claudio Zamitti	Physicist, USP	December 13, 1982
Marques, Ivan da Costa	Former member of CAPRE	July 9, 1980; December 1, 1982
Menezes, Luis Carlos de	Physicist, USP	December 14, 1982
Millan, Marilia Rosa	Cobra	December 2, 1982
Mônaco, Lourival C.	STI	July 14, 1980
Moreira, Manuel de Frota	CNPq	July 3, 1980
Moura Fé, José de Anchieta	CNPq	December 9, 1982
Nunes, Arthur Pereira	SEI	July 11, 1980; December 7, 1982
Pelúcio, José	UFRJ	June 30, 1980; November 29, 1982
Pena, Guilherme de la	Vice-President, CNPq	July 11, 1980; December 9, 1982
Pereira, José Lopes	President, INT	June 30, 1980

Name	Position or Organization	Date of Interview

Brazil (continued)

Name	Position or Organization	Date of Interview
Pinto, Ricardo Guedes	FINEP	November 29, 1982
Pirró e Longo, Waldimir	INT	June 24, 1980
Ripper, Mario	Doças de Santos	July 7, 1980; November 26, 1982
Rocha, José Ezil Veiga da	Strategic Activities Undersecretary, SEI	December 9, 1982
Rodrigues, Silvia Helena Vianna	Reporter	December 2, 1982
Sala, Oscar	Physicist, USP	December 13, 1982
Santos, Marcelo Dami de Souza	PUC; former President, National Nuclear Energy Commission	December 15, 1982
Santos, Maria Helena Castro	FINEP	June 25, 1980
Santos, Suelí Mendes dos	COPPE	July 3, 1980
Saur, Ricardo A. C.	Executive Secretary, ABICOMP	November 29, 1982
Schwartzman, Simon	IUPERJ	June 24, 1980; November 24, 1982
Soares, Manuel Fernando Lousada	CNPq	December 9, 1982
Soto, Juan Bautista	Physicist, COPPE	December 3, 1982
Tavares, José	INPES/IPEA	June 24, 1980
Teixeira, Descartes de Souza	Digibrás	December 10, 1982
Thome Filho, Zieli Dutra	Physicist, COPPE	December 3, 1982
Tigre, Paulo Bastos	Economist, UFRJ	December 3, 1982
Vargas, José Israel	Secretary, STI	July 10, 1980
Velho, Gilberto A.	Anthropologist, National Museum	November 30, 1982
Velloso, João Paulo dos Reis	Former Minister and Secretary, Planning Ministry and Secretariat	July 7, 1980
Zubieta, Roberto H.	Director Manager, Elebra SA; formerly of FATE	November 27, 1982

Name	Position or Organization	Date of Interview

Bolivia

Name	Position or Organization	Date of Interview
Aguirre, Carlos	Former Science and Technology Director, Planning Ministry	May 22, 1980
Ascarrunz Durán, René	Head Office of Norms and Technology	May 21, 1980
Bernal Yáñez, Gregorio	Director, Head Office of Norms and Technology	May 23, 1980
Cañipa, Lino	Andean Pact	May 21, 1980
Campos, José	Head Office of Science and Technology, Planning Ministry	May 21, 1980
Castaños, Arturo	Science and Technology Director, Planning Ministry	May 21, 1980
Garvizu, Carlos	Chief, Technical Information Service	May 21, 1980
Paredes, Carlos F.	Head Office of Science and Technology, Planning Ministry	May 20, 1980
Requena Suarez, Alberto	Director, Technology Transfer, Head Office of Norms and Technology	May 21, 1980
Schulczewski, Ramón	Head Office of Science and Technology, Planning Ministry	May 22, 1980
Suárez Morales, Ovidio	President, National Academy of Sciences	May 22, 1980
Velasco, J. Enrique	Chief of Scientific Research and Graduate Studies	May 22, 1980

Mexico

Name	Position or Organization	Date of Interview
Wionczek, Miguel S.	Colegio de Mexico, former Adviser to the President	July 17, 1980

Name	Position or Organization	Date of Interview

Peru

Alcalde, Arturo	Director, National Research Council	May 14, 1980
Dávila, Guillermo	National Planning Institute	May 19, 1980
Falconi, Jorge	National Commission of Foreign Investment and Technology	May 14, 1980
Flores, Gustavo	Andean Pact	May 13, 1980
Gese, Hugo	IBM	May 19, 1980
González Vigil, Fernando	Center for Development Studies and Incentives	May 13, 1980
Ñahui Ortíz, Alupio	ITINTEC	May 19, 1980
Prieto, Eduardo	National Planning Institute	May 19, 1980
Ruiz de Pardo, Carmen	Peruvian Institute of Nuclear Technology	May 19, 1980
Sagasti, Francisco R.	Analysis Group for Development	May 13, 1980
Samame Boggio, Mario	Director, Geological, Mining and Metallurgical Institute	May 15, 1980
Tola Pasquel, José	Dean, Catholic University; President, National Academy of Sciences	May 15, 1980
Vega, Jorge	Former Director, Technology Transfer, ITINTEC	May 14, 1980
Villagarcía, Carlos	Director, Technology Transfer, ITINTEC	May 15, 1980
Zevallos Muñíz, Marco Aurelio	National Research Council	May 14, 1980

United States

Daly, John	Agency for International Development	March 13, 1980

Name	Position or Organization	Date of Interview
United States (*continued*)		
Teitel, Simón	Inter-American Development Bank	March 13, 1980
Vernon, Raymond	Harvard University	March 10, 1980
Weiss, Charles	World Bank	March 12, 1980
Whiting, Van R., Jr.	Harvard University	March 10, 1980

Index

Designer: Leigh McLellan
Compositor: G&S Typesetters, Inc.
Text: Linotron 202 Baskerville
Display: Baskerville

Lightning Source UK Ltd.
Milton Keynes UK
UKHW010623270320
360939UK00004B/241